TELETEXT AND VIDEOTEX IN THE UNITED STATES

MARKET POTENTIAL
TECHNOLOGY
PUBLIC POLICY ISSUES

Data Communications Book Series

Basics of Data Communications. Edited by Harry R. Karp. 1976, 303 pp, softcover.

Data Communications Procurement Manual. By Gilbert Held. 1979, 150 pp, clothbound.

Practical Applications of Data Communications. Edited by Harry R. Karp. 1980, 424 pp, softcover.

Fiber Optics and Lightwave Communications Vocabulary. Edited by Dennis Bodson. 1981, 149 pp, softcover.

McGraw-Hill's Compilation of Data Communications Standards (Edition II). Edited by Harold C. Folts. 1982, 1923 pp, clothbound.

The Local Network Handbook. Edited by George R. Davis. 1982, 256 pp, softcover.

TELETEXT AND VIDEOTEX IN THE UNITED STATES

MARKET POTENTIAL
TECHNOLOGY
PUBLIC POLICY ISSUES

John Tydeman
Hubert Lipinski
Richard P. Adler
Michael Nyhan
Laurence Zwimpfer

Institute for the Future
Menlo Park, California

Data Communications

McGRAW-HILL PUBLICATIONS COMPANY

1221 Avenue of the Americas
New York, New York 10020

Production management by the Total Book.
The typesetter was Information Sciences Corporation.
The cover was designed by Cathy Canzani, Design Works.
Port City Press was printer and binder.

This book was prepared with the support of the National Science Foundation under grant no. PRA 8012731. Any opinions, findings, conclusions, and recommendations expressed herein are those of the authors and do not necessarily reflect the views of the National Science Foundation.

TELETEXT AND VIDEOTEX IN THE UNITED STATES

1234567890 PCP 8976543210

ISBN 0-07-000427-7

Library of Congress Cataloging in Publication Data
Main entry under title:

Teletext and videotex in the United States.

    Bibliography: p.
    Includes index.
    1. Teletext (Data transmission system)--United
States. 2. Videotex (Data transmission system)--
United States. I. Tydeman, John. II. Institute for
the Future.
TK7882.I6T43 1982   384   82-9970
ISBN 0-07-000427-7   AACR2

ABOUT THE AUTHORS

John Tydeman   Dr. Tydeman's activities at the Institute focus on the assessment of new communications technologies, such as teletext and videotex, and on the planning and evaluating of healthcare systems. The holder of a Ph.D. degree in systems engineering from the University of Lancaster (United Kingdom), he has served as a consultant on many statistical, economic forecasting, and corporate planning studies.

Hubert Lipinski   Dr. Lipinski's interests center on telecommunications, particularly computer-based group communications, and modeling methodologies and their applications to problems involving the interaction of technological, economic, and societal factors. He received a Ph.D. degree in theoretical physics from the University of California at San Diego.

Richard P. Adler   A communications specialist, Mr. Adler is particularly concerned with the social and cultural effects of media. He is coauthor of ''The Effects of Television Advertising on Children'' and has edited several volumes, including ''The Electronic Box Office'' and ''Understanding Television.''

Michael Nyhan   A communications and public policy analyst, Mr. Nyhan joined the Institute after serving five years with the Aspen Institute's Program on Communications and Society. He holds master's degrees in communications and public administration (public policy development) and is coeditor of ''The Aspen Handbook on the Media'' and ''The Future of Public Broadcasting.''

Laurence Zwimpfer   A telecommunications engineer with the New Zealand Post Office, Mr. Zwimpfer worked on the teletext and videotex technology assessment project during his six-month stay at the Institute. This work was part of a two-year Harkness Fellowship to examine new computer communication technologies and their societal implications.

CONTENTS

PREFACE   xi

1   DEFINING TELETEXT AND VIDEOTEX   1

What Are Teletext and Videotex?   2
How Do They Work?   4
What Is the Market for the Technology?   6

2   OVERVIEW OF STUDY METHODOLOGY   8

The Demand for Teletext and Videotex   9
Teletext and Videotex Technology   10
Policy Analysis   12

3   EXPERIENCE OUTSIDE THE UNITED STATES   14

Services   17
System Structure   21
Some Possible Policy Concerns for the United States   34
Conclusion   39

4   CURRENT STATE OF TELETEXT AND VIDEOTEX
    IN THE UNITED STATES   40

Current Trials   40
Analysis of Trials   45
Technological Attributes   47
Applications   52
Generic Classes of Information Services   56
Conclusion   59

5   FUTURE APPLICATIONS FOR TELETEXT AND VIDEOTEX   60

Insights about Future Applications from Current Trials   60
Future Applications   64
Potential Market for Home-Based Videotex Services   64

	The Business Market	80
	Growth of Teletext and Videotex	83
6	**TECHNOLOGY FORECAST: COMPUTER DATABASES**	**89**
	Network Structures	89
	Database Structure, Search Procedures and Storage	92
	Access Procedures and Accessibility	95
	Conclusion	98
7	**TECHNOLOGY FORECAST: COMMUNICATION NETWORKS**	**100**
	Communication Technologies	102
	Broadcast Television	102
	Multipoint Distribution Systems (MDS)	106
	Direct Broadcast Satellites (DBS)	107
	FM Radio	109
	Cable Television	110
	Switched Telephone Networks	114
	Packet-Switched Networks	118
	Conclusion	120
8	**TECHNOLOGY FORECAST: USER TERMINALS**	**122**
	Displaying Information	123
	Inputting Data	132
	Processing Data	135
	Storing Data	139
	Conclusion	144
9	**FUTURE ALTERNATIVES FOR TELETEXT AND VIDEOTEX**	**146**
	Basic Teletext and Videotex Services	147
	Comparison with Existing Teletext/Videotex Systems	155
	Enhanced Teletext and Videotex Services	157
	Conclusion	168
10	**POLICY ISSUE IDENTIFICATION**	**170**
	Background	171
	The Current Policy Environment for Teletext and Videotex	173
	Teletext and Videotex Policy Issue Identification	180
	Key Teletext and Videotex Policy Themes	184
	Policy Synthesis	186
11	**DEVELOPMENTAL POLICY ISSUES**	**189**
	Area 1: The Technological Path Standards	190
	Area 2: Structure of the Teletext and Videotex Marketplace	198
	Guarantee of Access	200

Competition   201
Area 3: The Range of Videotex Applications   204
Content Regulation   204
Copyright   207
Developmental Policy Option Profiles   209
Impact of Policy Option Profiles   211

12 CONSEQUENTIAL ISSUES ANALYSIS   214

Privacy and Security   214
Equity of Access   218
Consumer Protection   220
Industry Structure   223
Employment   227
International Trade and Communication   234
Conclusion   237

13 TELEBANKING AND TELESHOPPING—A VIDEOTEX
        POLICY CASE STUDY   238

Context   238
Structure of Policy Workshop   240
Policy Issue Identification   245
Policy Options for the Key Issues   246
Policy Option Profiles   249
Impacts of Widespread Penetration   249
Conclusion   252

14 TRANSFORMATIVE EFFECTS AND SOCIETAL IMPACTS   253

Teletext/Videotex Scenario: The Technology   254
Key Areas of Social Impact   255
The Home and Family Life   255
The Consumer Marketplace   258
The Business Office   262
The Political Arena   265

15 CONCLUSION AND FINDINGS   268

APPENDIXES

1   Examples of Character Sets   275

2   Selected Teletext and Videotex Applications and
    Associated Policy Issues   277

BIBLIOGRAPHY   287

INDEX   302

TO MIKE NYHAN

PREFACE

Teletext and videotex are gathering momentum in the United States and around the world. Although both technologies were invented outside the United States, there are now numerous technical and market trials under way in this country, trials that involve many of the major corporations—telephone companies, cable operators, banks, retailers, publishers, broadcasters, and computer hardware and software manufacturers. Although the direction and final consumer acceptance of the technology are still uncertain, the potential stakes are too great for companies to ignore developments in this rapidly changing field.

This book presents the findings of a technology assessment of teletext and vidotex, a study to assess the impact of teletext and videotex in the United States over the next 10 to 20 years. The project, which was conducted by a research team at the Institute for the Future, was sponsored by the National Science Foundation.

After providing an introduction to teletext and videotex and an overview of the methodology (Chapters 1 and 2), we then address three specific topics:

- The current state of teletext and videotex around the world and the likely market potential in the United States (Chapters 3, 4, and 5)
- The likely technological developments of teletext and videotex from the perspective of the user (Chapters 6, 7, 8, and 9)
- The public policy issues and societal impacts (Chapters 10 through 15)

The book has been written to appeal to a wide audience. There is a major focus on the public policy issues surrounding the penetration of teletext and videotex, and, as such, the book will be of value to corporate, trade, consumer, and government decision makers and advocates already in this field who wish to assess the potential of the technology. It will be of interest to those in industries and fields in which the technology will have a major impact, as well as those actively engaged in teletext and videotex software, hardware, field trials, or information services. For students of public policy and those outside the United States wishing to understand the complex character of the U.S. market and the regulatory framework for teletext and videotex, the book should also be of value.

ACKNOWLEDGMENTS

This book would not have been possible without the support and cooperation of a large number of people. Although their contributions have been invaluable in conducting our research, any failings or errors must remain the responsibility of the authors.

We wish to thank the many people who participated in the seven workshops we held in the United States and Canada, those who completed our futures questionnaire, and those who critiqued our early material.

We are indebted to those involved in various European teletext and videotex industries who generously gave their time to be interviewed on their respective systems. In particular, we owe special thanks to Hilary Thomas, CSP International, for her support in our data collection activities, and to Richard Hooper, British Telecom; Colin McIntyre, British Broadcasting Commission, Jacques Dunogue, Direction Generale des Telecommunications; Jacques Renevier, Sofratev; and Susan Collins, Infomart, who reviewed our chapter on non-U.S. teletext and videotex systems (Chapter 3).

We wish to thank the members of the oversight committee of our National Science Foundation project who offered valuable assistance and advice on the structure of the study and methods acutally used:

Dr. Walter Baer, Times Mirror
Professor Michael Botein, New York Law School
Professor Kan Chen, University of Michigan
Professor Paul David, Stanford University
Professor Donald Dunn, Stanford University
Dr. G. Patrick Johnson, National Science Foundation
Colonel Wayne Kay, Executive Office of the President
Professor Percy Tannenbaum, University of California at Berkeley

In particular, we wish to thank Pat Johnson for his criticism, support, and encouragement throughout the project.

Almost everyone at the Institute for the Future contributed in some way to the study. Robert Plummer helped in assessing the U.S. trials; Norman Klivans assisted in assessing the structure of the U.S. regulatory environment; Kathleen Vian assembled much of the material on societal impacts of teletext and videotex; and Mary Poulin assisted in the economic analysis. Our special thanks, however, go to Alexis Makarevich, who converted much of the scrawl from five authors into a legible manuscript; Ellen Margaret Silva, who provided the artwork; and Jim Storey, who edited the manuscript.

John Tydeman
Principal Investigator

CHAPTER 1

DEFINING TELETEXT AND VIDEOTEX

Computing and communication technologies have joined together to produce a new hybrid technology for delivering home-based information services. The distinctive feature of the technology is not the individual technical elements but the assembly of a total system comprising information banks, indexing structures, computer and communications hardware and software, system management, and billing. Through such an integrated system or systems, millions of people will have access to a wide array of information services—services that can be provided inexpensively, rapidly, and in environments chosen by the user. Because this evolving technology has the potential to change *how people use information and indirectly how they think,* it may well have an impact on many aspects of daily life as well as on the services currently provided by society.

This assessment is intended to contribute to a fuller understanding of the technological, economic, and social consequences of a widespread implementation of the technology. While our prime focus is on public policy issues, considerable emphasis has also been given to the underlying technological components and to the range of services that could be provided.

There is no single accepted name for this new technology. When referring to two-way information services to the home, terms such as videotex, viewdata, videotext, and interactive videotex are commonly used, while broadcast videotex, teletex, and teletext are often used interchangeably to describe the provision of one-way information services. As if that is not sufficiently confusing, the brand names of various systems themselves have taken on a definitional meaning. For example, Prestel, Telidon, Teletel, Antiope, Viewtron, QUBE, Ceefax, and Oracle are often used to describe the technology itself.

As a result, we have decided to use two generic terms to denote the technology. Following the International Telegraph and Telephone Consultative Committee (CCITT), we use the term *videotex* as the generic name to describe the provision of two-way information services and *teletext* for one-way services. In addition, we also use the term videotex to refer broadly to the class of systems that provide electronic information to the home.

WHAT ARE TELETEXT AND VIDEOTEX?

Tyler (1979*a*) has defined teletext/videotex as follows: "Systems for the widespread dissemination of textual and graphic information by wholly electronic means, for display on low-cost terminals (often suitably equipped television receivers), under the selective control of the recipient, and using control procedures easily understood by untrained users." It is significant to note that with the exception of the term "electronic," the definition is *medium free*.

Drawing on Tyler's definition, as well as those proposed by Fedida and Malik (1979), Madden (1979), and Winsbury (1979), we can describe the technology as follows:

- It is a broad class of systems, not a single distinct technology. In fact, there is a continuum of sophistication in database information transmission including:

> scrolling or recorded voice
> captioning
> teletext
> videotex—narrowband
> videotex—wideband
> personal computers
> personal computers with fully online database systems

- There is an orientation toward general use or widespread dissemination of textual and graphic information, which is remote from the user.
- Dissemination or transmission is by one or more electronic means, e.g., telecommunications links such as radio wave, TV signal, coaxial cable, copper wire, optical fiber, microwave, or satellite.
- The information is under user control. Although teletext and videotex have varying degrees of interactivity, neither is a passive medium. Information appears, and transactions are made or messages are sent (and even received) at the express command of the user.
- The system is easily understood by untrained users. That is, it is a system for noncomputing or nonelectronic specialists. This has implications for the way information is accessed, for the way information is packaged and for the way messages are sent and transactions are made.
- The array of services for users is limited only by the ingenuity of those providing them. Teletext and videotex are not just information-retrieval services. They pro-

vide an electronic window to a multitude of services in the information society. In subsequent chapters we identify five broad classes of videotex services.

information retrieval (e.g., news, weather, sports, advertising, directories, how-to guides)

transactions (e.g., reservations, teleshopping, telebanking)

messaging (e.g., electronic mail)

computing (e.g., interactive games, financial analysis)

telemonitoring (e.g., home security)

- The system is ''low'' cost. The whole concept from the original ''invention'' by Fedida has been aimed at a mass (home and business) market, at widespread implementation. The information services will achieve this status only if they are relatively low cost.
- It is not a new technology. It is a product of the merging of communication systems and computing. It represents a new combination of hardware, such as computers, databases, communication networks, television sets and terminal devices, and software, including frame design graphics and text, gateways, retrieval mechanisms, and billing.

These qualities do not, however, preclude videotex, or even teletext, from developing expensive information services for select user groups. While very few people may want to purchase a Mercedes-Benz on their videotex system or to have access to real-time information on the price of greasy wool at the Sydney stock exchange, these services can be made accessible via teletext and videotex. There will always be a market for specialized real-time information and for selective user-group databases, and videotex systems may provide suitable alternative means for accessing some or all of this information. The closed user-group facility on Prestel is recognition of this restricted market use.[1]

Teletext and videotex as a class of services compete with and could be said to replicate some existing computer-based timesharing services, word processing systems, electronic messaging, and even interactive games. A key element in the systems design is the focus on the household and the untrained user, not the professional computer person. Those services that require specialized training to execute are not included within the above definition, although a new software interface could put such systems into the videotex arena.

Throughout this project we will refer to teletext and videotex systems as a spectrum from simple one-way transmission, dumb-terminal interaction to sophisticated two-way transmission involving external data networks, intelligent terminals, and the potential for enhanced services.

---

[1]A closed user group is simply a number of users of a particular restricted service. Membership in the club is essential to access the service (i.e., a stockbroker investment service or a list of entertainment activities in Playboy clubs). On Prestel the term ''closed user group'' refers to intracompany restrictions on access to data; ''syndicated closed user group'' is the more general term for restricted access.

HOW DO THEY WORK?

Teletext

Information, consisting of alphanumeric characters or graphic images, is edited on a keyboard or generated from a computer-stored database. It is encoded in a bit stream of digital data at a transmission rate that is compatible with the color-TV system in use (PAL with 625-line pictures in most of Europe, and NTSC with 525 lines in North America).

The encoded data is multiplexed onto a video signal and transmitted with the TV signal on the unused lines in the vertical blanking interval—the period during which the scanning of the TV picture begins again (Roizen, 1980). The data in the TV signal are detected by a decoder attached to the TV set as an accessory connected to the radio frequency (RF) antenna socket or directly wired into the RGB beam circuits—the red, green, blue guns—of a color TV. The decoder accepts the digital data, stores one or more pages in a buffer memory, and displays these pages on the screen as directed by the user (Figure 1.1). When the viewer punches the number of the desired page (or screenful of information) on a keypad or keyboard, the buffer memory containing the page is kept in a "hold" condition. The page is then transferred to the TV screen via a character and graphic generator that is part of the decoder. The page remains on the screen until a replacement page is requested by the user or the system is switched off. The information may or may not take up the entire TV display. For example, the information may appear as an overlay to an ongoing TV program.

The information is "cycled" by the broadcast station. Access time, the delay between requesting a page and seeing it on the screen, depends on the number of pages being cycled, the rate of transmission (the bit rate), the number of TV lines

FIGURE 1.1 Broadcast Teletext.

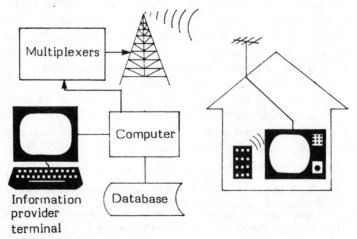

dedicated to carrying the information to the viewer, the amount of memory in the decoder, and the "importance" of the information. (It is possible to make every fifth or tenth page identical so that the waiting time for the information is virtually zero. The pages are slipped into the cycle more often but still exist in the computer in only one place. Alternatively, less popular information may be broadcast only at certain times.)

For example, using two lines of the vertical blanking interval, 100 pages of text with simple graphics may be cycled every 25 seconds. If a full TV channel were used for teletext transmission, more pages and quicker access could be achieved. In addition, other transmission media could be used to send large quantities of data to a dwelling or business for disc or tape storage. These could include cable TV, cellular radio, low-power TV, multipoint distribution systems, and direct broadcast satellites. These communication transmission networks are described in Chapter 7.

Videotex

As with teletext, pages of information are edited on a keyboard and stored in a computer database. The database design is such that it permits the accessing and rapid retrieval of specific items of information and the billing of customers using the system. The transmission lines between the user and the computer include the public telephone network (using appropriate modems to convert analog telephone signals into digital form for display) or a cable TV system with two-way capabilities.

Again a modified TV receiver with a decoder translates the data and builds up the video image (Figure 1.2). As with teletext, the decoder may be plugged into the antenna socket or directly wired into the RGB beam circuits. Pages for transmission are selected by the user on a numeric keypad or an alphanumeric keyboard. As the system has *full two-way capability*, the user may also send a message to the computer, the database, or another terminal. In telephone systems, the transmission of data to the user is usually at a higher speed than the transmission from the user to the

FIGURE 1.2 Videotex via Telephone Line.

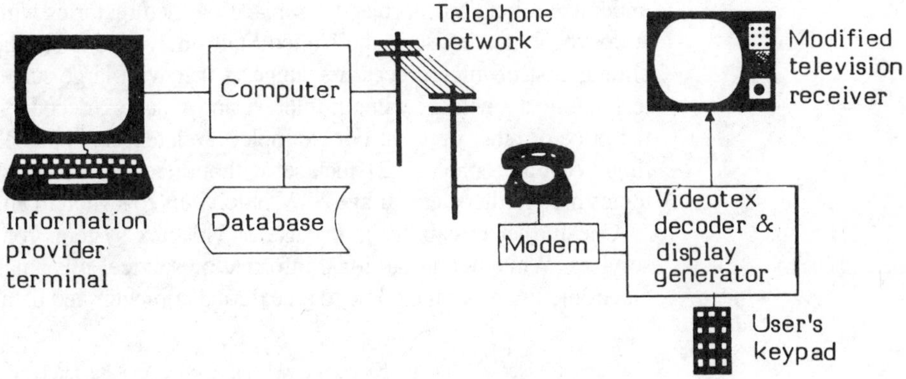

system (e.g., the British Prestel system uses 1,200 bits[2] per second from computer to user and 75 bits per second from user to computer).

The significant difference from teletext is that with a videotex system the data is not routinely cycled in a broadcast mode; instead, the individual users access the database as required. Thus, access time is now a function of the processing capacity of the host computer and the volume and pattern of usage.

From the above system descriptions it is evident that there are four general actors in any teletext/videotex system: the users of the service, the communications network provider (e.g., a common carrier or a broadcast station), the teletext or videotex system operator (e.g., the broadcast station or a private corporation that operates the teletext or videotex system), and the information service provider (e.g., any private corporation or public agency).

WHAT IS THE MARKET FOR THE TECHNOLOGY?

The key research question of this study—What are the public policy consequences of widespread implementation of teletext and videotex?—presumes the existence of a consumer and/or business market.

The potential consumer market is every household with a telephone or a television receiver—in excess of 98 percent of all U.S. households in 1980. As the technology utilizes existing home-based components, its "newness" has more to do with the way in which things may be done and the juxtaposition of the components than with the technology itself. Even so, American Telephone and Telegraph estimates that the market for telephone-driven videotex will not exceed 7 percent of U.S. households by 1990 (Sullivan, 1981). On the other hand, cable TV penetration (actual homes hooked to cable systems) is forecast to rise from the current 28 percent to more than 50 percent by 1990 and to "pass over" 75 percent of households. While virtually all existing cable systems are one-way only, most new franchise bidding includes provisions for two-way interactive services.

There may be other incentives to market teletext and videotex terminals. The economics of paper-based directories and manually operated switchboards, for example, have led the French Postal, Telegraph and Telecommunications Department to undertake a long-term project to replace paper directories with electronic directories. A cost/benefit exercise in the United States may lead to similar conclusions. The declining cost of microprocessors suggests that with large-scale production, equipping television sets with or without teletext and videotex decoders will be comparable to the price of other options. For example, a teletext-equipped 12-inch-screen color TV might cost the same as a 21-inch set without teletext. Advertising and marketing strategies may further "encourage" TV purchasers to switch to an enhanced set.

One further possibility is the use of videotex systems for fire and burglar alarm services. While not a traditional information service, it is a natural application for the electronic link from the home to an external computer and database.

---

[2]A bit is a binary digit (one that takes a value of 0 or 1). An 8-bit unit is referred to as a byte. A kilobyte, kbyte, is 1,000 bytes and a megabyte, Mbyte, is 1 million bytes.

At present the only commercial systems in operation in the United States provide teletext and videotex services directed at special groups such as captioning for the deaf, Dow Jones financial information, and home-based personal computer services (The Source, CompuServe). As teletext and videotex systems offering a variety of electronic services are introduced, and as user-friendly protocols are developed, the potential markets for the technology will expand. Our intention in this book is to examine the plausibility of widespread penetration within a 20-year time frame and to assess the public policy consequences of such developments.

CHAPTER 2

OVERVIEW OF STUDY METHODOLOGY

Technology assessment has been defined by Menkes (1979) as a process of examining "policy options and the decisions required to capture the benefits of technology while controlling its potentially harmful side effects." This requires constructing "a model of the interaction of technology, public policy, and society . . . so that the relationships between choice and institutional consequences can be elucidated and the effects demonstrated." The dimensions of a technology assessment include the public policy considerations of political, economic, social, institutional, technological, and legislative changes induced as a result of implementing a new technology.

The approach used in this book (Figure 2.1) may be summarized as follows. The widespread implementation of any new technology is a combination of market "demand-pull" and technology "supply-push" factors. The demand side involves identifying current applications of teletext and videotex, both in the United States and elsewhere in the world, and then searching for possible commercial and mass-market application areas that may arise over the next two decades. On the supply side the aim is to describe teletext and videotex in terms of their basic technological components, to forecast improvements of these components, and then describe alternative technological options for teletext and videotex systems based on these forecasts.

After developing an understanding of the technology, we conduct a detailed analysis of public policy issues and their likely impacts. For purposes of analysis, the policy issues are structured into developmental issues, those that shape the path of the emerging technology, and consequential issues, those that arise as a result of widespread penetration of teletext and videotex. Finally, some of the potential broad societal impacts of widespread penetration of the technology are discussed. The three major components of the study—market demand, teletext and videotex technology, and policy analysis—are now described in some detail.

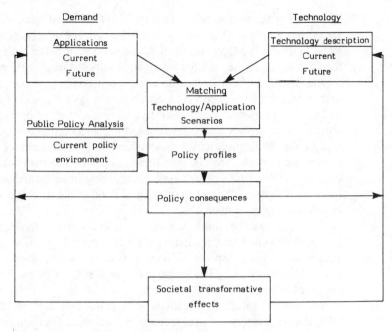

FIGURE 2.1   Framework for Technology Assessment.

THE DEMAND FOR TELETEXT AND VIDEOTEX

Teletext/videotex systems are anticipated to provide a wide range of information services. In many cases, for example, news, weather, and sports reports, these systems will be a direct *substitute* for current methods of information dissemination such as radio, TV news, telephone dial-it services, and parts of newspapers. In other instances the service is an *enhancement* of what is currently available, for example, teleshopping as compared to catalog shopping. And in yet other cases teletext/videotex offers the potential for an entirely *new* service, for example, electronic sensors for home energy management.

The environment in which teletext/videotex is emerging is itself turbulent. In addition to the existing technologies for delivering information services, there are many alternatives that will compete directly with these systems; for example, videodiscs, videocassettes, store and forward message services, and personal computers.

In this section of the study we address four questions:

- What are the current applications and uses of teletext and videotex?
- What generic classes of applications emerge?
- What are some likely future applications for the technology?
- What is the likely future market structure (size, demand) for these videotex information services?

As part of the review of the field trials in the United States, Europe, Canada, and Japan, our aim was to identify a comprehensive set of applications of videotex systems. In evaluating the existing trials and services it became clear that videotex systems are not merely electronic information retrieval media. By examining the fundamental structure of each application, e.g., a message is a link from user terminal to user terminal, a news report is a data flow from system operator to user terminal, and so on, it was possible to delineate five generic classes of applications—information retrieval, transactions, messages, computing, and telemonitoring. In some instances a particular application may be a combination of a number of the generic classes. Games, for example, may be considered to be both information retrieval and computing. The teletext and videotex applications being incorporated into the field trials are linked to the generic classes to get an indication of the current distribution of services as videotex begins to evolve.

To help answer the third question, we held a series of three "futures workshops" to identify likely future applications for the technology. The possible applications stretch across many aspects of modern western society, suggesting that videotex or videotexlike technology has the potential to alter the way we live our lives and conduct our businesses.

The likely future applications give one measure of the composition of teletext/videotex services over the next two decades. For each broad class of application we examined the size and nature of the existing demand and the likely future demand for these services, using personal consumption expenditure forecasts for the household and business expenditures on information services for the corporation. Three elements constitute a future market share projection: a natural growth (or decline) in the service due to the introduction of the new technology, a substitution effect (e.g., the introduction of the telephone caused a decline in personal mail), and a complementarity effect (the increased penetration of TV resulted in an increase in the sale of books).

An estimate of the likely size of the market was compiled from a market forecast questionnaire distributed to knowledgeable individuals in teletext- and videotex-related industries. These results were used to identify which of the services are most likely to evolve first for home and business use, the amount of time users will allocate to these systems, and the amount they will be prepared to pay for teletext and videotex services.

TELETEXT AND VIDEOTEX TECHNOLOGY

The technology is in an early developmental phase and is undergoing rapid change. New trials and services are being launched in North America and Europe, and most include some new technological enhancement. Our aim is not to monitor day-to-day changes in the technology but to focus on alternative future structures of teletext/videotex systems. Thus we describe the present state of the technology by examining current field trials, use these trials to develop a conceptual framework in which to represent the changing technology, describe the technological components of a videotex system, identify alternative components (enhancements or substitutes to present components) likely to be introduced over the next 20 years, and develop forecasts of all of these components.

As the specific elements of a teletext or videotex system will change over the next decade it is important for a policy study that a framework be established that is relatively stable and in which any trial, now or in the future, may be analyzed. Thus, for selected teletext/videotex systems now in use or under trial in Europe and North America, this general framework is used to define the characteristics of the systems from the viewpoint of one of the major actors—*the users*—and to compare the various trials in terms of their technological configurations.

Further, we examined the organizational configuration of the trials to see what industry alignments are emerging. In this way we determined what groups are currently the system operators, the communications network providers, and the information providers, and then identified similarities and differences in methods of system operation, transmission, terminal design, and display. This provides a starting point for examining the likely directions for the technology.

We have used the term "system characteristic" to cover the basic functions of the system such as display, input, storage, processing, billing, etc., and the term "technological option" to describe the range of technological possibilities for each characteristic—the display may take place on a TV, CRT, and so on. We have also outlined the basic attributes of each characteristic.

The attributes are key elements in that they not only describe the nature of the technological option but also help in linking applications of teletext/videotex service to various technological configurations. For example, an electronic messaging service may be defined for a user in its most basic form—a black-and-white display unit with an alphanumeric keyboard and either a telephone or two-way cable TV connection. The attributes are thus pivotal in linking alternative technologies for teletext and videotex to classes of applications as illustrated below:

Technological display options	Display attributes	Applications
TV	Color	Teleshopping
CRT monitor	Text	Telebanking
	Resolution	Messaging
Printer	Graphics quality	Electronic newspaper
Plotter		

Having established a framework for analysis and reviewed the state of the "technological art," the focus of the study then narrows to that of the user. Each of the key actors in a teletext or videotex system has a different perspective of the system. The information service provider is concerned with data management, the system operator is concerned with the design and functioning of the teletext or videotex system, and the communications network provider is responsible for getting the information from the database to and from the user. The user must consider the system characteristics and modes of accessing information; in particular, cost, transmission speed, one-way and two-way capabilities, and terminal requirements. This narrowing of focus to the user is essential for the technological component forecasting or else the information and electronic industries in their entirety would be under examination.

The basic research questions are:

- What technological components are likely to constitute teletext/videotex systems over the next 20 years?
- What are their characteristics?
- How much will they cost?

Because there is no unique set of technological components that categorically determines a teletext/videotex system, our scope at this stage is intentionally wide. Further, there are a variety of systems and possible attachments or add-ons to the basic technological configurations. Areas of rapid technological breakthrough such as local storage, microprocessing power and voice recognition and synthesis could radically alter the currently "acceptable" designs of teletext and videotex systems. Thus, our aim is to describe the field of technological options in terms of the attributes identified earlier without trying to prejudge any one final design at this stage.

Twenty-year forecasting is at best a risky exercise. In this study, forecasts were based on assumptions of widespread penetration of the technology. From this set of forecasts of the various technological components, the reader can postulate any system design and, using linear cost summation, compare it in terms of cost and attributes with any other system.

Finally, the technology and the market material are integrated into a number of scenarios for teletext and for videotex. These scenarios represent an effort to match the alternative technological configurations (databases, networks, terminals) for teletext and videotex with the market demand as represented by the forecasts for each generic application. We illustrate this matching with a number of *basic*, or present-day, teletext and videotex systems, and a number of *enhancements* to these basic systems for each generic class of application. By using the attributes of the technology, these scenarios also highlight important differences between teletext and videotex.

POLICY ANALYSIS

This is the prime focus of the technology assessment. The aim of the policy analysis is to provide a context for policymakers to assess their role in the emergence of teletext and videotex services. The major inputs to the analysis are the technology/application scenarios just described and the public policy arena that currently exists, or is evolving, in the United States.

There are three sections to the policy analysis:

- Identifying the significant policy issues surrounding teletext and videotex technologies and applications
- Describing policy profiles as the technology evolves over the next two decades
- Analyzing the impacts of widespread implementation of the technology for alternative policy profiles

Because of the hybrid nature of teletext/videotex systems, policy issues arise from the areas of telecommunications and computerization as well as from the applications themselves. We identify 10 existing bodies of law or regulatory mechanisms which potentially affect the development of the technology. These are: the First Amendment (do electronic newspapers have the same protections as print newspapers?); the Federal Communications Act (the carriage-content debate); antitrust laws (anticompetitive practices); the Federal Communications Commission (FCC) common-carrier regulation (role of the dominant common carrier, AT&T); FCC broadcasting regulation (are the regulations of normal TV broadcasting applicable?); federal and state banking laws (procedures for electronic funds transfers); the Federal Trade Commission (FTC) regulation of sales and advertising (consumer protection laws); copyright law (the protection of intellectual property); state utility regulation (rates setting and monitoring services); and cable franchising and regulation (provisions for access). In functional terms these regulations relate to editorial content and control, centralization/monopolization of services, access for information providers, economic dislocation, privacy, technical standards, equity of access, control/ownership of information, legal liability, job substitution/protectionism, security of data transmission, confidentiality, and consumer protection.

It is clear that many different policy issues and public interest questions arise as the technology is being implemented. The standards debate, for example, is probably of no concern once there is a mass market. Standards, however, can shape the direction of the new technology and influence the rate of implementation. Issues such as this are called *developmental issues*. A range of policy options is specified for each issue and then, by aggregating them across all developmental issues, alternative policy profiles are described. Two specific profiles were constructed to contrast likely developments in regulatory policy: a ''deregulatory'' profile and a ''proregulatory'' profile.

Consequential issues are defined as issues that arise as a result of widespread implementation of the technology. These issues represent the combined effects of developmental public policies and market demand for the technology. They include employment and labor market considerations; new work patterns; emergence of new industry structure; as well as individual concerns about privacy, consumer protection, and equity of access.

To complete the technology assessment we broadened the analysis to the general societal context by asking the question: What are the *transformative effects* or societal impacts of widespread implementation of teletext and videotex? These effects have been divided into four main areas in which teletext and videotex systems may make substantial impacts, namely, the home, the marketplace (for financial transactions and purchase of goods), the office, and the political arena.

CHAPTER 3

EXPERIENCE OUTSIDE THE UNITED STATES

Unlike many new computer communication technologies, teletext and videotex first emerged in Europe rather than in the United States. A British research engineer, Sam Fedida, is generally recognized as the inventor of videotex, or viewdata as it was then known (Fedida, 1975). With the support of the British Post Office (now British Telecommunications), Fedida was able to demonstrate a working model of viewdata as early as 1974. By 1976 the British Post Office had mounted a public viewdata trial, and a full commercial service known as Prestel began in 1979 (Ford, 1979*a*). In parallel with the development of viewdata, the British broadcasting authorities, British Broadcasting Corporation (BBC) and Independent Broadcasting Authority (IBA), were developing broadcast teletext services (Ceefax and Oracle). In 1976, both the BBC and the IBA received full government support to begin nationwide public teletext services (Ferrarini, 1981).

Teletext and videotex also emerged in France during the 1970s. In the early 1970s, the French Center for the Study of Telecommunications and Television (CCETT) was established in Rennes as a joint research center to develop new technologies for the French PTT and for Telediffusion de France (TDF), the French broadcasting authority. The outcome of this work was Antiope, a system incorporating both broadcast teletext services and interactive videotex services. In 1978, France adopted a national plan based on the recommendations of Simon Nora and Alain Minc (1980) to develop an integrated information services network using the Antiope system. The first major implementation was the French PTT's Teletel videotex service, which began in 1981 as a home-based trial in Velizy, a suburb of Paris. In this trial, households were equipped with Antiope-adapted color television sets connected to the telephone lines via a modem. In 1982, a start was made on a 10-year program to equip all telephone subscribers in the Ille et Vilaine region with a black-and-white directory assistance terminal as a substitute for paper directories.

Other developments in Europe have been based largely on the British teletext and videotex systems. By the end of 1981, West Germany, Switzerland, Sweden, Finland, Norway, Austria, the Netherlands, Belgium, and Italy were all conducting Prestel-based videotex trials, and Sweden, Austria, Belgium, the Netherlands, Finland, and West Germany all had British-standard teletext services under way.

During the mid-1970s, the Department of Communications in Canada became interested in teletext/videotex developments. By the end of 1978, an integrated teletext/videotex system called Telidon had been developed. In 1979 the Canadian government approved a $9-million four-year program for Telidon field trials and further development of the system's components (Ferrarini, 1981). By the end of 1980, however, the total number of participants in these trials remained quite small. Early in 1981 the Canadian government reaffirmed its commitment to Telidon by approving $27.5 million for further development and marketing efforts (*International Videotex/Teletext News*, 1981*a*). In addition to government support, private industry in Canada has been responsible for much of the Telidon software (Infomart) and hardware (Norpak) developments.

Teletext/videotex developments in Japan have occurred even more recently. During the late 1970s, a number of wideband interactive video information service experiments were conducted. These included CCIS (Coaxial Cable Information System), HI-OVIS (Highly Interactive Optical Video Information System) and VRS (Video Response System) (Takasaki and Kitamura, 1980). These systems were found to be too expensive using current technology for wide-scale home-subscriber use, and therefore the Japanese PTT (Nippon Telegraph and Telephone Public Corporation) developed a telephone-based videotex system called Captain (Kumamoto and Kitamura, 1980). Captain is similar to the British Prestel system, but because of the need to generate up to 3,000 different characters to accommodate the use of Chinese characters (Kanji), the character/pattern generator is located at the videotex service center rather than in the decoder unit in users' terminals. The first public trial of Captain began in December 1979 and involved about 1,000 terminals (Yasuda, 1980).

Tables 3.1 and 3.2 summarize the major teletext and videotex systems, respectively, in operation outside the United States at the end of 1981. A common element in the United Kingdom, France, Canada, and Japan has been the provision of substantial government support in teletext/videotex developments. Furthermore, the PTT (Postal Telephone and Telegraph administrations) in Japan and Europe and Bell Canada, the dominant telecommunications carrier in Canada, have taken leading roles in promoting the development of new information technologies.

In the United States, neither of these elements is present in teletext/videotex developments. There is no government mechanism for developing a nationally coordinated teletext/videotex system, and AT&T, the dominant telecommunications carrier, has moved slowly in this area because of regulatory uncertainty about its role in providing information services.

Teletext and videotex applications in many countries have concentrated on provid-

TABLE 3.1   TELETEXT SYSTEMS IN OPERATION

Country	Generic system	System operator	Extent of service as of December 1981*
Australia	Ceefax	Private TV network	20,000
Austria	Ceefax	ORF (Austrian TV Authority)	125,000
Belgium	Antiope	BRT (French speaking)	50
	Ceefax	BRT (Flemish speaking)	8,000
Canada	Telidon	TV Ontario	55
Finland	Ceefax	Finnish Radio	20,000
France	Antiope	Telediffusion de France	1,500
Holland	Ceefax	NOS (Dutch television)	170,000
Sweden	Ceefax	Sveriges Radio	80,000
United Kingdom	Ceefax	BBC	
	Oracle	IBA	350,000
West Germany	Ceefax	ARD/ZDP (Associations of German Radio Broadcasters)	140,000

*Sources: International Videotex Teletext News, 1982; Teletext, 1982.

TABLE 3.2   VIDEOTEX SYSTEMS IN OPERATION

Country	Generic system	System operator	Extent of service as of December 1981*
Canada	Telidon	Bell Canada and provincial telephone companies	2,000
		Infomart	300
		Department of Supply and Services	200
Denmark	Prestel	PTT	200
Finland	Prestel	Company consortium	200
France	Teletel	PTT (1) Videotex	3,000
		(2) Electronic directory	2,000
Holland	Prestel	PTT	4,000
Japan	Captain	PTT	1,000
Spain	Prestel	PTT	200
Sweden	Prestel	PTT	100
Switzerland	Prestel	PTT	150
United Kingdom	Prestel	British Telecom	15,000
Venezuela	Telidon	OCEI	30
West Germany	Prestel	PTT	6,000

*Source: International Videotex Teletext News, 1982.

ing information retrieval services. In the United States, however, both business and home consumers are exposed to a much wider range of information services. For example, computer timesharing, personal computers, videodiscs, videocassettes, and computer games have all attained a much higher penetration in the United States than in other countries. In addition, some 80 percent of the world's databases are stored in U.S. computer systems.

Because of the greater diversity in both information needs and information industry structure, it is unlikely that the development of a national teletext/videotex system in the United States will follow the same evolutionary path as in other countries. However, the systems developed in the United Kingdom, France, Canada, and Japan provide an indication of possible alternative services, structures, and public policy issues that may have relevance for the United States.

These four systems are therefore analyzed to determine:

- The classes of service being provided or proposed
- The broad structure of each system
- Any technical or public policy considerations that may be of relevance to U.S developments

SERVICES

The applications of information services currently being offered are as much a function of the interests of the information providers that are participating as they are of any market demand. Teletext mainly serves the public or residential sectors, including specific user groups such as the hearing impaired. The information services include stock market reports, weather reports, general news, news flashes, sports results, subtitles, leisure information, jokes, games, recipes, and quizzes. Videotex services are predominantly business-oriented in the United Kingdom, Finland, the Netherlands, and Switzerland, and residentially oriented in Germany, France, Canada, and Japan.

Teletext

The United Kingdom is by far the largest residential market for teletext. The services offered on Ceefax include news, food-guide, news flashes, finance, an entertainment guide, an alarm clock, sports, weather and travel, and some subtitling on BBC 1; and news background, fun and games, TV and radio programs, a finance spectrum, sport features, consumer reviews, and numerous other pages such as Christian comment and public organization listings on BBC 2. The Independent Television (ITV) service, which is now heavily supported by advertisers, also offers regional teletext services around the United Kingdom.

The three networks (BBC 1, BBC 2, and ITV) offer approximately 100-page magazines that are frequently updated and changed during the day (Figure 3.1). Both the BBC and ITV recently increased the number of vertical blanking interval lines providing their teletext services from two to four. This reduced the total time to transmit a 100-page magazine to 15 seconds. The BBC is now experimenting with

FIGURE 3.1 A BBC Ceefax Index Page.

''live'' exchanges of teletext pages such as news, weather, and finance between Austria and the United Kingdom.

In France, teletext is business-oriented. The French offer closing quotations of the stock exchange on TF1 and special weather forecasts (small-boat and light-plane), news, and various information services on the other two stations (TF2 and TF3) (Figure 3.2). Some are pay services, e.g., weather and stock exchange information, and selective access decoders are being developed for use with magnetic or microprocessor cards. This enables an access cost to be fixed for certain services at particular times. The developments in France reflect the fact that the cost of semiconductor memory is decreasing more rapidly than the cost of data transmission. Cheap local storage and local processing provide the user with the opportunity of storing a catalog or ''database'' once and then receiving updates from the broadcaster.

As Antiope software is similar for both the teletext and videotex systems, the use of hybrid systems of broadcast and telephone to provide an interactive capability is being explored. Further, the Didon packet data broadcasting network allows the broadcaster to send a number of different services on each channel. Whereas with Ceefax, for example, there is one information provider and one set of pages per channel, with Antiope there are a number of sets of information available simultaneously on each channel.

Both the French and British teletext systems are aimed at providing rapidly changing information to a large volume of users. The British have selected the home market and the French have explored, in addition to home services, specific large user groups

FIGURE 3.2 Teletext Information from the French Stock Exchange.

that require such information, e.g., corporate information that must be available to all employees. The French are also embarking on a small-scale trial to place teletext systems in public locations.

A major public use of teletext has been for the deaf. In Sweden, the Ceefax-based system operated by Sveriges Radio AB was introduced initially on an experimental basis for the deaf and hearing-impaired. It has about 100 pages of news, weather, and general information as well as limited subtitling services for popular Swedish and foreign programs. The British Ceefax system carries a special ''page'' for the deaf, which consists of 30 cycling subpages, and the three stations (BBC 1, BBC 2, and ITV) offer a total of about seven hours of subtitling per week.

Videotex

There are three classes of general videotex services being offered. First, there is the public database service. In the case of Prestel, for instance, there is both a general-interest database and a business database. Users typically have a paper-based directory to determine the page number of various information services as well as the access cost per page, except where a flat fee is charged for information services. In addition there are autodial, coin-operated public access videotex terminals that are programmed to access certain pages on the database. The range of public videotex services for various national systems is shown in Table 3.3.

The second class of applications is closed-user group services. In this case a syndicate has exclusive access to a database. Membership can be on a fixed fee or a fee

TABLE 3.3 VIDEOTEX SERVICES OFFERED IN SOME NATIONAL SYSTEMS

Applications	United Kingdom	Canada	Finland	Germany	Japan	Holland	France
Information retrieval	×	×	×	×	×	×	×
Messages (a) Directional (to particular users)			×				
(b) Systemwide (to all users)	×	×	×	×			×
Transactions with direct payment				×			×
Transactions without payment	×	×		×	×	×	×
Data processing (a) Local (b) Remote		×					
Games/ entertainment	×	×	×	×	×		×
Education	×	×			×		
Electronic directory							×

plus pay-as-you-use basis. The information is itself either proprietary, for example a stockbroker's services or market newsletters for a set of clients, or of value to a particular paying clientele.

The third class of applications is in-house applications or private systems. These may be information services or message services for national or international corporations. One example is General Electric's videotex system introduced for Whitbreads in the United Kingdom. A second example, on Prestel's database, is an in-house system used by International Computers Limited (ICL) to communicate information to their international subsidiaries on software faults and update communication protocols.

There are three main types of private videotex systems emerging: (1) a stand-alone computer system, which may range from a small microcomputer with four to five access ports and a database of around 5,000 pages to a minicomputer with 16 to 120 access ports and a database of around 180,000 pages; (2) a front-end minicomputer that operates off a private (or public) mainframe computer; and (3) a mainframe computer system itself. The personal computer market has also begun offering in-house videotex possibilities. Norpak has announced a Telidon board for the Apple II computer to be available in mid-1982; IBM personal computers now include a Prestel in-house videotex system.

There is no clear boundary between existing teletext services and videotex serv-

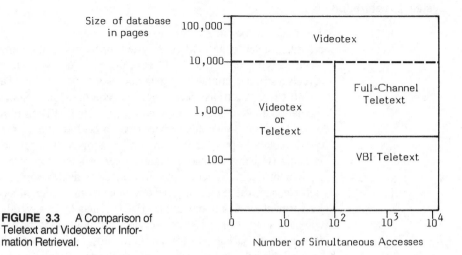

Size of database
in pages

100,000

10,000

1,000

100

Videotex

Full-Channel
Teletext

Videotex
or
Teletext

VBI Teletext

0   10 10^2 10^3 10^4

Number of Simultaneous Accesses

FIGURE 3.3 A Comparison of Teletext and Videotex for Information Retrieval.

ices. Many of the information retrieval services being provided are the same on both systems. Drawing from the European experience, it is possible to classify teletext and videotex systems in terms of the size of the database and the volume of simultaneous accesses (Figure 3.3). Teletext has a comparative advantage when there are a large number of users for a limited amount of frequently changing information, such as news bulletins.

SYSTEM STRUCTURE

In some countries, teletext and videotex have evolved as separate developments; in others, their development has been integrated. In the United Kingdom the Ceefax and Oracle teletext systems developed independently of the Prestel videotex system, but both systems share the same display technology, i.e., the integration is at the receiver end. In France and Canada, the Antiope and Telidon systems, respectively, have emerged as integrated developments, providing a common operating system for both teletext and videotex. These teletext and videotex systems are all being marketed in the United States.

The structure of teletext and videotex systems may be described in terms of a number of basic elements:

- System organization
- Database arrangements
- Communication networks
- Character sets
- User terminals
- Methods of accessing information
- Billing arrangements

We examine each of these seven system elements in turn.

System Organization

The major actors in existing teletext/videotex systems include information service providers, teletext and videotex system operators, and communications network providers. A comparison of the Ceefax/Oracle,[1] Antiope, and Telidon *teletext* systems reveals a striking similarity in general system organization (Table 3.4).

Most *videotex* trials and services outside the United States are being provided by the national telecommunication carrier, i.e., the PTTs in Europe and Japan. Finland is an exception. Videotex services in Finland are being provided by a tricompany conglomerate: a publishing company (Sanoma Oy), a private telephone company in Helsinki (Helsingen), and a manufacturer of television receivers and other electrical equipment (Nokia Electronics). In Canada, videotex services are being provided by Bell Canada, the dominant telecommunications carrier, under the jurisdiction of the Department of Communications (DOC), and by many other provincial telephone companies. In some cases the services are now being provided by cable companies (Cablecom in Saskatchewan), and Infomart, a private corporation, is a system operator on both cable and telephone. Like Prestel, Telidon was developed with government support; unlike Prestel, however, implementation has been left up to the private sector. A comparison of operational videotex systems emphasizes the dominant role of the telephone network in videotex development (Table 3.5).

[1]Ceefax/Oracle: Ceefax (BBC) and Oracle (ITV) are compatible in that signals from both systems can be received using the same decoder.

TABLE 3.4   EXISTING TELETEXT SYSTEMS: COMPARISON OF ATTRIBUTES

Characteristics	Ceefax/ Oracle	Antiope	Telidon
Information provider	Broadcasting authorities	Companies, newspapers	TV Ontario
System operator	Broadcasting authorities	Broadcasting authorities	TV Ontario
Database structure	Centralized (some regional services)	Regional	Centralized
Communications network (System operator to user)	TV broadcast (VBI)	TV broadcast (full channel and VBI)	TV broadcast, satellite and cable broadcast
Interface device	Decoder	Decoder	Decoder
Display	TV	TV	TV
Input	Numeric pad	Numeric pad	Numeric pad
Processing	One page	One to 100 pages	One page
Storage	Nil	Nil	Nil

TABLE 3.5   EXISTING VIDEOTEX SYSTEMS: COMPARISON OF ATTRIBUTES

Characteristics	Prestel	Teletel	Telidon*	Captain
Information provider	Various	Various	Various	Various
System operator	Telephone carrier (PTT)	PTT	Telephone carrier,cable company, electronic publishing company	PTT
Database structure	Centralized, replicated databases	Distributed systems with gateways to independent third party databases through public data networks	Centralized, replicated databases; local independent databases	Centralized, replicated databases
Communications network	Telephone	Telephone	Telephone cable	Telephone cable
Interface device	Modem	Modem	Modem	Modem
Display	TV or CRT	Dedicated terminal	TV or CRT terminal	TV terminal
Input	Keypad or keyboard	Keyboard	Keyboard	Keypad or keyboard
Processing	One page	One page	Multipages	One page
Storage	Nil	Nil	Character	Nil

*A number of Telidon trials are in progress; not all are identical.

A key structural element in the development of videotex services is the design of the system operator's "service center." The service center is responsible for administrative and control activities including passwords, user access, billing, central indexing, messaging, and any form of user monitoring. Separate service centers, or as the British call them, information retrieval centers, have been established in Germany, Canada, and the United Kingdom, the ultimate aim being to allow all consumers, or at least a good majority, access to videotex services with no more than a local exchange call. The French, however, are exploring the possibility of a completely decentralized system.

Database Arrangements

In current *teletext* systems, the service center provides the database and computing facilities. Service centers are located at the broadcast station, e.g., the BBC in the United Kingdom, or Sveriges Radio in Sweden, and they may integrate information

FIGURE 3.4   Database Arrangement for Prestel Videotex System.

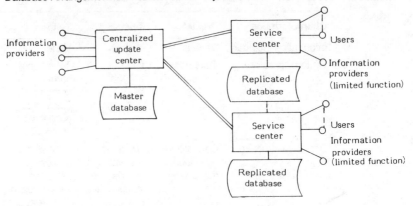

from other sources. For example, the results of the French stock exchange are continually updated and disseminated via Antiope.

In current public *videotex* systems, three different database arrangements have been used. The British Prestel system maintains a master database at a centralized update center and a replicated database at each information retrieval center (Figure 3.4).

The Prestel database structure ensures uniformity in the arrangement of data, thus simplifying access, but it has the major disadvantage of requiring large duplicated databases. Hence, as the amount of information available on Prestel expands, the amount of computer storage increases by a factor equal to the number of databases in the system. However, as a result of the establishment of a public packet-switched network in the United Kingdom in 1980, the Prestel database is being modified to allow access to remote databases, similar to the existing German Bildschirmtext structure (Figure 3.5).

FIGURE 3.5   Database Arrangement for Bildschirmtext Videotex System.

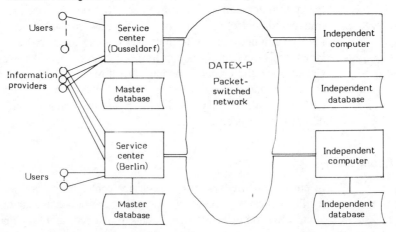

A key application of videotex in Germany is expected to be catalog shopping and reservations, much of which is already computerized by companies offering these services. The need for external databases, without centralized duplication, was seen to be of paramount importance. A major limitation of the initial Berlin/Dusseldorf trial was the absence of a single update center (as in the Prestel system). In order to maintain duplicate files in both service center databases, information providers must repeat the update process for each database.

The French Teletel videotex system extends the distributed database concept one step further. Unlike the British and German systems, Teletel has no databases provided by the service provider (the French PTT).[2] Subscribers are linked via information retrieval centers (or service centers) and the public packet-switched data network (Transpac) to remote databases (Figure 3.6). Information providers are responsible for all billing and database access procedures.

Communication Networks

All three *teletext* systems (Ceefax/Oracle, Antiope, and Telidon) use the broadcast television network to provide the communication link between the system operator and the user. Existing services in Europe use from two to four lines in the vertical blanking interval on each television channel, although the French have also experimented with regional dedicated full-channel teletext services.

Existing *videotex* systems typically involve three communications network components: information provider to database, database to service center (or system

---

[2]For the initial Velizy trial, however, the French PTT is providing a limited database at the service center.

FIGURE 3.6 Database Arrangement for Teletel Videotex System.

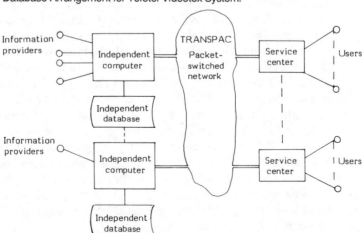

TABLE 3.6   COMMUNICATION NETWORKS IN EXISTING VIDEOTEX SYSTEMS

System (Country)	Info. Provider to Database	Remote Database to Service Center	Service Center to User
Prestel (U.K.)	Telephone	Packet	Telephone
Bildschirmtext (West Germany)	Telephone	Packet	Telephone
Teletel (France)	Telephone or packet	Packet	Telephone
Telidon (Canada)	Telephone	Packet	Telephone/cable
Captain (Japan)	Telephone	Telephone	Telephone

operator), and system operator to user. The communication networks used in existing videotex services and trials are shown in Table 3.6.

Systems employing the telephone network to link users to the service center require data modems to convert digital data into an analog form for transmission over the telephone network. Typically, these modems transmit information *to* the user at 1,200 bits/sec and *from* the user at 75 bits/sec. In many countries outside the United States, modems are supplied on a rental basis by the telephone carrier and are frequently integrated into the other terminal equipment.

Communication links between service centers and remote databases are typically provided by high-speed leased lines or packet-switched networks. In the Prestel system, information retrieval centers are linked to the centralized update center using dedicated telephone network lines operating at 4,800 bits/sec.

Character Sets

The International Organization for Standardization (ISO) and the International Telephone and Telegraph Consultative Committee (CCITT) of the International Telecommunication Union (ITU) have adopted a common 7-bit code for transmitting digital information over the public switched telephone network. The standard is known as ISO 646 or CCITT Recommendation V.3. This code provides 2^7 (128) unique combinations, which is sufficient to define the English alphabet, control characters (e.g., delete, escape, line feed), and some elementary graphics (e.g., punctuation marks). All existing videotex systems use a code based on this standard (see Appendix 1).

The 7-bit code is partitioned into two areas, the control set, which is composed of 32 codes, and the graphics set, which is composed of the remaining 96 codes. The basic graphic set (G0) is not able to satisfy the character requirements of all European, Asian, Scandinavian, and Roman languages, and therefore supplementary character sets have been defined, e.g., character set G2 is used to provide the additional symbols required in the French and German languages. Graphics set G1 is used for simple graphics displays (see Appendix 1).

User Terminals

Teletext/videotex business and home-based user terminals typically include three main components (Figure 3.7):

1 An input device, for selecting "pages" of information
2 A decoder, for translating and assembling the digital teletext/videotex signal into a video display
3 A display device for presenting the information to the user

The most common input device for all existing teletext and many videotex systems is a simple numeric keypad (Figure 3.8). In teletext systems, the keypad has local significance only, i.e., it is used to identify the number of the information page and instruct the decoder to seize and display the appropriate frame from the "magazine" of pages being broadcast. In videotex systems, numbers entered on the keypad are transmitted to the videotex database where the required page is selected. Originally, most videotex systems that involved mainly information retrieval and preformatted messages used a keypad as a basic user input device. However, 30 percent of residential Prestel users, and all business terminals, have alphanumeric keyboards. The Canadians use an alphanumeric keyboard, and in France, where a more extensive service is proposed (electronic telephone directories), specially designed user terminals incorporating a full alphanumeric keyboard are being used (Figure 3.9).

There is no single design or keyboard layout in user terminals, but a noticeable feature of some of the designs is the nonconventional arrangement of the alphabetic keys. The keys in a Teletel terminal, for example, are arranged in alphabetic order rather than in typewriter keyboard format.

The information providers in both teletext and videotex systems almost invariably use a full alphanumeric keyboard as the basic input device. In-house systems currently being developed and commercial services for information providers include a wide range of terminal facilities, e.g., bulk update, multiple page editing, black-and-white as well as color displays, and advanced editing facilities.

FIGURE 3.7   Teletext/Videotex Terminal Components.

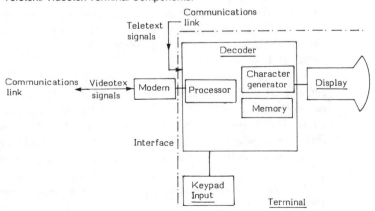

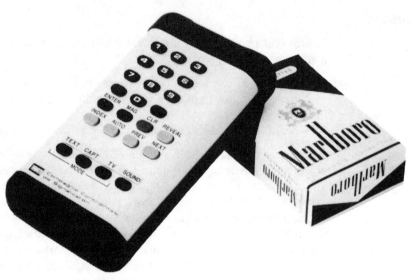

FIGURE 3.8   Teletel Keypad. (*Source: Sofratev.*)

Teletext/videotex decoders include a processor, memory, and generally a character/pattern generator.[3] Decoders may be connected externally to the TV, i.e., plugged into the RF antenna socket (set-top decoders) or wired directly into the beam circuits of the RGB guns (built-in decoders). A fundamental technical difference between Antiope and the other three systems (Prestel, Telidon, and Captain) is the method used to transmit display attributes, e.g., character color and background color. The Antiope system uses parallel picture attributes, whereas other systems use serial picture attributes. In the case of serial attributes, the attribute code is sent within the data stream and is normally displayed as a space in the appropriate background color; this problem is avoided using the parallel attributes method. However, approximately twice as much memory is required in the decoder for parallel attribute systems (16-bit storage for each display location—8 bits for the information character and 8 bits for the control character—compared with 8-bit storage for serial systems). A European standard announced by CEPT (European Conference of Postal and Telecommunications Authorities) incorporates both serial and parallel attributes within the basic standard.

A more fundamental difference between existing teletext systems is the method of signal transmission used: synchronous or asynchronous. The complexity of the decoder (and hence the cost) depends on the method selected. The British Ceefax/Oracle systems use synchronous (or fixed format) transmission, whereas the French Antiope and Canadian Telidon systems use asynchronous (or variable format) transmission. In the fixed format system, there is a fixed relation between the

---

[3]The Japanese Captain system is the exception. The character/pattern generator is located on the service provider's premises.

FIGURE 3.9 French Teletel Terminal with Full Alphanumeric Keyboard. (*Source: Matra*)

transmitted data and the position of the corresponding characters displayed on the television screen. This simplifies the construction of the decoder since the address on each data stream gives adequate information as to the nature and location to display the rest of the data in the bit stream. This is a distinct advantage when there is also considerable interference to the broadcast signal (Crowther, 1979).

In contrast, the variable format system involves the continuous transmission of a stream of data (information characters and control characters) in which there is no inherent relation to the television display. The advantage of the variable format system is its greater flexibility in coping with future changes in teletext technology (CBS, 1980).

All existing teletext/videotex terminals, with the exception of the French Teletel terminal, use either a standard color-television receiver with a set-top decoder or a modified color-television receiver with a built-in decoder. At present, the built-in decoder generally provides a better-quality text display. However, for displaying graphics, the quality achievable depends on the underlying method of picture construction. The British and French teletext/videotex systems use an *alphamosaic* approach, whereas the Canadian Telidon system uses an *alphageometric* construction. A third alternative, known as *alphaphotographic*, is being planned for all systems to provide a greater degree of refinement.

The alphamosaic approach divides each video frame into a number of blocks, namely, 24 rows of 40 blocks in European systems. Within each block an alphamosaic terminal generates an 8×10 matrix of pixels (picture elements) for alphanumerics and, because of ISO limitations on the number of bit string combinations, a 3×2 matrix for graphic elements. Thus the total number of pixels for a full-frame alphanumeric (text) display would be 76,800, and 5,760 for graphic displays (Prestel system). Clearly, a binary representation of each pixel would require a very large memory in the terminal decoder to assemble each video frame. In practical systems, the blocks provide a displayable character that is derived from the basic character code (Figure 3.10).

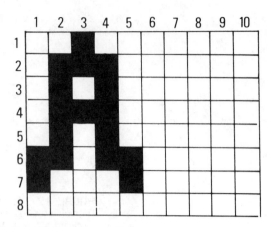

FIGURE 3.10 Character Construction from Pixels.

Development of the alphamosaic system by the French has led to the notion of down-loaded alphabets, i.e., a new alphabet (which may be graphics) is down-loaded to the user's decoder and stored for the appropriate number of frames. A specific graphic refinement is the dynamically redefinable character set (DRCS) concept, which allows higher resolution graphics by creating graphics dynamically in much the same way as the down-loaded alphabet creates nonstandard letters. In essence, before transmitting a page with a special character the operating system transmits the codes for the character to a random-access memory (RAM) in the terminal. The decoder then reads the standard characters in the read-only memory (ROM) of the terminal, and then looks for special characters in the RAM.

The Canadian Telidon system is based on an *alphageometric* approach. Unlike the alphamosaic systems, the Telidon display is compatible with any level of resolution (e.g., 525, 625, or 1,125 lines per frame). Telidon uses picture description instructions (PDIs) to define basic shapes such as a point, a line, an arc of a circle, a rectangular area, or a polygon. The PDIs, which are down-loaded and decoded in the terminal, create the shapes for each video frame between X-Y coordinates. The picture quality depends on the degree of resolution selected, e.g., 256 × 256 pixels, 512 × 512 pixels, or 1,024 × 1,024 pixels. The higher the resolution is, the more decoder memory is required, which increases the cost of the equipment. The Telidon approach is not radically different from the DRCS systems, and compatibility seems likely in future systems.

Figure 3.11*a* and *b* illustrates the different levels of graphic quality achievable with alphamosaic and alphageometric systems.

With the third method, the *alphaphotographic approach*, it is possible to achieve full-definition video (of the same quality as broadcast television pictures) using the telephone network as the communications link. With current technology and a transmission speed of 1,200 bits/sec over the public switched telephone network it takes about 60 seconds for a picture image covering approximately one-ninth of the screen to appear, e.g., using the British picture service Prestel (Smith, 1980). The

FIGURE 3.11*a* and *b* Alphageometric
and Alphamosaic Displays.

Telidon, Canada

Sofratev, France

British Telecom, United Kingdom

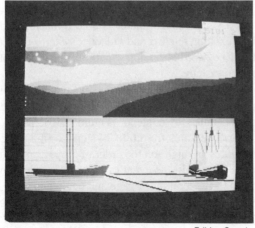

Telidon, Canada

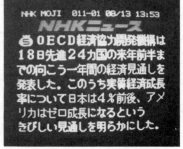

NHK, Japan

(*a*) Alphageometric   (*b*) Alphamosaic

FIGURE 3.12   Alphaphotographic Display.

British Telecom, United Kingdom

Canadian approach has involved sending the characteristics of the pictures independently of the resolution, and then recreating them in the terminal (similar to the PDI
method for text and graphics) (Figure 3.12).

Methods of Accessing Information

A fundamental design characteristic of all first-generation teletext and videotex systems is the adherence to hierarchical menu searches, i.e., tree-structured databases.[4]
The aim of the tree structure is to provide the user with a simple "menu" choice that
does not require skill with computer languages or extensive training to be understood,
and can be accessed with a simple keypad. As the computer power for such a search is
small compared with that needed for keyword searches, considerable cost savings can
be achieved since a small minicomputer is able to accommodate a large number of
access ports. Most existing teletext and videotex systems use tree-structured searches.
A certain degree of keyword accessing capability for groups of pages has been introduced in the Telidon system.

The usual procedure, once a telephone connection has been made for videotex, is
to specify a user identification number (to log-on) and then to sequentially negotiate
increasingly specific tables of contents (Figure 3.13). It has been estimated that a
"typical" Prestel user requires from 6 to 14 steps to find a specific piece of information and that error rates are high (on the order of 50 percent of users fail to get through
four nodes without error!). In the event of an error it is usually necessary in existing
systems to return to the general index page and begin again.

In all cases the option exists to go straight to a particular page by entering the page
number, for example 4163289. In the case of the French electronic directory a particular name may be typed in or a joint keyword-tree search may be undertaken using
geographic locations, names, or businesses. Comprehensive paper-based support

---

[4]For a detailed discussion of tree structures, see Thomas (1980).

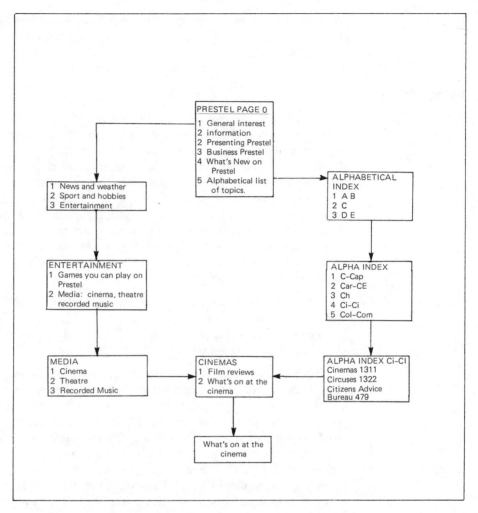

FIGURE 3.13   A Tree Structure Method of Accessing Information: Prestel Example. (*Based on Prestel Users Guide, April 1979.*)

directories have emerged in both the United Kingdom and Germany to help videotex users directly access the pages of interest. For teletext systems, the limited number of frames makes access somewhat easier and the paper directory simpler.

The Swedish system Datavision, the Finnish system Telset, and Telidon have included limited "high-level" keyword searching where the keywords are available within certain databases and are unique to particular databases, for example, TWA may use a different set than Amtrak or Pan Am.

In closed-user groups and in-house systems, the searching mechanisms have sometimes been streamlined by the information provider. Fintel, a business/finance database on Prestel, for example, offers a service whereby a corporation name, industry code, product, or other description may be entered, and in return the user receives references to all recent *Financial Times* material on that subject. This has its parallel in full-text searching.

Billing Arrangements

The billing arrangements differ markedly between teletext and videotex. In current systems the user is charged for the decoder in the TV set, either in the purchase price of a set or in the leasing arrangements. For teletext the cost of the service is absorbed by the system operator. In the French system, a smart card has been developed to facilitate pay teletext by allowing selective access to the system for a price. In the United Kingdom, the BBC's teletext service, Ceefax, is subsumed under the television license fee, and the ITV service, Oracle, is paid for by advertisers.

For videotex there are numerous billing arrangements. First, in Europe the modem may be leased by the user and information provider or included in the purchase price of the videotex terminal. Second, users are typically billed for three services each time they access the system:

A system connect cost, which may vary according to the time of day, and thus reflect peak loading

A cost for the specific page of information or a flat fee for the information service

A local zone or long-distance telephone call charge

This billing structure has been developed for basic information retrieval and does not reflect enhanced and more general videotex services such as data processing.

The Swedes are proposing to base charges on cost per access time rather than cost per page viewed. (This issue has already caused some concern in the United States, where it has been observed that high-speed transmission rates, even 9,600 bits/sec, with intelligent terminals can reduce access time to a few seconds.) In Switzerland, for example, there is some concern as to the feasibility of an external database network with a centralized PTT billing service unless there is monitoring of every access to every external database, something neither the PTT nor information providers want to occur.

In this area the experience from pay services in the United States would seem to be as relevant as the European experiences.

SOME POSSIBLE PUBLIC POLICY CONCERNS FOR THE UNITED STATES

The teletext/videotex public policy debate—both nationally and internationally—has already begun. It is the all-pervasiveness of the technology that has aroused such interest in its potential impact.

As part of our data collection, a number of European organizations were visited to gain insights into the structure of the teletext/videotex networks and to identify public policy issues that may have relevance to the U.S. debate. Seven countries were visited—Denmark, Finland, France, Sweden, Switzerland, United Kingdom, and

West Germany—and interviews were conducted with members of the PTTs, the broadcasting authorities, research laboratories, organizations testing teletext/videotex, and various commissions on mass media and new technology.

Although some of the European public policy concerns are local and thus country specific, a number of themes emerged that may have bearing on the public policy debate in the United States. The following issues and approaches to resolving issues, while not exhaustive, are indicative of the public policy concerns in the European videotex trials and operations. The European country associated with each issue is the country in which the issue or policy was given most prominence during discussions. It is not meant to suggest that it is the only country in which the issue is of importance or that it is the country in which the issue has most importance.

Definition of Teletext/Videotex

1 There must be a search for the relevant code of behaviors. Are the laws of the press on responsibility and reliability valid? Is videotex in fact text transmission? Is teletext/videotex print or broadcast? (Denmark, Sweden, U.K.).

2 If videotex is defined as a closed-user group, it is not subject to the broadcasting regulation. Can all videotex be defined in this way? (Sweden).

3 Who owns the vertical blanking interval—the PTT, which owns the network, or Sveriges Radio, which broadcasts the programs? (Sweden).

Broad Societal Impacts

1 New technology and an increasing trend toward monopolistic ownership of the media could result if unchecked in an even more centralized and powerful information elite within the society (Denmark).

2 The combined effects of teletext and videotex create the environment for home-based work, increased productivity, and energy savings (France).

3 Electronic transactions change the consumer decision-making process—the way in which products are bought and sold (France).

4 Prestel changes the traditional union demarcation debate in the United Kingdom. Abolition of the need for double-key strokes (journalists and compositors) has widespread professional implications (U.K.).

5 Videotex "demassifies" information—i.e., it allows information to be individually packaged (France).

6 It affects thinking patterns—what are the implications of learning and decision making using information presented nonlinearly and organized on screens of 75 to 100 words? There are similar problems for preparation of material for display (prose, style, grammar, sentence construction, use of parenthesis) (U.K.).

7 Videotex pages have no room for nuances. This could affect the depth to which material is collected (U.K.).

8 Videotex will not obey the "rules" of a mass media. As it will be individualistic, there will be pressure by community groups to have their own information services. "Who gets what?" and "What are the societal consequences?" are two key issues (Denmark).

9 As videotex is interactive it may not be suitable for viewers to use modified 26-inch TV screens. Concerns have been expressed as to the physical effects on the eyes of continual close-up monitoring of text on CRT screens (Denmark).

10 The current generation of videotex systems requires the development of cognitive skills such as typing, and logical skills such as hierarchical tree searching. The need for these skills is a major impediment to widespread adoption (U.K.).

11 There is no payment for BBC teletext services in the U.K. as it is funded entirely by license fees (U.K.).

Content-Carriage

1 Printed-matter laws and regulations apply to videotex services. There is no messaging service available for general users, although presumably in-house and even closed-user groups will be able to offer internal electronic mail services (Finland, West Germany, Sweden, Denmark).

2 PTT is to remain the sole carrier and solely a carrier. There will be no role for PTT as a value-added carrier. Information providers are responsible for content (France).

3 There are different regulations for publishing, broadcasting, telephone, and databases. Strict laws have been passed forbidding integration and then interrogation of databases—a violation of civil liberties. The relevance of these laws for videotex will emerge during trials (France).

4 The relation of videotex to traditional publishing media is to be resolved. Newspapers are the major information providers now. Does freedom of the press (editorial freedom, control, and content issues) apply? (U.K.).

5 The greater the number of channels available, the more liberal the uses of the system. Scarcity of channels of communication implies a need for regulation, intervention, quality monitoring (U.K.).

6 Deutsche Bundespost will operate all billing arrangements for information providers, including access to external databases. User statistics are maintained by Bundespost, but they are all confidential. Only aggregate information is divulged (West Germany).

7 But Deutsche Bundespost is not allowed to offer data-processing services. Storage of information is allowable provided that the messages are not changed (West Germany).

Privacy and Security

1 Telset bills all users and information providers but does not record what pages were accessed or the number of pages accessed. Each page has a specified cost. This cost, along with modem rental charges and connect charges, is billed monthly by Telset (Finland).

2 A microprocessor memory card is being tested to guarantee individual security (France).

3 Collection of money and invoicing of services is the sole responsibility of the information provider during trials (France).

4 Individual pages accessed by users are known by British Telecom but are not available for internal or external use. Aggregate information only is provided (U.K.).

5 The freedom of speech issue carries over to reading material. Is the state, or anyone else, entitled to know what I view on a videotex system? (U.K.).

6 Privacy law states that it is illegal to store information about people or organizations without special approval. This also applies to storing names of users for message files (Sweden).

Consumer Protection and Equity of Access

1 All goods displayed in stores must show a price and certain details on labels. This also applies to videotex. All videotex frames must show a price, and users are allowed to specify dollar limits on the value of pages they want to look at. Each family member will be given a number and an upper limit for spending (Finland).

2 Videotex offers protection against unsolicited mail. A mailbox can be edited without the contents being displayed. (Finland).

3 Laws are in place to prevent information about individuals being made public. Further laws prevent public or private registries and databases on individuals from merging. Videotex may threaten this aspect of individual security (Denmark).

4 An electronic purchase requires a definite action by the purchaser. No default options (U.K., Finland).

5 Market researchers may want to find out detailed consumer behavior patterns. Unless strict codes are enforced it is going to be possible to monitor national, regional, and local usage patterns (U.K.).

6 Mail orders via Bildschirmtext can be cancelled even after the item has been received (West Germany).

7 PTT subsidizes the cost of information services to give minority groups access to videotex (France).

8 There are explicit policy objectives to ensure that the information-rich are not subsidized. The broad Swedish policy is to guarantee access to public information, e.g., rights and obligations, basic statistics, and energy conservation (Sweden).

Copyright

1 Newspapers are not allowed to print weekly TV schedules, only daily TV schedules, as the BBC holds copyright for all of its program scheduling. Weekly programming guides are the domain of the BBC. In Germany radio and TV guides are produced by individual broadcasters up to six weeks in advance. This puts pressure on the networks to maintain a program sequence (U.K., Germany).

2 Authors of talks on TV and radio retain the copyright. Any further "creative" use of material could require original owners to be repaid (U.K.).

3 Does the copyright for information published in one medium carry over to another (especially an electronic medium)? What are the implications of taking hard copy from an electronic medium? (U.K.).

4 With down-loading it is technically possible to copy and modify frames (U.K.).

National Standards

1 There are no national laws to ensure uniform videotex standards (Finland).

2 Sockets are being included in TV set manufacture to incorporate video tape recorder, videotex, teletext, etc., even though these services not yet available (France).

3 A common national standard is being developed for telex, teletex, teletext, videotex, and facsimile (home and work) (France).

4 Videotex is compatible with teletext, but there are technical problems if compatibility is to be maintained with teletex and telex (U.K.).

5 Who is responsible for the quality of the database—the information providers or the system provider? This issue is especially ticklish in international/transborder data flows (U.K.).

International Communications

1 The adoption of a common European videotex standard (CEPT) leaves a major issue. Should fixed format transmission or variable format transmission be used for teletext? (France, U.K.).

2 In the case of transborder flows of information:

With whom does the responsibility lie?

What are the international copyright implications?

What happens with messages, as it is forbidden to use multiplex lines to send data in some countries?

Where is the line between corporate information and advertising (which is illegal on TV in some countries)?

What are the consequences of defining videotex as broadcast, print, or telecommunications? (it makes a major difference in some European countries) (Denmark, France, West Germany, U.K., Sweden).

Advertising-Liability

1 There are restrictions on TV and radio advertising of cigarettes and alcohol. These are believed to carry over to videotex. Advertising is carefully monitored and all the regulations for newspapers are believed to apply, e.g., misleading information, such as nonstatistically valid comparisons of products, are forbidden and the "negative option" (if you don't reply then. . .) is also prohibited (Finland).

2 Advertisements have to be labeled, although they can be integrated with information services (France, West Germany).

3 Types of advertisements may be restricted (U.K., Denmark, Sweden).

4 It will be illegal to send unsolicited circulars over telephone lines (U.K.).

5 The Consumer Board is concerned with rules for content, i.e., the quality, reliability, and variety of information on the system. These will be enforced if information providers are to be allowed to advertise (Sweden).

6 The information providers in most countries have prepared a code of practice relating to rights of parties, malpractice, contractual responsibilities of involved parties, advertising, and content (U.K., Switzerland).

Content-Impartiality

1 Videotex, like newspapers, need not be impartial (Finland).

2 The charter of the BBC requires balance and objectivity for services including teletext. This doesn't carry over to antisocial organizations (e.g., terrorists, anti-Semitic groups) (U.K.). A similar charter exists for Sveriges Radio, which is responsible for content (Sweden).

CONCLUSION

This brief overview of the teletext and videotex developments outside the United States provides a backdrop for analyzing the systems within America. By focusing on generic classes or types of applications, it is clear that teletext and videotex have the potential to offer more than information retrieval services. This immediately broadens the scope to include transactions, electronic messaging, data processing, home management—services that are still very much in their infancy in Europe.

The examination of the structure of teletext and videotex systems provides guidelines for building a framework to analyze the trials in place in the United States and to forecast the technological components of teletext and videotex systems. It also places the various technical debates on display and transmission standards within a context that can be adapted to the United States.

Finally, the preliminary work on policy issues suggests some directions that the United States policy debate may take. It also highlights, by implication, some of the solutions being proposed in Europe. In that sense it provides an early warning for issues that are likely to crystallize once the trials and services have become operational in the United States.

CHAPTER 4

CURRENT STATE OF TELETEXT AND VIDEOTEX IN THE UNITED STATES

An increasing number of teletext and videotex trials and services have been announced or are operating in the United States. These trials are important in that their outcomes will influence the future directions of teletext and videotex in this country. For selected systems now in commercial use or under trial, we have investigated the following:

- key participants in the trials and the functional roles and relationships among them
- the technological characteristics of the system, from the viewpoint of the users
- the specific services included in the trials

Analysis of this information provides an initial framework from which to draw inferences about likely development of teletext/videotex in the United States and a context in which to identify a number of potential public policy issues.

CURRENT TRIALS

For the purpose of our analysis, we examined in detail more than two dozen teletext and videotex trials that illustrate the wide range of actors and services involved in the development of this new technology.

Teletext

The teletext trials included in our sample are as follows:

1 *KSL.* The first United States broadcast teletext trial was launched in 1978 in Salt Lake City by Bonneville International. A two-way system using a push-button phone for upstream requests was tested as part of the trial.

2 *NCI/closed captioning*. This is a service for hearing-impaired viewers in which subtitles are added to regular television programs. The service can be accessed by purchasing an add-on freestanding adapter unit for any TV set or a 19-inch color-TV set equipped with a special decoder. Initially developed by the Public Broadcasting Service, closed captioning is now carried on selected ABC programs as well as PBS. Textual captioning information is transmitted on line 21 of the vertical blanking interval and does not interfere with regular TV programs. The system is passive, in that the viewer exercises no control over the displayed text. As of late 1981, approximately 40,000 decoders had been sold at $250 through Sears Roebuck stores.

3 *WFLD*. Beginning in April 1981, the Chicago television station WFLD began broadcasting a 100-page teletext magazine called "Keyfax" produced by Field Electronic Publishing Inc. (Figure 4.1). In the fall of 1981, WFLD began an additional service called "Nite Owl," in which selected Keyfax pages are displayed sequentially over the entire channel during early morning hours. In early 1982 the service was taken over by a consortium of Field, Honeywell and Centel (Keycom) for distribution on both cable and broadcast TV.

4 *KNXT/KCET/KNBC*. During 1981, these three Los Angeles television stations began separate but coordinated field tests using Antiope software and hardware. The test was initiated by CBS through its owned and operated affiliate KNXT, along with public TV station KCET, then was joined by NBC affiliate KNBC. The two commercial stations' services included advertising as well as news and other features (Figure 4.2).

5 *WETA*. The most elaborate publicly funded teletext trial began in Washington, D.C., in 1981. Sponsored by the Corporation for Public Broadcasting, the National Science Foundation, the National Telecommunications and Information Administration, and the United States Department of Education and organized by the Alternate Media Center at New York University, the test involved placing Telidon decoders in 40 homes and 10 public places in the Washington area. Several newspapers and a number of public institutions served as information providers.

6 *KPIX*. This field trial was launched in early 1982 in San Francisco by the CBS affiliate owned and operated by Westinghouse Broadcasting. Like the Los Angeles stations, KPIX used French-made Antiope equipment for its teletext service, which it called "Direct Vision."

FIGURE 4.1 Producing a Teletext Magazine, WFLD. (*Source: Bill Horne, Field Electronic Publishing.*)

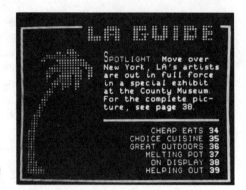

FIGURE 4.2   A KCET Teletext Page.

7 *Cabletext.* This service, offered by Southern Satellite Systems, uses the vertical blanking interval of Atlanta "Superstation" Channel 17 (WTBS) to carry the Reuters News Service. Decoders are rented to cable operators who, in turn, select the information pages they wish to display on their cable news channels. The viewer at home has no selection capability.

8 *Time Inc.* Using a full satellite transponder to provide national distribution, Time Inc. created a number of "electronic magazines" consisting of 4,000 to 5,000 pages. In its initial test in 1982, the service will be delivered over a full cable channel to subscribers on Southwestern Cable TV, a Time-owned system in San Diego.

Videotex

An even larger number of videotex projects have been launched to date. While most of them are simply field tests, several are full-fledged commercial services:

1 *QUBE.* This service, which Warner Amex Cable Communications has offered on its Columbus, Ohio, system since 1977, represents the first two-way cable service in the United States. QUBE is more properly described, however, as a "polling-response" system than true videotex. In conjunction with live video programming, the system permits viewers to register answers to questions, participate in game shows, and order merchandise. It also includes pay-per-view programming and a home security service. Although QUBE does provide text-on-demand, a small trial involving access to The Source (see below) through home computer terminals was run over the system. There were approximately 50,000 QUBE subscribers in Columbus at the end of 1981, and QUBE systems were scheduled to be included in new systems being built by Warner-Amex.

2 *The Source.* The first commercially available videotexlike service in the United States, The Source is now owned by *Readers' Digest.* News, financial data, games, electronic mail, and other services are offered through a timesharing system accessible nationwide to home and personal computer users through the Telenet and Tymnet packet-switched networks.

3 *CompuServe*. This is also a computer timesharing service for personal computer owners. CompuServe is owned by H&R Block and offers news, games, and quizzes, want ads, and other services. CompuServe is the system operator for a group of 13 United States newspapers that are testing "electronic newspaper" delivery in an experiment coordinated by the Associated Press. As of late 1981, CompuServe had approximately 17,000 subscribers nationwide.

4 *Comp-U-Star*. This discount buying service (unrelated to CompuServe) allows a home computer user to access a database and search for the product and price desired. Products can be ordered through the system and charged to a credit card. The service is offered by Comp-U-Card.

5 *BISON*. This test, conducted by A. H. Belo, involved home terminals provided by Texas Instruments connected to a two-way cable system in a Dallas suburb. Information has come from the Dallas Morning News, the Associated Press, and Dow Jones, and the service costs $40 per month.

6 *OCLC*. The On-Line Computer Library Center (OCLC) and Banc One of Columbus, Ohio, jointly organized this test. It involved providing an electronic library catalog, a video encyclopedia, a community calendar, educational programming, and home banking to 100 randomly selected households over a three-month period in early 1981.

7 *Green Thumb*. Under sponsorship of the United States Department of Agriculture, farmers in two Kentucky counties received information on weather, agriculture, commodities markets, and related subjects. A special terminal, the "Green Thumb Box," received selected pages of information in bursts over a phone line, then stored it for retrieval by the farmer.

8 *Dow Jones*. Beginning in 1980, Dow Jones has offered a news retrieval service over telephone lines to personal computer users on a subscription basis. The service provides up-to-the-minute content from the organization's domestic and international news services, as well as from Dow Jones's two major newspapers, *The Wall Street Journal* and *Barron's*. It also provides price quotations on 6,000 stocks and other securities listed on four major U.S. exchanges. Dow Jones has entered into an agreement with Radio Shack so that owners of TRS-80 computers can purchase a software package to access the News Retrieval Service. In late 1981, Dow Jones also began offering a one-way noninteractive news service to cable systems distributed in the vertical blanking interval of one satellite transponder.

9 *EIS*. AT&T's first test of the feasibility of an electronic information service (EIS) took place in Albany, New York, in 1979. The test involved 75 households that were provided with terminals to access white- and yellow-pages information and public service announcements. A follow-up test of an enhanced directory service planned for Austin, Texas, was cancelled by AT&T after opposition developed from newspaper publishers.

10 *Viewtron*. This test, jointly sponsored by Knight-Ridder Newspapers and AT&T, began in the second half of 1980 in Coral Gables, Florida. It was designed to test consumer reaction to an electronic home information medium and linked modified TV sets in homes to a central computer via telephone lines. The services offered include news, shopping, travel, banking services, games, learning aids, consumer advice, etc. The initial phase, involving some 700 sites, has been completed. A larger market test, involving as many as 5,000 users, is scheduled for 1983.

11 *INDAX*. This interactive service is being developed by Cox Cable Communications for use on its systems around the country. Two-way cable is being used to supply interactive shopping, banking, educational, and information services (Figure 4.3). After testing on the Mission Cable System in San Diego, INDAX will be available on other Cox systems, including those in Omaha and New Orleans.

12 *Times Mirror*. In 1982, this multimedia company began a test involving Canadian Telidon equipment and some 350 sites in the Los Angeles area. A portion of the homes in the test were linked via telephone lines, while other homes accessed the service via two-way cable in order to compare the advantages and disadvantages of the two transmission media. As the trial got under way, Times Mirror announced the creation of a joint venture with Infomart, a Canadian software marketing company, to distribute Telidon systems in the United States.

13 *CBS/AT&T*. During 1982, CBS undertook its first test of an interactive videotex service in cooperation with AT&T. The initial test, which involved 200 families in Ridgewood, New Jersey, allows people to obtain information from CBS-owned publications (e.g., *Woman's Day*, *Mechanix Illustrated*), shop, bank, and make travel reservations. The test employs both computer terminals and modified TV sets.

14 *Continental Telephone*. This large independent telephone company began a videotex trial in 1982 in Manassas, Virginia, using French-made Antiope equipment. The trial involves 100 terminals and a total of 7,000 frames of information.

15 *Express Information*. This test of bank-at-home services was conducted in Knoxville, Tennessee, in 1980–81. A joint venture of United American Bank, CompuServe, and Radio Shack, this service gave users access to the CompuServe database and to a variety of bank services, including reviewing accounts and bill paying.

There have also been a number of other trials of home-banking services. Among the banks that have conducted or participated in such tests are Security Pacific, Bank of America, and First Interstate Bank, Los Angeles; Citibank, Chemical Bank, and

FIGURE 4.3   Banking Service on Indax.

Chase Manhattan Bank, New York; and First Bank System, Minneapolis. These trials are reviewed in the discussion on videotex transaction services in Chapter 13.

ANALYSIS OF TRIALS

The trials and services emerging in the United States reflect a variety of institutional arrangements and technological configurations. From a functional standpoint, there are four broad roles that are found in all trials:

1 *Information is assembled.*

Teletext/videotex systems rely on information providers to collect and assemble information into a usable or salable form. We use the term *information service provider* to refer to the organizations responsible for the following tasks: collection of raw or transformed data, processing or reprocessing of the data, reformatting into frame size, and editing of changes. In some videotex network configurations, information service providers may also supply their own computer database, which can be linked to the videotex system. In this case, the information provider is also responsible for storing and indexing the data as well as possibly billing users for the rights to access the product. It is important to note that the term information service provider is used to encompass the provision of all of the information-related services on teletext and videotex—not just pure information such as a news service.

2 *Information is managed.*

Information management functions include the storage and indexing of data as well as user billing. As noted above, the role of the information provider may include these functions for the data he or she provides. However, there are likely to be many information providers who do not have their own computer storage facilities. The management function for these organizations must then be provided by the teletext/videotex *system operator*.

3 *Information is transmitted.*

The transmission medium and the supporting facilities differ both between videotex and teletext systems and among the various market trials. In fact, more than one carrier may be involved in a single teletext/videotex system. Communication links are required to connect users to their local teletext/videotex service center (e.g., a local telephone network or a cable television distribution system) as well as connect remote databases to the local system (e.g., packet-switching networks or satellite links). The term *communications network provider* will be used to describe the organization or organizations responsible for the provision of communications links in the teletext/videotex systems.

4 *Information is received (or at least receivable).*

As with any public or private good, there must be a demand for the good; in this case, someone must receive the information or information services. The *user*, or consumer, may be a public or private corporation, a private individual, or a special group and may either pay for the service according to usual market-derived principles, receive the information without direct cost (for example, public goods or advertisements), or be subsidized in the cost of the service.

TABLE 4.1   ROLES AND TASKS IN TELETEXT/VIDEOTEX SYSTEMS

Principal roles	Prime task
Information service provider	Data collection and assembly
System operator	Management
Communications network provider	Transmission
User	Consumption

In actual teletext/videotex systems the boundaries may not always be so distinct as Table 4.1 suggests. We have already noted that one organization may be involved in more than one activity. In addition, the information service provider role can have a number of variations. First, some organizations may offer an "umbrella" service, e.g., access to storage, editing, and even transmission for those wishing simply to provide a limited information service—subinformation providers. These "packages" of information services may be formatted for distribution by system operators on various teletext and videotex systems. However, the four broad roles—collecting, managing, transmitting, and receiving information—are always required and, as such, may be seen to constitute the core or basic subsystems. The general relation between these subsystems is illustrated in Figure 4.4.

In Europe, as described in Chapter 3, there has tended to be a single arrangement for these systems in each country. This is because these systems have been developed with substantial government support and direction. Typically, private companies are allowed to act only as information providers, while the government-run PTT serves as both system operator and communications network provider. In this country, by contrast, there has been virtually no central planning or development. As a result, emergence of the technology in the U.S. has been characterized by great diversity in terms of the organizations that have played the key roles (Tables 4.2*a* and 4.2*b*).

It is evident from this summary that no clear pattern has yet emerged in this country regarding the distribution of functions among the parties interested in providing teletext/videotex services. In some instances, a single organization plays all three roles of information provider, system operator, and communications network provider. In other trials, different groups play each of these roles. In the Viewtron trial,

FIGURE 4.4   Teletext/Videotex System Structure.

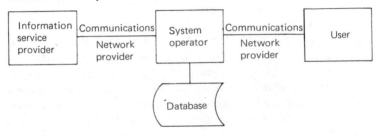

TABLE 4.2(a)   U.S. TELETEXT TRIALS AND SERVICES: THE PARTICIPANTS

Teletext Trial	Information service provider	System operator	Communication network provider	Target user group
KSL	TV station	TV station	TV station	TV viewers
NCI captioning	Nat'l Captioning Institute	Nat'l Caption-ing Institute	TV station	Hearing-impaired
WFLD	Publisher	Private corp.	TV station	TV viewers
KNXT/ KCET/ KNBC	TV stations, others	TV network/ public TV station	TV stations	TV viewers
WETA	Various	Research organization	TV station	TV viewers
KPIX	TV station	TV station	TV station	TV viewers
Cabletext	Publisher	Satellite carrier	Satellite carriers, cable company	Cable subscribers
Time Inc.	Publisher	Publisher	Satellite carriers, cable company	Cable subscribers

for example, AT&T provided only the communications network, while Knight-Ridder acted as system operator and information provider (along with other information service providers). In the CBS trial in New Jersey, AT&T will act as both network provider and system operator, with CBS acting only as information provider. In its EIS trial in Albany, AT&T played all three roles. While the diversity of arrangements is greatest in the area of videotex, even the teletext trials reflect a number of different functional relations.

Whether a single dominant pattern *will* emerge in the United States is also not apparent. It is possible that the technology and the market will support several different kinds of arrangements. How the teletext/videotex market is structured is, in part, a public policy concern. The policy implications of this issue are considered in Chapter 11.

TECHNOLOGICAL ATTRIBUTES

The second aspect of the trials concerns the technology itself. The four fundamental roles provide different viewpoints from which to consider this technology. The user is concerned with system characteristics such as how information is received, how the system displays information, how selections are made, and whether there are local storage or processing capabilities. The system operator is concerned with characteristics such as how information is received from information providers, how it is stored and accessed for transmission, and how users are charged. The communications net-

TABLE 4.2(b) U.S. VIDEOTEX TRIALS AND SERVICES: THE PARTICIPANTS

Videotex Trial	Information service provider	System operator	Communication network provider	Target user group
QUBE	Various	Cable co.	Cable co.	Cable subscribers
The Source	Various	Private corp.	Telco, Packetnets	Home computer users
CompuServe	Publishers, others	Private corp.	Telco, Packetnets	Home computer users
Comp-U-Star	Private corp.	Private corp.	Telco	Consumers
BISON	Publishers	Publisher	Telco	Residential
OCLC	Library, bank, others	Private corp.	Telco	Residential
Green Thumb	Various	University	Telco	Farmers
Dow Jones	Publisher	Publisher	Telco, satellite carrier, cable company	Investors, businesspeople
EIS	Telco	Telco	Telco	Residential
Viewtron	Various	Publisher	Telco	Residential
INDAX	Various	Cable co.	Cable co.	Cable subscribers
Times Mirror	Various	Publisher	Telco, cable co.	Residential
CBS/AT&T	Publisher	Telco	Telco	Residential
Continental Telephone	Various	Telco	Telco	Residential
Express Information	Bank, others	Bank	Telco	Bank customers

work provider is concerned with alternative network structures for delivering the information, while the information provider sees the system in terms of its data management capabilities: how information is input, edited, updated, and transmitted to the system operator. It is around these four sets of characteristics that our framework for describing teletext/videotex systems is organized, with particular emphasis on the user.

Each system characteristic is described in terms of a number of performance *attributes*. For example, when describing a transmission interface from a system operator to a user, the attributes to be considered include speed of transmission, physical device used, cost of interface device, cost of communications link, ease of access, and whether the link is one-way or two-way. There are a number of *technological options* available to provide each characteristic, and each option contains some or all of the attributes to a greater or lesser degree. For example, a TV set, which is one option for displaying information, does not offer the user hard-copy output and can have other resolution and display limitations (e.g., color versus black-and-white).

A variety of technological options can be used in designing a teletext/videotex system, and the choices made determine the ultimate quality of the performance attributes of the system. Table 4.3 gives a detailed breakdown of system characteristics and pairs each set of characteristics with the performance attributes and the technological options that affect it from the viewpoint of the user. As illustrated in the table the number of technological options for the components of teletext/videotex systems is very large. Any attempt to rigorously consider all combinations of system components would not only be very confusing but also of doubtful value. An alternative approach, which we have adopted, is to classify the various U.S. trials and services in terms of the technological components actually being used. This provides a concise summary of the current state of teletext/videotex developments.

Consider first a classification of trials and services in terms of the communications network used to deliver information to the end user. Three primary networks have

TABLE 4.3 SYSTEM CHARACTERISTICS, ATTRIBUTES, AND
TECHNOLOGICAL OPTIONS FROM A USER VIEWPOINT

System characteristic	Interface to communications network provider	Display	Input	User processing capability	User storage
Attributes	Transmission speed 1 or 2 way Ease of access Cost	Text format Color Graphics Speech Hard copy Video-frame Motion Cost	Numeric Alphabetic Positional Sensory Cost	User-programmable Built-in programs Cost	Size/capacity Access time Volatile/nonvolatile Read only/read-write Cost
Technological Options	Phone modem Decoder for broadcast TV Decoder for cable TV Decoder for FM subcarrier Decoder for multipoint distribution system Satellite dish for direct broadcast satellite	TV set TV monitor Computer terminal Speech synthesizer Printer Plotter	Keypad Keyboard Joystick Light pen Touch-sensitive display Graphics tablet Physical sensors	"Dumb" decoder Preprogrammed decoder Down-loadable decoder User-programmable decoder User-programmable host Preprogrammed host	Floppy disc Audio cassette Video disc Video cassette Hard disc Random access memory (RAM) Read only memory (ROM)

TABLE 4.4 U.S. TELETEXT/VIDEOTEX TRIALS AND SERVICES:
COMMUNICATION NETWORKS

Broadcast TV	Telephone	Cable TV system
KSL	The Source	Cabletext
NCI captioning	CompuServe	Time Inc.
WFLD	Comp-U-Star	QUBE
KNXT/KCET/KNBC	BISON	INDAX
WETA	OCLC	Times Mirror
KPIX	Green Thumb	Dow Jones
	Dow Jones	
	EIS	
	Viewtron	
	Times Mirror	
	CBS	
	Continental Telephone	
	Express Information	

been used to date (Table 4.4). In addition to the technical differences among them, these three networks have developed under quite different regulatory schemes. This raises the possibility that the same service to the same users might be regulated differently according to the communications network chosen.

In terms of the effective transmission speed[1] the various approaches are as follows:

Broadcast TV: 3,840 char/sec[2] (5.69 Mbits/sec for two lines of vertical blanking interval).

Telephone: Usually 30 char/sec. (AT&T trials provide 120 char/sec transmission. The Source provides an optional 120 char/sec service at about twice the normal cost.)

Cable TV: 1,320 char/sec per channel in teletext mode (1.7 Mbits/sec for 2 lines of vertical blanking interval). 173,000 char/sec in full data channel mode.

From this comparison we see that the telephone-based system, while the most widely available, is the slowest at delivering information to the user. Thus the provider of a system faces significant trade-offs between speed and availability in selecting a transmission medium.

Categorizing some of the current systems, this time by display format, produces the breakdown shown in Table 4.5.

The most notable feature of the breakdown in Table 4.5 is the lack of agreement. Certainly no single display format has yet emerged as standard in the United States.

---

[1] By "effective transmission speed" we mean the sustained rate at which the user receives characters. A broadcast teletext service might transmit at a very high rate but only during two lines of the vertical blanking interval. The effective rate would be number of characters per line × number of lines used per second.

[2] Typically, each character is represented by 10 bits, including a parity check.

TABLE 4.5   U.S. TELETEXT/VIDEOTEX TRIALS AND SERVICES: DISPLAY FORMATS
(number of columns × number of rows)

32 × 16	32 × 15	32 × 20	40 × 20	40 × 24	80 × 24
CompuServe*	NCI	KSL	BISON	WFLD	The Source
Green Thumb			Viewtron		
OCLC*			KMOX		
INDAX					
QUBE					
Express Information					

Note: If a paper printing terminal is used, the number of rows becomes irrelevant.

The tendency toward narrow (32 column) formats reflects the belief that the average TV receiver cannot cope with wider lines due to the bandwidth limitation or overscan. However, this reduces the utility of a page of information: a 32 × 16 page contains only 64 percent as many characters as a 40 × 20 page.

Another way of categorizing the systems is by display device (Table 4.6). The debate over display format extends to the display device itself. It is not yet clear whether the American consumer would prefer to receive text-based services on the TV set used for entertainment viewing or on a separate device.

A final breakdown of interest concerns the user's input device—keypad, keyboard, or both (Table 4.7). What is reflected in this table is an uncertainty about access techniques and about desirable services. For menu-driven access, a simple keypad is sufficient; an alphanumeric keyboard allows more flexible, but possibly more compli-

TABLE 4.6   U.S. TELETEXT/VIDEOTEX TRIALS
AND SERVICES: DISPLAY DEVICES

Home TV set	Computer terminal or home computer*
All teletext trials	The Source
Viewtron	CompuServe
INDAX	Comp-U-Star
Green Thumb	Dow Jones
QUBE	Express Information
OCLC	
Times Mirror	

Note: Home computers may, of course, connect to the home TV set.

TABLE 4.7 U.S. TELETEXT/VIDEOTEX TRIALS AND SERVICES: INPUT DEVICES

Numeric keypad	Alphanumeric keyboard	Both keypad and keyboard
All teletext	The Source	Viewtron
INDAX	EIS	
QUBE*	CompuServe	
Green Thumb	Express Information	
WETA		
OCLC*		
Times Mirror		

*Note: More extensive than a standard keypad but not a full keyboard.

cated, types of searching. Those who believe that user-to-user messaging (or other types of user input) may play a key role in the acceptance of videotex are, of course, committed to the full keyboard.

An analysis of the early developments of teletext and videotex in the United States points out an obvious sorting-out process. Because so much is unknown, and because there are so many potential actors in the evolution of this technology, the early years are bound to appear like a disparate collection of tests and applications. Clearly, there is little, if any, compatibility in the systems and trials proposed thus far. The various teletext and videotex systems not only are incompatible with systems within their own generic class but also are incompatible from one class to another. It is not clear at this point, for example, if a decoder designed for a teletext test will have any use in a videotex application. Standards setting, whether by industry or government, has yet to penetrate many dimensions of this technology.

APPLICATIONS

The most important aspect of these systems, at least from the user's viewpoint, is the services that they provide. As is often the case in the introduction of a new technology, initial applications tend to be derived from established uses of other technologies or, in this case, of other media. It is not surprising that many early applications of teletext and videotex are derivatives of print and broadcast media.

This section begins by examining current applications and develops a broad framework for considering both existing and future applications.

Table 4.8 summarizes the applications being tested in U.S. teletext and videotex trials as of early 1982 (including tests that have been completed). Details of these applications are as follows:

News, Weather, Sports, and the Electronic Newspaper Almost every trial offers some news, weather, and sports information. Newspapers and wire services

TABLE 4.8 U.S. TELETEXT/VIDEOTEX TRIALS AND SERVICES: APPLICATIONS

	Teletext								Videotex														
	KSL	NCI Captioning	WFLD	KNXT/KCET/KNBC	WETA	KPIX	Cabletext	Time Inc.	QUBE	The Source	CompuServe	Comp-U-Star	Bison	OCLC	Green Thumb	Dow Jones	EIS	Viewtron	INDAX	Times Mirror	CBS/AT&T	Continental Tel.	Express Info.
News, weather and sports	■		■	■	■	■	■		■	■	■	■			■	■		■	■	■	■	■	■
Electronic newspaper				■			■		■	■	■					■		■		■	■		
Advertising	■		■	■	■		■		■		■	■			■			■	■	■	■	■	
Program captioning	■	■	■	■	■	■																	
Education			■	■	■		■		■		■			■	■			■		■	■		
Financial information				■			■		■	■	■					■	■	■	■	■	■		■
Library information							■				■			■				■		■			
Electronic directories							■			■	■	■						■	■	■	■	■	
Teleshopping							■		■		■	■						■	■	■	■	■	
Home banking							■		■		■							■	■	■	■		
Electronic messaging							■			■	■		■				■	■		■	■		
Games/entertainment	■		■	■			■		■	■	■							■	■	■	■		
Home computer support										■	■						■				■		
Personal files										■	■						■				■		
Home security																		■		■	■		

have become both information providers and system operators. In some instances, only selected features or stories are offered. Because of the limited number of pages they offer, broadcast teletext systems have tended to provide only a headline-type service. In other cases, the entire editorial content of the paper is available for electronic retrieval.

Most major U.S. newspapers have or are converting to electronic processing of information. In practical terms this means that their product, news, is packaged for digital storage and retrieval. As such, it is available for uses other than simply printing. Facing rising energy costs, paper costs, and delivery costs and changing reader needs, U.S. newspapers have begun to look at alternative means for news distribution. A consortium of 13 newspapers and the Associated Press has launched a test of electronically delivered news over CompuServe. While this test is aimed initially at home computer users, other teletext and videotex users could easily follow.

Several large U.S. newspaper companies have begun developing their role as information providers, either through participating in trials or sponsoring their own tests (e.g., Knight-Ridder, Times Mirror, and the *Dallas Morning News*). Currently, the user can receive news by way of passive scroll on a cable TV system, as pages in a teletext or videotex system, or through a home computer.

Advertising Advertising on teletext and videotex systems may be of several varieties:

Display (e.g., food market specials of the week)
Classified (e.g., help wanted, houses for sale)
Promotional (e.g., ''This weather information brought to you by _____ .'')

In teletext and videotex the advertising may be more information oriented (i.e., "infomercials") than sales oriented and may be tied to advertising messages in other media. Thus, the traditional distinction between advertising and editorial content may tend to become blurred.

Because there is no developed audience of videotex and teletext users, advertisers have yet to develop advertising strategies for use in these media. In several tests a number of large corporations have become information providers, blending their roles as advertiser and information provider. In the Viewtron test, the at-home user has the chance not only to see the supermarket's specials of the week but also to order them.

Program Captioning   Program captioning is the provision of simultaneous text to what is being aurally presented on a television program. The service has a direct, obvious benefit to persons with hearing impairments. Captioning can also be a valuable service to multilingual audiences. The captioned material can be added to the program either at the time of production or afterward. In the United States, closed captioning has been authorized by the FCC and has been operational since 1979. Decoders are purchased through a retail chain, Sears Roebuck, and programs are captioned for two commercial networks and the Public Broadcasting Service by the National Captioning Institute.

Closed captioning might be thought of as a forerunner to teletext. It is technologically more limited in that only captioning can be provided with each program; with teletext more applications are possible in one system. For example, teletext could provide captioning for the deaf and Spanish captioning as well. Experimental program captioning has been included as part of KNXT's teletext trial in Los Angeles.

Education   The Viewtron trial includes features such as health tips. The Green Thumb test in Kentucky provides extension/education information to farmers in two rural counties. Cox's INDAX System includes an "Electronic Schoolhouse," which offers math drills and other educational material. The WETA test has included scientific information from the Smithsonian Institution and consumer tips from several government agencies. Mathematical games, part of the Channel 2000 test (OCLC), have reportedly been quite popular. KCET is experimenting with elementary-level instructional material for in-school use to supplement instructional television broadcasting.

Financial Information   Almost every teletext trial has included summaries of stock market activity and a selection of other quotations. In 1980 Dow Jones announced a financial information service oriented toward the home computer user. Current-day quotes in stocks, bonds, options, mutual funds, and U.S. Treasury issues are available (historical quotes are available for stocks). In addition, the user can access current news from the pages of *The Wall Street Journal, Barron's*, and the Dow Jones News Service. Both The Source and CompuServe have financial information available as well.

Library Information   Specialized libraries and information banks are already providing electronic retrieval services. It is likely some of these will participate in videotex trials to see if further markets for those products can be developed. The New York Times Information Bank, for example, provides a limited version of its information service to The Source, which in turn provides it to its subscribers.

OCLC, a nationwide online library cataloging service, has tested the viability of videotex for library service. In its Columbus, Ohio, test, participants could access the card catalog of the Columbus library system, request a desired book, and receive it in the mail.

Electronic Directories   According to the early wisdom, a videotex system seems to have greater utility when used as an index, a guide, or a directory service. Electronic directories could be provided for any institution or corporation with frequently changing names and listings. Of course, the white pages of the telephone directory would be a prime candidate for an electronic delivery system. Such an application has been tested by AT&T. A major limitation of the yellow pages for advertisers has been its yearlong time frame between updating. An electronic yellow page (EYP) service could be updated almost instantaneously. In addition to its updating capability, an EYP service could also give the advertiser a direct link to the consumer. In other words, browsing in the EYP could lead to ordering via the system.

Teleshopping   All major videotex trials include some teleshopping applications. For example, Comp-U-Card offers a discount buying service. It also offers the service through The Source. Several cable systems are testing advertiser-supported video-based shopping services.

Electronic Banking   The emergence of automatic teller machines (ATMs), bill paying by telephone, and the computerization of bank records have brought the possibility of home banking closer to reality. Several banks have begun tests through videotex like services. Banc One, a financial holding company headquartered in Columbus, Ohio, conducted a test as part of OCLC's Channel 2000 trial. Some banks have tested home-banking services via personal computers. Several banks are participating in Cox's INDAX service in San Diego. At the simplest level, a user can review the status of his or her accounts. At a more advanced level, he or she can manipulate funds and pay bills over these systems.

Electronic Messaging   Both The Source and CompuServe offer electronic mail and messaging capabilities at a relatively low cost ($4 to $5 per hour). Messaging has become a very popular feature on The Source. In other videotex systems, however, messaging capabilities are still limited. With hand-held keypads, the possibility of sending preformatted messages to other people who access the system is there. Alternatively, users might be able to access a community bulletin board much as personal computer users do now for receiving information from other users. A bulletin board has been a feature of the Viewtron test in Coral Gables.

Games/Entertainment   Although teletext and videotex are often thought of as inexpensive means of delivering ''pages'' of information to the home, both have entertainment and game potential as well. Entertainment functions such as cartoons, quizzes, and horoscopes have been quite popular in early trials. The degree of interactive involvement in the game or entertainment source is usually a function of the sophistication of the system. In teletext, for example, a viewer's choice might be as simple as selecting a question on one page and then viewing the answer on another page. At the other end of the spectrum, a cable service might send game software to the home computer user, allowing him or her to interact with it or with others who might also wish to play.

Home Computer Support   Several current services provide software to home computer users. The Source, CompuServe, and Dow Jones are offering various services of potential interest to the home computer hobbyist. Packages providing financial or tax analysis are obvious examples. Through The Source and CompuServe, the home user can take advantage of much greater computing power as well as a wide range of other services.

Personal Files   Videotex systems could provide users with remote (as opposed to local) storage of material supplied by the user. Examples might include the source codes of user-written programs, documents, or personal directories. The home-computer-oriented timesharing services already offer this service.

Home Security   The first two-way service to be offered widely on cable systems is likely to be some form of home security protection. This application involves only a small amount of data being returned upstream from the home to a central computer. A home security service is reported to be one of the most popular options offered on Warner-Amex's QUBE system in Columbus.

GENERIC CLASSES OF INFORMATION SERVICES

From the preceding discussion, five generic classes of information services can be identified (Figure 4.5):

Information retrieval. In this category, the user gains access to information in a computer's database. The information may include general news bulletins, financial quotes, entertainment listings, basic directory services, the provision of subtitles for TV programs, or special interest services such as encyclopedia information that does not change rapidly. Information can be delivered in either a one-way teletext or a two-way videotex system.

Transactions. This area involves user interaction in order to make a reservation, purchase a product, register an opinion, or manipulate financial accounts (teleshopping and telebanking). Again, various levels of sophistication are possible. They range from simple "yes" or "no" responses to full catalog perusal capabilities with ordering and payment systems built in. The service requires a two-way videotex system.

Messaging. This service application allows the user to communicate with others on the system, or possibly with others on other systems, or even with those not on any

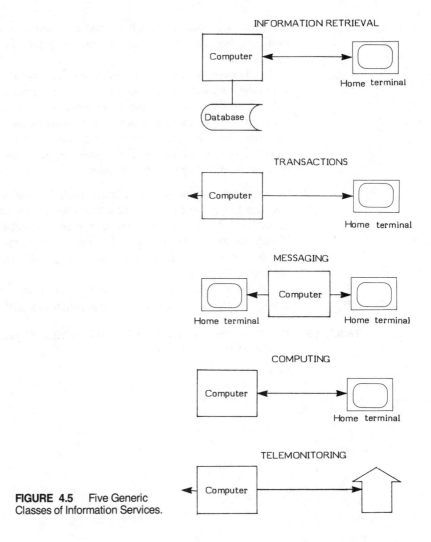

FIGURE 4.5   Five Generic
Classes of Information Services.

system (via down-loading messages for mail or print distribution). The sophistication of the messaging capability is dependent upon the type of input device the user has. A keypad limits the user who wants to send a message to a preformatted message. The user who has a full keyboard terminal, on the other hand, can send a full-text message. Full message service requires a videotex system, but general messages may be down-loaded on one-way systems.

Computing. The ability to enhance a user's terminal with additional processing or storage power could be an attractive service area for users with "intelligent" terminals. Indeed, this service could allow a user to preprogram his or her terminal to capture certain items of information, games, etc., much as a videocassette recorder can be preprogrammed to record TV programs at specific times. A two-way system is necessary for interaction with a remote computer, but a one-way system, plus an intelligent terminal, is sufficient for down-loading software for localized computing (telesoftware).

Telemonitoring. Emergency—fire, medical, and burglar—protection for the home via some two-way electronic means such as polling is already being developed independently. A videotex computer could serve as the central monitoring point for these services. In addition, controlling or monitoring home and business energy consumption via some electronic two-way service is technologically feasible. Such a service could be tied to a videotex package, particularly where homes and buildings were designed for such control devices.

What is important to keep in mind in all of these application areas is that teletext and videotex are not so much new technologies—some argue that they are quite the opposite—but rather are more properly seen as potentially new mass-market services. If they are to become widespread services, it is likely that the preceding categories will be part of a service package, but it is also likely that new, as yet undeveloped services will emerge.

Table 4.9 summarizes the existing U.S. trials and services in terms of these generic classes of information services. It demonstrates the different stages of development of

TABLE 4.9 TELETEXT AND VIDEOTEX IN THE US: TRIALS BY GENERIC CLASSES OF APPLICATIONS

	INFORMATION RETRIEVAL	TRANSACTIONS	MESSAGING	COMPUTING	TELEMONITORING
KSL	■				
NCI Captioning	■				
WFLD	■				
KNXT/KCET/KNBC	■				
WETA	■				
KPIX	■				
Cabletext	■				
Time Inc.	■				
QUBE		■	■	■	
The Source	■	■	■	■	
CompuServe	■	■	■	■	
Comp-U-Star	■	■	■		
Bison	■		■		
OCLC	■				
Green Thumb	■			■	
Dow Jones	■				
EIS	■				
Viewtron	■				
INDAX	■	■			
Times Mirror	■	■	■		
CBS/AT&T	■				
Continental Tel.	■				
Express Info.	■				

these classes: virtually every trial includes information retrieval in one or more forms. A smaller number of trials include transactions, and fewer still offer messaging or computing. Only one system reviewed here includes telemonitoring, though this service is developing quickly on other cable systems.

CONCLUSION

This brief overview of teletext and videotex trials in the United States indicates a fragmented market approach to technological development—a variety of communications networks, the lack of a standard display format, and the various combinations of services offered. It is as if these trials are searching for the right ''packages'' that will appeal to home consumers.

While the results of most of these trials are not public, a categorization such as we have presented can illuminate what people perceive to be the most promising teletext/videotex packages. Moreover, by tracking a broad range of such trials over time in a common framework, one can try to discern the key issues that affect the viability of teletext/videotex systems.

CHAPTER 5

FUTURE APPLICATIONS FOR TELETEXT AND VIDEOTEX

It has become conventional wisdom that "applications, not technology" will determine the success or failure of videotex. However, it is possible to forecast the evolution of technological components with greater confidence than to forecast future developments in applications. At this early stage there is relatively little data on which to base conclusions about which particular services, or combination of services, will prove most popular or on how much people will be willing to pay for them. The purpose of this chapter is to identify the fullest range of potential videotex and teletext applications and to provide an indication of the magnitude of the markets for these applications.

INSIGHTS ABOUT FUTURE APPLICATIONS FROM CURRENT TRIALS

Experience with videotex and teletext trials and services to date may at least offer some clues as to the pattern of future uses. Table 5.1 lists applications in terms of the number of U.S. trials in which they have been included.

This census shows that news, weather, and sports information has been included in virtually every trial; personal files and electronic messaging have been included in fewer trials. These statistics may simply reflect the fact that a news, weather, and sports service is relatively easy to provide, while videotex-based personal files and messaging not only are more difficult to provide but also can raise more complicated policy issues.

Table 5.2 lists the categories of information that have been accessed most frequently on Prestel in its first year of operation. This listing suggests that videotex may turn out to be as much a source of entertainment (e.g., games, quizzes, sports, jokes, horoscopes) as a channel for obtaining "serious" information. On the other hand,

TABLE 5-1   TELETEXT AND VIDEOTEX
APPLICATIONS BY NUMBER
OF U.S. TRIALS, 1981–82

News, weather, and sports	27
Banking transactions	15
Advertising	10
Teleshopping	10
Games/entertainment	10
Electronic newspaper	9
Financial information	9
Education	7
Electronic messaging	5
Program captioning	5
Home computer support	3
Personal storage files	3
Electronic directories	1

these rankings are based on initial use of the system and may reflect the limitation imposed by Prestel's tree-search database structure, which tends to make accessing large-scale databases slow and cumbersome. It is also important to note that Prestel was designed as an information-retrieval service and did not initially include more sophisticated services such as teleshopping or messaging.

Analysis of usage of The Source, which offers an array of data processing, transactional, and messaging options in addition to information retrieval, provides yet a different picture. As Table 5.3 indicates, services involving messaging were most popular, followed by news-oriented information retrieval. However, subscribers to The Source, who are mostly computer hobbyists, are not necessarily representative of the broader public, who are the intended audience for videotex.

A much more diverse population was included in a survey of consumer interest in various interactive services conducted by the Benton & Bowles advertising agency (Table 5.4). The results indicate that the services that evoked the greatest interest

TABLE 5.2   MOST POPULAR TOPICS ON PRESTEL, 1978–79

Games	Travel	Holidays
Quizzes	Business news	Company information
Stock market	Restaurants	What's on
Sports	National news	Consumer advice
Jokes	Horoscopes	Cars

Source: Winsbury, 1981.

TABLE 5.3   MOST POPULAR SERVICES ON
THE SOURCE, JANUARY 1980

1. Electronic mail
2. Bulletin board (classified advertising)
3. Chatting (terminal-to-terminal communication)
4. User directory associated with items 1-3
5. UPI (news and sports)
6. New York Times Consumer Data Base
7. Unistox (financial data)

Source: Plummer, 1980.

were those that offer clear financial benefits. However, the survey did not include entertainment-oriented interactive services such as games and quizzes. The survey also made no mention of costs and therefore provided no information on how much (or how little) consumers would be willing to pay for these services.

While these results are interesting, none of them is conclusive. Taken as a whole, they suggest that the success of any videotex service will depend not only on its content but on such factors as convenience, ease of use, reliability, and cost. How a service is packaged and priced may be as significant as *what* the service is, especially in an environment in which an ever-increasing array of information and entertainment media are competing for a share of the consumer's time and money. As Table 5.5 demonstrates, there is a variety of alternative media—some old, some new—available to perform each of the categories of videotex functions.

TABLE 5.4   INTEREST IN INTERACTIVE SERVICES

	% Responding	
Service	**Very interested**	**Somewhat interested**
Fire/burglar alarm	49	32
Shopping information (prices)	35	35
News, weather, sports	26	39
Meter reading	26	29
Home banking	23	29
Ticket reservations	18	25
Travel reservations	17	25
Opinion polls	16	28
Travel schedules	15	30
Home shopping	10	27
Financial news	10	22

Source: Cablevision, 1981*b.*

TABLE 5.5   COMPETITIVE
MEDIA/TECHNOLOGIES

		Home	**Business**

1.	Information retrieval	Newspapers
		Magazines
		Books
		Television
		  Broadcast
		  Cable
		  STV*
		Audio cassettes
		Records
		VCR and videodisc
		Radio
		Telephone
		Yellow pages
			Newspapers
			Magazines and newsletters
			Books and reports
			Audio cassettes
			VCR and videodisc
			Telephone
			Yellow pages
			Online databases
			Paper files

2.	Transactions	Checks Credit cards Catalog shopping Telephone bill paying Telephone shopping	Checks Purchase orders Telephone transfer Computerized billing and payments
3.	Messaging	Telephone Mail Telegraph	Telephone Mail and private carriers Telegraph Teleconferencing Facsimile Electronic mail Specialized common carriers
4.	Computing	Calculators Video games Electronic games Home computers	Calculators Personal computers In-house data processing Timesharing services
5.	Telemonitoring	Stand-alone alarms Autodial alarm systems Security patrols	Stand-alone alarms Autodial alarm systems Security patrols

*Subscription television.

Timing will also be important. Certain more specialized videotex applications may not become available commercially until relatively widespread penetration is achieved on the basis of more general interest services. An example of this pattern can be seen in the cable field, where the development of cultural pay programming services was made possible by the prior success of general appeal pay channels such as Home Box Office.

Finally, technology plays a role in determining applications. The services a system can provide are limited by the technical characteristics of the system. Such factors as transmission speed (upstream and downstream), error rate, graphic capabilities, and terminal characteristics play a part in how a service is implemented. The severe restriction in the number of pages that can easily be accessed in a VBI-based broadcast teletext system is an obvious example. Another limitation on the potential of videotex is imposed by the resolution of the 525-line standard for U.S. television—a standard that was established with no reference to the capabilities needed for displaying pages of text. The development of a new high-resolution (1,125-line) TV standard would greatly increase the range of potential videotex services, but movement toward such a standard will be influenced by many factors other than videotex (e.g., the development of wall-size flat TV screens or the introduction of digital techniques for the storage and transmission of television programming). As the technological components of videotex systems evolve, so will the range of services these systems carry.

What these considerations mean is that experience with videotex systems to date does not provide an adequate basis for predicting which specific videotex applications will be successful in the future and which will not. However, it is possible to identify a much larger number of *potential* applications than have been included in videotex service or trials to date.

FUTURE APPLICATIONS

Assuming that teletext and videotex do achieve widespread penetration over the next 20 years, what might these systems be like? What services might they offer?

A series of three "futures" workshops was held as part of this technology assessment. These workshops included an exercise designed to elicit a broad range of potential applications for teletext and videotex. More than 50 potential applications generated at the workshops are listed in Table 5.6.

Applications are categorized according to their primary functional classification—either as information retrieval, transactions, messaging, computing, or telemonitoring—along with a brief identification of likely providers and users of the service. Many of the applications listed could be implemented in either a one-way teletext system or in a two-way videotex system. For example, many one-way information retrieval services could be enhanced by interactive transactions and/or messaging capabilities. An electronic version of the Official Airline Guide would be classified as an information retrieval service; addition of a trip-planning service would involve computing, while inclusion of the capacity to make reservations would require a transaction capability. A few applications (e.g., computer-assisted instruction, which utilizes both information retrieval and computing) are true hybrids requiring two or more functions.

POTENTIAL MARKET FOR HOME-BASED VIDEOTEX SERVICES

Very few of the applications in Table 5.6 are wholly new. The absence of novel uses may simply be a reflection of the fact that it is easier to extrapolate from current uses than to foresee true innovations. Still, it is likely that the success of videotex systems will depend on their ability to satisfy existing needs more effectively or more efficiently than to create new needs. It is possible, therefore, to get a sense of the magnitude of the potential market for videotex services by examining the size and nature of

TABLE 5.6   FUTURE TELETEXT/VIDEOTEX APPLICATIONS

Application	Description	Providers	Users
I Information retrieval			
Application area: electronic publishing			
Electronic newspapers	Text from local papers and wire services. Ability to retrieve background stories.	Local newspapers, national news services	Everyone
Specialized newsletters	High-cost newsletters with timely information.	Small and large publishers	Investors, collectors, professionals
Electronic encyclopedia	Online reference service with frequent updating to include new developments.	Encyclopedia publishers	Students
Application area: library and reference service			
Database access	Opportunity to access specialty databases from home.	Specialized database providers, libraries	Professionals, researchers
Catalog review	Listing of all library materials. Capability of reserving/requesting items from library.	Public libraries	Researchers, writers
Application area: community services			
Community bulletin board	Listing of local events accessible by subject type, place, etc. Also proposed as a public information utility. Electronic referral services.	City government, local newspaper, community groups	Local residents
Transit/travel information	Bus, train, and airline route schedules plus intercity connections. In enhanced version, trip planning capabilities are available.	Transit authority, airlines, travel agents, AAA	Travelers
Emergency information	Latest reports on accidents, road conditions, weather, air pollution, etc.	Local government, highway patrol	Local residents
Government information	Listing of meetings and hearings. Meeting agenda. Notice of reports, regulation changes, etc.	Local government	Local residents
Housing availability	Multiple listing service. House, apartment, condominium sales and rentals. Hotel and motel space availability.	Landlords, realtors, hotels and motels	House hunters, travelers
Comparison shopping	Umbrella group collects information on current prices for particular products.	Nonprofit community group	Consumers

(Continued)

TABLE 5.6 FUTURE TELETEXT/VIDEOTEX APPLICATIONS

Application	Description	Providers	Users
Electronic hotlines	Match requests to information (e.g., poison control information—type of poison entered and recommended action immediately displayed).	Special interest groups (e.g., poison control problems center)	Residents with emergencies
Foreign language service	Translations of community announcements, emergency, and other information.	Local government, ethnic community groups	Non-English speakers
Captioning	Subtitling of TV programs, news, etc.	Caption center, broadcasters	Hearing-impaired
Electronic directories	Open or closed systems for providing listings of employees, buildings, stores, hours of service, etc. Telephone white pages.	Companies, associations, phone company	Employees, customers, group members, phone users
Application area: education			
Course listings	Extension courses, night school classes, private school offerings available by subject, location, fee, etc.	Schools, adult education programs	Students
Computer-assisted instruction/computer-managed instruction	Course material programmed to move with individual learner's speed and capability.	Schools, specialized companies	Students
Special services for home-bound students	Interactive "correspondence" courses.	Schools	Students
Supplemental materials for education TV programs	Online "Sesame Street" type materials to practice lessons/ideas from broadcast program.	CTW, local school districts, public TV	Young children
Do-it-yourself training	Step-by-step instruction for home repair, car repair, cooking, etc.	School systems, product manufacturers, specialized companies	Do-it-yourselfers
Literacy	Basic language and mathematical skills.	School districts, training community colleges	Dropouts, older students, new arrivals
Retraining	Tutorial programs including linking new job interest and positions available.	Unions, companies, community colleges	Unemployed

	Application area: health		
Storefront medicine	Systems with details on diagnosis, treatment, drugs, health risks, costs, etc.	Hospitals, medical associations, specialized publishers	Professionals, paraprofessionals, consumers
First aid	What to do in case of . . ., could include follow-up numbers to call.	Red Cross, drug companies	Anyone with emergency

	Application area: entertainment		
Electronic jukebox	Access to large library of recorded music, lectures, old radio shows, etc.	Radio, TV stations, libraries	General community
On-demand TV	Dial-up program selection of movies, TV shows.	Cable, broadcast companies	Pay-TV subscribers

	Application area: advertising		
Electronic yellow pages	Directory-type information on products and services updated regularly, accessed by categories. In enhanced version, could be combined with order-taking capability.	Telephone companies, merchants	Consumers
Supplement to TV advertising	National ads with local dealers and price information. More detailed description on product. Could have order-taking capability.	Advertisers	Consumers
Classified advertising	Similar to newspaper classified advertising but with more capabilities for indexing and retrieval.	Individuals, retailers	Consumers
Display advertising	Show weekly specials and product information. Could be used in conjunction with other ads. May include ordering capability.	Retail merchants, chains, department stores, supermarkets	Consumers

	II Transactions		
Electronic checkbook	Purchase made over system; purchase amount automatically deducted from purchaser's checking account.	Banks, savings and loans	Bank customers
Electronic funds transfer	Bank customer has access to all accounts via home terminals. Can manipulate accounts and make transactions.	Banks, savings and loans	Bank customers
Electronic credit cards	Account number is entered in system for purchases, bill paying, etc.	Banks, savings and loans, credit card companies, retail stores	Bank customers

(Continued)

TABLE 5.6   FUTURE TELETEXT/VIDEOTEX APPLICATIONS

Application	Description	Providers	Users
Electronic catalogs	Online access to catalogs with order-taking capability.	Catalog merchants (Sears, Consumer Distributors)	Consumers
Gambling	Quotations, pool size, risk information. Potential extension of off-track betting offices.	State agencies or private companies	Adults
Entertainment options	Electronic box office service. Leisure-time planning. Vacation options. Could include reviews, ratings, guidebook information.	Ticketron, travel agents	Everyone
III Messaging			
Electronic mail	Point-to-point messages for personal correspondence, bill paying, advertising, etc.	System operators, post office, specialized carriers	Anyone
"Videotexgram"	Efficient way of sending same message to multiple locations.	Businesses	Customers, suppliers
Conferencing	Shared textual space by group.	Videotex system operators	Special interest groups
Serendipity machine	Unplanned meeting with people of like interests via system.	Videotex system operators	Anyone
Electronic welcome wagon	Information on community and neighbors provided by online welcomer.	Local chamber of commerce	Newcomers
Referenda or quasi referenda	Citizen input to government officials on controversial issues.	Local government	Citizens
Consumer action	Complaints and reactions to products or services are channeled through consumer ombudsman service with feedback to participants.	Nonprofit community group, Better Business Bureau	Consumers
Electronic gossip	Links residents with similar interests to share information, opinions, and ideas.	Community groups	Local residents
Internal business communication	Intraoffice memos; closed-user groups.	Companies	Employees
Market research	Use of new product is tracked through online polling of consumers.	Retailers, manufacturers	Consumers
IV Computing			
Video games	Interactive games with potential of adding multiple players. Could be combined with TV programs so that participants at home influence course of program.	Teletext/videotex system operator	Anyone
Home computing service	Provide additional computing power to home terminals.	CompuServe, The Source, etc.	Personal computer owners

TABLE 5.6   FUTURE TELETEXT/VIDEOTEX APPLICATIONS

Application	Description	Providers	Users
Personal information storage	Private electronic files for home-related information	Videotex service provider	Anyone
Financial management	Personal finance management.	H&R Block, CPAs, etc.	Anyone
Telework	Text editing, file maintenance, data entry, and analysis.	Companies, service bureaus	Employees
Extensions of corporate management systems	Available to workers at home and at other convenient locations (e.g., hotel rooms).	Companies	Employees
Inventory/stock control	Extension of present computerized systems to smaller and more decentralized companies and franchise organizations.	Companies	Branch offices, local dealers or distributors
V Telemonitoring			
Home security	Remote sensors. Emergency fire, police notification.	Cable companies, phone companies, security companies	Residents
Health and safety monitoring systems	Assist in at-home care (e.g., ECG, blood pressure readings).	Public health agencies, medical groups	Chronically ill, elderly
Energy management	Control and regulation of household energy use. Meter reading.	Gas and electric utilities	Residents

the existing demand for these services. In the following sections, we will review overall consumer expenditures for major applications under information retrieval, transactions, messaging, computing, and telemonitoring.

In these sections, Tables 5.7 to 5.10 are derived from the Institute for the Future's Forecast of Consumer Expenditures. The figures for average monthly expenditure per household are based on a 1980 total of 80 million U.S. households. By 1990 the number of U.S. households is projected to grow to 94.3 million. In addition, total real expenditures are projected to grow at an average rate of 3 percent from 1980 to 1990. The tables include estimated growth rates for individual products and services during the 1980s.

Information Retrieval

The largest number of applications identified in Table 5.6 are related to information retrieval. This is not surprising given the fact that videotex systems were originally developed in Europe primarily as information retrieval systems. The incorporation of additional applications has been a more recent development.

Table 5.7 lists 1980 consumer expenditures for print and nonprint media, first in terms of total annual expenditures, and second in terms of average expenditure per

TABLE 5.7   INFORMATION RETRIEVAL EXPENDITURES

	Total 1980 expenditures ($ billions)	Average household expenditure/ month	Percent estimated annual growth 1980-1990
Print media			
Newspaper subscriptions	2.20	$ 2.30	3.0
Magazine subscriptions	1.10	1.15	3.4
Encyclopedias and bookclubs	1.15	1.20	2.7
Other nonsubscription print purchases	2.50	2.60	3.3
Total print expenditures	6.95	$ 7.25	
Nonprint media			
TV sets	10.60	$11.10	2.7
Radios	0.85	0.90	3.3
Phonographs and tape recorders	2.40	2.50	3.0
Records and tapes	3.10	3.25	3.1
Cable TV	2.30	2.40	8.0
Total nonprint expenditures	19.25	$20.15	

household per month. It is generally assumed that information retrieval via videotex would compete most directly with information currently obtained from print media such as newspapers, magazines, and perhaps encyclopedias. Average monthly expenditure per household for print in all forms in 1980 was $7.25. However, the monthly expenditure for electronic media was more than 2½ times as great—$20.15 per month, the largest item being the purchase and repair of television sets. If videotex does achieve widespread use, it is likely that expenditures for television sets will increase. On the other hand, a videotex service that delivered music on demand would compete directly with the sale of phonograph records and tapes.

Education   A special but potentially significant information retrieval subcategory involves educational applications of videotex. Several current trials have included educational services, and the futures workshops identified specific applications ranging from simple listings of available courses to supplemental educational materials to TV programs to the provision of computer-aided instruction.

Table 5.8 lists 1980 consumer expenditures for education, which amounted to $14.4 billion. This figure both overstates and understates the potential of this area for videotex. More than half this figure, or nearly $9 billion, is made up of school tuition. It is highly unlikely that a videotex-based service would ever substitute for formal school-based education.

At the same time, expenditures by consumers account for only a fraction of overall U.S. expenditures on education. In 1979, this figure totaled $151 billion, or roughly 7 percent of the year's GNP, most of which comes from government tax dollars.

TABLE 5.8   CONSUMER EXPENDITURES FOR EDUCATION

	Total 1980 expenditures ($ billions)	Average household expenditure/ month	Percent estimated annual growth 1980-1990
Education			
Elementary-high school			
Tuition	2.22	$2.31	3.8
Books	0.36	0.37	2.8
College			
Tuition	6.17	6.42	2.2
Books	1.24	1.29	2.4
Business/secretarial			
Tuition	0.54	0.56	3.6
Books	0.15	0.16	3.3
Miscellaneous education expenditures	3.75	3.87	2.8
Total education expenditures	14.43	$14.98	

Schools on all levels represent a very large market in their own right. If videotex systems do become effective educational tools, it is likely that they would be utilized within schools, much as personal computers are now being rapidly adopted. A number of attributes of videotex—ease of use, low cost, emphasis on color and graphics—make these systems attractive as an educational aid, particularly with younger students.

From the consumer's viewpoint, the areas of do-it-yourself instructional and educational enrichment seem particularly promising. These areas have already been targeted as markets for recorded video programming and personal computer software. Within the business market, discussed below, there may be other educational applications in the area of job-skill training and professional development.

Advertising   Another important potential source of videotex revenues is advertising. Table 5.9 lists 1980 expenditures on advertising. These figures indicate that annual expenditures for advertising are more than twice as great as total personal expenditures for both print and nonprint media. If videotex does become a widely used medium in this country, it will inevitably become of interest to advertisers. In this regard, it is important to note the economic significance of advertising in supporting existing media. As of 1980, advertising accounted for approximately 50 percent of total magazine income, 75 percent of newspaper income, and 100 percent of commercial radio and television income. The loss of even a modest portion of this income could have significant adverse consequences for these media. At the same time, if videotex is able to attract advertising, it could represent a subsidy that would make it unnecessary for individual consumers to support the entire cost of the service.

Which media are most likely to be affected by competition from videotex for advertising? Most advertising applications of videotex proposed to date—such as classified real estate and help-wanted ads—have been locally based. Figure 5.1 shows the breakdown between local and national advertising expenditures. The chart indicates that the primary media for national advertising are television, magazines, and direct mail, while local advertising is particularly important for the yellow pages

TABLE 5.9   ADVERTISING EXPENDITURES

	Total 1980 expenditures ($ billions)
Newspapers	15.9
Radio and TV	15.3
Magazines	3.3
Yellow pages	2.8
Other media	17.7
Total advertising expenditures	55.0

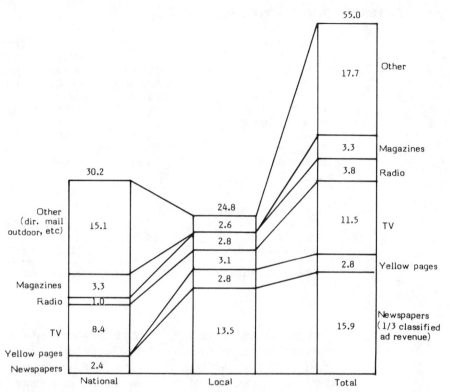

FIGURE 5.1   Local and National Advertising Expenditures by Medium—1980. ($ billions)
(Source: McCann-Erickson, 1981)

(which is almost entirely local), radio (74 percent local), and newspapers (85 percent local). The chart also indicates that newspapers account for more than half of all local advertising expenditures. The interest in videotex from the newspaper industry has been motivated as much by a desire to protect its advertising revenue base as by its desire to explore new methods of delivering news to its readers.

Transactions

Transaction applications of videotex fall primarily into three broad categories: home-banking and bill-paying services, teleshopping, and remote reservations. Already, all three of these are rapidly growing areas, and the emergence of videotex is likely to accelerate their development further. Table 5.10 lists consumer expenditures for some of the transactions that are likely candidates for being carried out in videotex systems. The first category comprises expenditures connected with banking and bill paying. Banking fees include charges for various bank services such as check writing, over-drafts, and safe-deposit boxes; they do not include interest paid on loans. Most relevant to a videotex transaction service are expenditures for check writing and credit

TABLE 5.10 TRANSACTION EXPENDITURES

	Total 1980 expenditures ($ billions)	Average household expenditure/ month	Percent estimated annual growth 1980-1990
Banking and bill paying			
Banking fees	1.95	$ 2.05	3.6
Postage for payment of bills	1.20	1.25	2.8
Teleshopping			
Direct marketing sales	36.00*	37.50	10.0
Services requiring reservations			
Airline travel	6.60*	6.90	3.9
Vacation lodging	4.95*	5.15	3.8
Entertainment admissions	14.65*	15.25	3.7

*These figures represent the total values of goods and services purchased.

card transactions. In fact, bank charges to customers reflect only about one-fourth of the actual cost of processing checks. Although banks have been raising their service charges, they have not been able to increase them as quickly as their own costs. In 1982, the average cost of processing a check at a major bank was 41 cents, an increase of 36 percent in two years (*Business Week*, 1982). With individuals writing a total of some 20 billion checks annually, actual costs of check processing by banks could be approximately $8 billion (*Bankers Desk Reference*, 1980). Converting some of these transactions to electronic form (including automated-teller machines and point-of-sale terminals as well as electronic home banking) offers banks the opportunity to reduce their labor costs. A 1980 survey by the Electronic Money Council found more than half the banks surveyed offered some form of telephone bill paying for accounts that are to be paid on a regular basis such as utilities, credit card accounts, or home mortgages (Mier, 1981). Almost all these current services are semiautomated; that is, they involve a human intermediary at the bank who completes the customer's transaction. In a fully automated videotex system, the customer would presumably carry out the entire transaction on his or her own.

The ability to move funds electronically becomes more significant when it is linked to a teleshopping service. In this case, a home user would be able to peruse an electronic catalog, select an item for purchase, and arrange for electronic payment via either an electronic check or an electronic credit card transaction.

Nonstore retail sales are currently growing at 10 to 15 percent annually, more than twice as fast as the overall rate for retail sales, and total nonstore sales are in excess of $100 billion annually (Cooney, 1981). The Direct Mail Marketing Association has estimated that in 1979 a total of $36 billion was spent by consumers on products sold through all forms of direct mail marketing. In that same year, more than 85 million catalogs were sent through the mail to American consumers. If even a fraction of this enormous market were to be captured by videotex, teleshopping could become a major application area.

The final category of potential videotex transactions is reservations. Here again, the magnitude of transactions that potentially could be conducted via videotex is very large. The entries in Table 5.10 are illustrative of the services requiring reservations. Although travel agents will undoubtedly continue to play a role in arranging vacations and booking air travel, there are many routine travel reservations that could be handled directly by individuals via videotex. The same is true in terms of purchasing tickets for plays or concerts or other reserved-seat performances that are now provided by Ticketron-type services. (It should be noted that only part of entertainment admissions are reserved-ticket attractions.)

In all these applications, the chief attraction of videotex is convenience. It can provide a service at home and on demand that allows consumers to examine a variety of products or services, determine their price, place an order, and arrange for payment in a single transaction.

Messaging

The speed and convenience of electronic messaging are likely to make it a major application area for videotex. The use of electronic mail is already growing rapidly for business communications. Videotex systems will bring the technology necessary for electronic messaging into the home.

Table 5.11 lists consumer expenditures for mail and telephone services. As in the case of information retrieval, total consumer expenditures for electronic media (telephone and telegraph) are several times greater than spending on print-based media (mail). However, growing use of videotex for whatever purposes is likely to increase rather than decrease overall telephone usage.

The average monthly expenditure per household of $2.80 for postage does not suggest a large potential for home-based electronic mail. A somewhat different picture

TABLE 5.11   MESSAGING EXPENDITURES

	Total 1980 expenditures ($ billions)	Average household expenditure/ month	Percent estimated annual growth 1980–1990
Postage	2.65	$ 2.80	2.8
Stationery	0.40	0.45	2.4
Telephone and telegraph	22.20	23.00	2.6
Total messaging expenditures	25.25	$26.25	

emerges if we look at the total volume of mail. In 1979, the United States Postal Service handled a total of 99.8 billion pieces of mail. Mail sent *from* households totaled only 17.5 billion pieces—or about 17.5 percent of the total mail volume. Mail sent *to* households from nonhouseholds (i.e., business and government) equaled 55.5 percent of the total. Thus mail to and from households combined represented 73.0 percent of all mail. The business sector, including business correspondence to and from households, accounts for 84 percent of all mail (see Table 12.8). Videotex-based electronic mail might offer an efficient means of carrying much routine correspondence, such as notices, statements, and bills.

Electronic messaging and electronic mail are not likely to become significant applications until videotex systems achieve relatively widespread penetration. The great advantage of the current mail and telephone systems is their universal access. It will be many years until videotex reaches sufficiently high penetration to rival them directly. Nonetheless, electronic videotex messaging may well begin to be made available early on as an option for certain routine communications as an alternative to conventional channels.

Computing

The computing-oriented applications for videotex fall into two broad categories: games and home computer support. The current markets for these applications are still relatively small, as Table 5.12 indicates.

Computer-based video games represent a rapidly growing fraction of the $4.5 billion annual market for toys and games. By 1981, total expenditures were in excess of $1.2 billion, and they are estimated to reach $3 billion in 1982. The current generation of video games is "programmable"—that is, the games can be changed by purchasing additional cassettes. Master game units presently cost $150 to $300 and individual game cassettes are sold for $10 to $35. Sustaining interest in video games depends on the ability to provide a variety of different games. As of 1981, Mattel offered 25 different games for its Intellivision system, while Atari offered 43 different cassettes. Sales of game cassettes alone were more than $600 million in 1981.

Because the cost of buying cassettes can exceed the cost of the master unit, a videotex system could rent game software as an alternative to purchasing cassettes. In 1981, Mattel launched its Playcable service, which provides a rotating selection of interactive games on cable television systems for a fee of $6 to $10 per month to sub-

TABLE 5.12   COMPUTING EXPENDITURES

	Total 1980 expenditures ($ billions)	1980 penetration (% of all households)
Home video games	0.70	3.5
Home computers	0.18	1.0

scribers who own an Intellivision unit. Since virtually all video games involve moving an image, videotex terminals capable of offering games would have to include a joystick or another such device in addition to a keypad or keyboard.

The home computer market is slower in maturing than the video-game market. Sales to individual consumers amounted to only 20 percent of the total personal computer sales of $750 million in 1980. These sales are projected to grow rapidly over the next two decades (Figure 5.2). A technology assessment of personal computers estimated that computers would be in from 10 million to 40 million households by 1990 (Nilles, 1981).

If home computers do develop into a mass market, videotex services would be likely to provide a similar kind of support for them as Playcable provides for home video games. Assuming that actual growth of personal computers falls midway between Nilles's high- and low-growth estimates, there will be 28 million home computers in 1990. If 20 percent of this group subscribed to a videotex service, there would be 5.6 million subscribers in 1990. The same assumptions would yield 11.4 million subscribers in the year 2000.

Telemonitoring

The final category of potential videotex applications includes the automatic monitoring of homes for burglary and fire protection and for energy management. Table 5.13 shows estimated 1980 expenditures for home security. Like computers, the security market has been oriented almost entirely toward business applications, with sales for home protection amounting to only 5 percent of total 1980 sales of approximately $8 billion. However, as awareness of and concern about crime have increased, so has

FIGURE 5.2 Home Computer Cumulative Sales to First-Time Buyers. (*Source: Nilles, 1981.*)

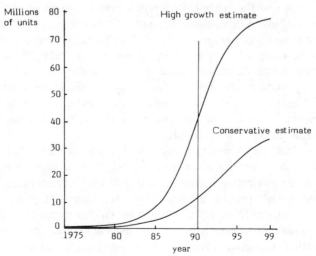

TABLE 5.13 TELEMONITORING EXPENDITURES

	Total 1980 expenditures ($ billions)	1980 penetration (% of all households)	Average monthly expenditure for security-equipped households
Home security equipment and services	0.4	3.0	$13.90

interest in home security systems. The total market in 1990 has been forecast to grow to $23 billion, of which $2.5 billion would be for home protection.

Only a small fraction of homes that have security systems today have telecommunications capabilities. Typically, these systems include an autodialer that alerts police when an alarm is tripped either by fire or intrusion. The chief obstacle to increased penetration of these sophisticated systems is their high cost—typically several thousand dollars for installation plus a monthly maintenance fee.

The most aggressive marketing of remote security monitoring systems at present is being done by the cable television operators, who have concluded that a security service is the easiest interactive service to sell their customers. (The Benton & Bowles study summarized in Table 5.4 showed that a fire/burglar alarm system evoked the highest level of consumer interest.) A 1980 NCTA survey found some 70 cable systems operating or developing some form of home security service. Installation fees ranged from $500 to $1,000, with service charges running from $4.00 to $18.00 per month (*Cablevision*, 1981*b*). One persuasive argument operators are able to make in selling this service is that many insurance companies are offering discounts of 15 to 35 percent on homeowner's insurance for people who have installed cable security systems.

These systems can be quite sophisticated. For example, Pioneer's VIP security system combines a fire-sensing capability that can detect both heat and smoke, a medical alert system, an emergency "panic" alert, and a burglar alarm. The burglar alarm can be equipped with a variety of sensors, including door and window contacts, motion detectors, pressure pads, and strobe and infrared beams (*Cablevision*, 1981*b*). Perhaps most noteworthy, the security system can be installed on its own or as part of the "interactive VIP system with complete entertainment and data communications structure." Thus, it may be that an interactive videotexlike system will develop on cable beginning with a security service.

A final telemonitoring application is automatic meter reading and energy management. These systems would permit an electric utility to issue load-shedding instructions to participating homes in order to disconnect certain appliances (e.g., air conditioners) when peak load limits are reached. Since the primary benefits of these systems are for utilities, it will be up to the utilities to market such systems to individual customers. To be cost-effective, this application will almost certainly be piggybacked on other interactive services. A few large commercial and public buildings are now equipped with such systems, but no homes have yet been wired for energy management. However, another sharp rise in energy costs could stimulate more rapid development of these systems.

Residential Markets: Summary

Some general conclusions can be drawn from the preceding analyses. As the summary of expenditures in Table 5.14 shows, applications related to teleshopping and telereservations, electronic mail, and advertising represent the largest potential markets for videotex. Home computing and video games and home security are small markets at present, but they are growing very rapidly and may emerge as significant application areas. Also, it appears likely that certain applications, such as messaging and advertising, will not develop fully until videotex systems reach relatively widespread penetration.

Of course, other media will continue to compete for a share of these markets. If videotex becomes significant, other media are likely to evolve to meet its competition. Videotex systems offer a number of advantages, including speed and convenience. The critical factor in determining its acceptance for any given application is likely to be cost and ease of use compared with those of the media with which it competes.

TABLE 5.14   SUMMARY OF POTENTIAL VIDEOTEX CONSUMER MARKETS

	Total 1980 expenditures ($ billions)
1. Information retrieval	
Print media	6.95
Nonprint media	19.25
Education	14.40
Advertising	55.00
2. Messaging	
Postage and stationery	3.05
Telephone and telegraph	22.20
3. Transactions	
Banking and bill paying	3.15
Direct mail sales	36.00*
Reservations (airline, hotel, entertainment)	26.20*
4. Computing	
Home video games and home computers	0.88
5. Telemonitoring	
Home security	0.40

*Represents total value of goods and services purchased.

THE BUSINESS MARKET

Although the consumer market is assumed to hold the greatest long-term potential for videotex, many claim that in the short run it is the business market that is likely to develop most rapidly. As evidence, observers cite the experience in England, where more than 85 percent of the early subscribers to Prestel have been businesses (Chapman, 1981). Similarly, approximately four-fifths of current sales of personal computers are to the business market (Nilles, 1981).

According to this view, businesses, unlike consumers, have been accustomed to making substantial investments for information. As corporations have grown larger, more complex, and more decentralized, their information needs have become more acute. The expansion of computer applications from simply performing numerical calculations to supporting a wide range of information processing functions has also made business more conscious of its information needs. Perhaps the most significant current manifestation of this trend is the movement toward office automation, which involves applying computer-based technologies to a broad range of clerical and managerial tasks. Among the systems and devices generally included under this rubric are word processing systems, personal computers, and intelligent copier/printers. If this movement succeeds, it is possible to foresee the day when the job of every white-collar employee (and many blue-collar employees) involves some sort of interaction with a computer terminal.

What role is there for videotex in this environment? The problem of identifying specific business-oriented applications for videotex is somewhat different than that of identifying home-oriented applications. In the latter case, there are a number of potential information services for consumers that would be novel and distinctive (e.g., electronic yellow pages, electronic mail, telemonitoring). The question is whether the general public wants these services badly enough to pay for them. On the other hand, businesses are accustomed to paying for information; the problem here, however, is to define a distinctive role for videotex vis-à-vis the numerous alternative information technologies and services with which it must compete. Thus, a consumer at home might look at a decision about subscribing to a videotex service as a choice between an information/transaction service and, say, the purchase of a videocassette recorder or a subscription to a pay TV network. In the business environment, the choice is more likely to be between adopting a videotex system or subscribing to an online database service or investing in a distributed word processing system.

There is one additional difference between the home and business environments that may be significant to the growth of videotex. A key element of systems conceptualized to date has been the role of a television set as the display device for videotex information. The reason for this is that television sets are ubiquitous in the home: TV sets are in 98 percent of U.S. households; 85 percent of households have color sets and more than 50 percent have two or more sets (Nielsen, 1981). Teletext and videotex systems have been developed as add-on services that would allow consumers to make greater use of their television sets.

In the business environment, by contrast, television sets are relatively rare. Because television evolved as a medium of entertainment rather than of information, its place has been in the home rather than in the office. Of the more than 100 million

TV sets in the United States, fewer than 1 percent are likely to be found in offices. Thus, if videotex is to succeed in the business market, it will require the development of turnkey videotex.

Business Information Acquiring, managing, and disseminating information have come to represent major business activities. According to one widely cited study, half of the entire U.S. work force is primarily engaged in dealing with information (Porat, 1977). A study conducted by AT&T found that business expenditures for information in 1978 totaled $200 billion—$104 billion for personnel, $34 billion for telecommunications, $24 billion for word processing and mail, and $16 billion for electronic data processing (J. White, 1981). AT&T estimates that business expenditures in these categories are currently $250 to $300 billion and are projected to reach $800 to $900 billion by 1990. Obviously, only a small portion of these expenditures needs to be captured by videotex-type services to represent a sizable new market.

Several studies have more closely examined the kinds of information services likely to be provided by videotex. One survey of the "information industry" found that its 1979 revenues amounted to $9.4 billion (Information Industry Association, 1980). Included in the survey were more than 1,000 firms engaged in providing primary information, secondary information, computer services, retail information services, information support services, and seminars and conferences. Excluded from the survey were "traditional" information services such as data processing, business newspapers, periodicals, and books.

The existing business that is closest to a videotex service is the online database service industry. As Table 5.15 indicates, total 1981 revenues for this comparatively new industry were approximately $1.25 billion (International Resource Development, 1981). These services are provided by timesharing companies, publishing firms, and specialized online service suppliers. The industry as a whole is expected to grow fourfold by 1991 to total revenues of $5.6 billion. As of 1981, the three largest categories—credit, financial/economic, and financial transactional databases—accounted for more than 75 percent of total database revenues. These categories consist largely of highly specific numerical information (e.g., an individual's or a company's credit rating) for which the attributes of a videotex system would not offer major advantages.

Because of the availability of computing services, it may be that business applications of videotex will have to find a niche not currently serviced by available alternatives. In England, for example, the most successful business-oriented videotex service has been the provision of travel information such as transportation timetables and fare schedules, and hotel rates and availabilities. These services, which have accounted for more than half of all accesses by business, were utilized by small- to medium-size English travel agencies which, unlike those in the United States, did not have access to affordable computerized information services (Chapman, 1981). It is also interesting to note that business videotex users in England have made substantial use of the portion of the videotex database *not* specifically intended for businesses. While only 13 percent of videotex terminals were in households as of 1981, more than 40 percent of total page accesses were to the so-called residential databases. Part of this usage

TABLE 5.15 EXPENDITURES FOR ON-LINE
DATABASE SERVICES

Category	1981 expenditures ($ millions)
Credit	450
Financial/economic	342
Financial transactional	160
Marketing and demographics	145
Legal	55
Scholarly	52
News	28
Government watch	10
Patents	7
Other transactions	5
Lists	3
	1,257

Source: International Resource Development, 1981.

may be accounted for by employees making personal use of a system not otherwise available to them. But, in addition, it is likely that there are legitimate business uses for a considerable amount of general interest information.

Closed User Groups   A closed user group (CUG) refers to a user group in a videotex system with controlled access to information or other services. Although CUGs have not been included in the early U.S. trials, they are common in the British Prestel system. An example is the "in-house" service established by British Leyland (BL) to improve communication between itself and its dealer network. The network consists of approximately 1,800 outlets, which range from large companies to small repair shops located "from Land's End to John O'Groats, plus some in Ulster and the offshore islands" (Hutt, 1981). BL's Stock Locator system now provides this network with details on all unsold BL automobiles throughout the United Kingdom. The system, which provides information updated daily, replaced a mail-based system that provided less information and was as much as two weeks out of date. Because the system is interactive, dealers are able to use it to enter requests and to report on changes in their inventories of vehicles.

A similar system is being implemented in France by Viniprix, a large chain of food stores. A private Teletel system is being used to control inventory and communicate information on sales and to offer promotions to 250 shops in the Paris area. Store managers enter the system once a day to receive bulletins and to enter orders for merchandise (*Cablevision,* 1981a).

An extension of the closed-user group on a public videotex system is a private videotex system for a single organization or a small group of organizations. In addition to information retrieval and messaging, the system could be used for company marketing presentations, company directory services, office memos, in-house training, demonstrations of procedures such as fire emergencies, and resource use monitoring. In most cases it will be possible to link into a public videotex system for additional information and services.

Telework

Perhaps the least developed market is "telework" services supporting individuals who work from home. As of 1981, an estimated 20 thousand to 30 thousand workers conducted their business from home via electronic means (Zientara, 1981). Videotex and related systems will offer additional facilities to permit telework. A real constraint, however, will continue to be the conventions and traditions of the workplace. In the absence of a dramatic shift in social or economic conditions—e.g., recurrence of severe gasoline shortages over an extended period of time—telework seems unlikely to become a major videotex application in the short term. For specific types of employment, however, it holds great promise.

Business Market: Summary

The business market already includes a variety of electronic communication and computing technologies offering information services similar to those envisioned for business videotex. What videotex offers is the integration of a number of these services within a single system. In addition, it offers a graphic capability that can move the general business world beyond text-based applications.

The ultimate acceptance of videotex systems by U.S. business may depend less on the specific information available in these systems than on their "user friendly" characteristics. In terms of word processors and personal computers, ease of use has become a critical factor as markets spread beyond specialized users to a broader range of employees.

GROWTH OF TELETEXT AND VIDEOTEX

The preceding analyses suggest that the potential residential and business markets for teletext and (particularly) videotex are very large. If videotex succeeds in capturing even a fraction of these expenditures, it will almost certainly emerge as a major medium. But will this happen? And if so, when? These questions remain unanswered.

Historical data indicates that different media have grown at very different rates over the past century (Table 5.16). AM radio and television spread very rapidly once they were introduced. They reached more than one-quarter of all U.S. homes in less than a decade. Television, the most successful of all media, was in three-quarters of all households in just 11 years and reached 90 percent in just 16 years. While pay television is still relatively new, its growth to date has been almost as rapid as television in its early years. (Videocassette recorders, which are even newer, are also matching TV's early pace.) What all three of these media have in common is that they

TABLE 5.16   HISTORICAL GROWTH OF MEDIA IN THE UNITED STATES

Medium	Year invented	Year of commercial introduction	Number of years to reach X% of all U.S. households						
			1%	5%	10%	25%	50%	75%	90%
Telephone	1876	1877	20	N/A	N/A	N/A	69	80	93
AM radio	1895	1920	3	4	5	8	11	18	27
Television	1927	1946	3	4	5	6	8	11	16
Cable TV	1949	1949	11	19	23	32	—	—	—
Color TV		1955	7	10	12	14	17	22	—
Pay TV		1974	2	4	6	—	—	—	—
Videocassette		1978	2	4	—	—	—	—	—

N/A = Not available
Source: Sterling and Haight, 1978.

are essentially entertainment-oriented. And the only direct expense involved in gaining access to television and radio programming was the initial outlay for purchase of a receiver.

At the other extreme is the telephone. Although we take for granted virtually universal telephone service, its initial growth was extremely slow. It took two decades for even 1 percent of U.S. households to get telephone service, and it was nearly 70 years before half of all households had telephones. Similarly, cable television grew gradually at first, taking some 23 years before it reached 10 percent of households. And even though cable is growing rapidly today, it is not clear when its penetration will reach the 75 percent level.

On which end of this spectrum will teletext and videotex fall? The experience to date in the United Kingdom, which has been a pioneer in developing these media, suggests slow growth for videotex. After nearly three years of full-scale commercial service, fewer than 2,000 British households have subscribed to Prestel. However, service is relatively expensive, has been limited to information returned based applications, and the United Kingdom is not necessarily a reliable guide to what will happen in this country.

At the same time, results from current field trials and consumer surveys described at the beginning of this chapter are also of limited use. Field trials have almost all been small and of short duration and have seldom involved actual consumer payments. Even where results have been obtained, they have generally not been publicly available. Results from consumer surveys that involve asking for responses to new and unfamiliar products should always be treated with caution.

Despite these limitations we have attempted to estimate the likely growth of videotex over the next two decades. We decided that the best approach would be to solicit the views of professionals who were active in and knowledgeable about the development of videotex in the United States. We therefore prepared a questionnaire, which we circulated to a list of interested individuals that had been compiled during the project. The 77 people who replied represented a variety of different industries and disciplines (Table 5.17). The respondents were also asked to assess their own level of expertise in teletext/videotex on a five-point scale. Virtually all of them described themselves as either "familiar," "very familiar," or "expert" in terms of their knowledge of teletext and videotex.

The remainder of the questionnaire focused on the respondents' views of the likely status of teletext and videotex in the years 1985 and 2000.

Most Popular Services

The first question was concerned with identifying which of the generic classes of services will be utilized most frequently. Table 5.18 lists the aggregate results of the rankings in terms of the total number of accesses for the home market and the business market. Note that a sixth generic classification, "games/entertainment," is included along with the other five described in this chapter. Subsequent to preparation of the questionnaire, it was decided that this category, though probably distinctive from a user's perspective, was not distinctive in terms of the functional requirements of the applications. From this perspective, games and entertainment services would represent either information retrieval (e.g., jokes and riddles) or computing (e.g., video games).

Perhaps the most noteworthy result in this table is the identification of "transactions" as the leading videotex application in the year 2000. This suggests that future videotex systems may be considerably different than present-day systems.

TABLE 5.17 INDUSTRY AFFILIATION AND EXPERTISE OF SURVEY RESPONDENTS

Industry affiliations		Expertise in teletext/videotex	
Publishing	13	1. Unfamiliar	0
Research/consulting	13	2. Casually acquainted	3
Government agency	11	3. Familiar	22
University	7	4. Very familiar	30
Cable TV	6	5. Expert	22
Other commercial	5		
Telephone	3	Total	77
Communication/computing	3		
Equipment manufacturing	3		
Broadcasting	3		
Electronic publishing	2		
Financial	2		
Other legal	1		
Union	1		
Paper manufacturing	1		
Educational associations	1		
Medical library	1		
Trade associations	1		
Total	77		

TABLE 5.18   POPULARITY OF TELETEXT/VIDEOTEX APPLICATIONS

	Home market			
	Aggregate results of rankings			
	Teletext		Videotex	
Class of service	1985	2000	1985	2000
Information retrieval	1	1	1	2
Transactions	—	—	3	1
Messaging	—	—	4	3
Computing	3	3	5	5
Games/entertainment	2	2	2	4
Telemonitoring/home management	—	—	6	5

	Business market			
	Teletext		Videotex	
Class of service	1985	2000	1985	2000
Information retrieval	1	1	1	1
Transactions	—	—	3	3
Messaging	—	—	2	2
Computing	2	2	4	4
Games/entertainment	3	3	6	6
Telemonitoring/home management	—	—	5	5

Penetration Levels

Perhaps the most intense interest about the future of teletext and videotex concerns the rate at which these systems will be adopted by the public. We therefore asked our respondents to estimate the size of the consumer market for these systems in 1985 and in 2000 assuming that command systems were available in 1982. As Table 5.19 indicates, teletext is expected to grow more rapidly initially, but videotex is expected to be almost as widespread by the end of the century. Assuming that there will be some 85 million households in the United States in 1985 and 100 million in the year 2000, our estimates indicate that there will be between 1.7 million and 4.3 million videotex households in the United States in 1985, and that the number will grow to between 30 million and 38 million households by the year 2000.

TABLE 5.19   PENETRATION OF TELETEXT AND VIDEOTEX

	% of households				
	1980	1985		2000	
		Mean	Median	Mean	Median
Teletext	<0.1	7.2	5	39.2	35
Videotex	<0.1	5.1	2	38.3	30

TABLE 5.20 IMPACT OF TELETEXT/VIDEOTEX ON THE BUSINESS MARKET

	1985		2000	
	Number of responses	**%**	**Number of responses**	**%**
Little or no impact	38	56%	3	4%
Moderate impact	29	43%	20	30%
Large impact	1	1%	45	66%
Totals	68	100%	68	100%

Impact on the Business Market

Respondents were also asked to indicate the level of impact that teletext/videotex services are likely to have on the business market. The results are shown in Table 5.20. The majority of respondents expect that teletext/videotex will have had little or no impact on the business market by 1985, while just one respondent predicted a large impact. By the year 2000, in contrast, virtually all respondents predict a moderate or large impact, with only a small minority anticipating little or no impact two decades from now.

Usage of Teletext and Videotex

A final series of questions dealt with how much time people will devote to using these systems and the amount of money they will be willing to pay for them. Table 5.21 indicates that both usage and expenditures per household are expected to grow substantially as these systems evolve.

Using the figures on household expenditure along with the estimates of penetration, it is possible to calculate forecasts of overall industry revenues (Figure 5.3). Using the more conservative estimates of 2 percent penetration and $5 per week per household ($260 per year) expenditure yields total annual consumer expenditure of

TABLE 5.21 USAGE OF TELETEXT AND VIDEOTEX

	1985		2000	
	Mean	**Median**	**Mean**	**Median**
Household consumption of all teletext and videotex information services (hours per week)	5.0	3.0	13.8	10.0
Household expenditure on all teletext and videotex services (1980 $ per week)	6.2	5	14.6	10.0

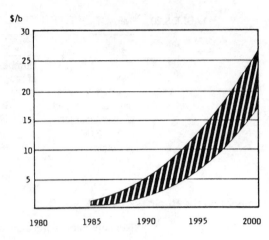

FIGURE 5.3 Total Consumer
Expenditures on All Videotex
Services.

$443 million for videotex service in 1985. The more optimistic estimates of 5.1 percent penetration and $6.20 average weekly expenditure yield total 1985 expenditure of $1.4 billion. The more conservative and more optimistic estimates for total consumer expenditures in the year 2000 (in 1980 dollars) are $15 billion and $29.1 billion respectively. And these figures do not include expenditures for videotex hardware, consumer-oriented advertising expenditures, or business expenditures for videotex.

CHAPTER 6

TECHNOLOGY FORECAST: COMPUTER DATABASES

For the home or business user, the three key technological elements of a videotex system are the computer database (discussed in this chapter), the communications network (reviewed in Chapter 7), and the user terminal (described in Chapter 8). The computer database is the heart of teletext and videotex systems. It stores the information, transaction programs, messages, and computer programs that constitute the teletext or videotex information services.

Remote access to such computer databases is by no means unique to teletext and videotex. During the last decade, many computer system applications have utilized this characteristic, e.g., computer timesharing, bibliographic search services, electronic messaging, point-of-sale terminals, and automatic bank tellers. The economics of these applications and the high skill level required have generally prohibited their extension to the home user. Teletext and videotex overcome these limitations and allow such services to be offered to business and home users alike.

The specific database arrangements for Prestel, Teletel, and Bildschirmtext were reviewed in Chapter 3. The purpose of this chapter is to identify the database requirements for general teletext and videotex systems and forecast their likely development during the next two decades. The key aspects of the database for the user are the supporting network structures, the database structure and search procedures, and the accessibility of the system.

NETWORK STRUCTURES

Centralized database arrangements are suitable for sytems that have a limited number of frequently changing pages, such as most vertical blanking interval teletext systems, and for small microcomputer and minicomputer turnkey videotex systems designed

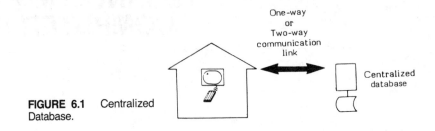

FIGURE 6.1 Centralized Database.

for corporate use (Figure 6.1). One such turnkey system using Antiope equipment is offered in the United States by Videodial. The microcomputer version has eight user ports and a database of 5,000 pages of information, and the minicomputer system has 70 user ports and a database of 180,000 pages of information. The system is suitable for specialized closed-user groups either on a public videotex system such as The Source or on private videotex systems.

Systems with centralized databases are also able to take advantage of overhead efficiencies by concentrating all their resources at one physical location. However, they are unsuitable for transaction-type services or information services that involve external information service providers. They are also limited in terms of the total number of interactive users that can be connected, e.g., it has been estimated that when the number of users subscribing to The Source exceeds 30,000, the number of simultaneous accesses may become too great for the computer hardware to handle.

Furthermore, national videotex services with centralized databases such as The Source or CompuServe also incur the cost of long-haul communications for each access. With the introduction of packet-switching networks, this communications cost has been reduced to a relatively small amount (less than $5 per connect-hour during off-peak times). Nevertheless, this is still sufficiently high to provide a deterrent to wide-scale penetration. Although communication costs are expected to continue to decline during the next two decades (Chapter 7) database costs are expected to decline even more rapidly. For both teletext and videotex systems, there are likely to be substantial economies of scale for the information providers if they are able to link their source databases directly with system operators' databases or with a videotex center. Such a center, which is also the link to the user terminal, could provide indexing, a system directory, log-on, error detection, accounting, and statistical services. It could also reformat and repackage software services from third-party databases and thus present a standard virtual terminal protocol to the different services independent of the home terminal (Ball, 1980). Therefore, economic considerations, the mix of local and national databases, and the problem of dealing with a large volume of simultaneous accesses suggest that distributed databases are an alternative arrangement for videotex systems.

Distributed databases are essentially one of two types: master/replicated databases and external databases.

In a *master/replicated database* arrangement, each database contains the same information; users typically access the database that is closest to them (Figure 6.2*a*). One database is designated the "master"; information providers update this database only. All other replicated databases are then automatically updated using high-speed data communications links. This database arrangement was implemented in the original Prestel system. The key advantage of this arrangement is that, because the databases are under the control of one organization, access procedures and page formats

FIGURE 6.2 Distributed Databases: Alternative Network Arrangements.

(a)   Master/replicated

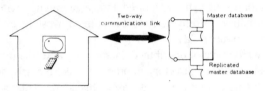

(b)   Master/replicated + external

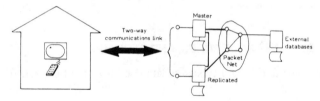

(c)   Fully distributed

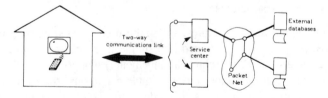

(d)   Database hierarchy

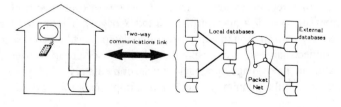

can be readily standardized. The arrangement is particularly suitable for general purpose information retrieval applications. However, for information targeted at special interest groups, replication in all databases is likely to be extremely inefficient. Furthermore, enhanced videotex applications such as teleshopping and telebanking would not be possible since access to external databases is required.

In an *external database* arrangement, information is stored in many different physically separate computer databases. Frequently accessed information or services may be duplicated to keep down the cost of communications, but in general, information requested will be obtained directly from the source database and transmitted via a high-speed data communications network (typically a packet-switching network) to a master database (Figure 6.2*b*) or to the user's local videotex system operator (Figure 6.2*c*) and thence to the user. The local system operator may or may not operate a local database; in the latter case, the system operator simply provides "gateway" functions, e.g., indexing, and possibly billing, to independent databases. Furthermore, databases may be structured into a hierarchy, with the most frequently accessed information information stored at the lowest level (possibly even in the user's terminal) (Figure 6.2*d*).

No matter which type of distributed database is used, the overriding objective is to include a "meta-service" directory in order to make the network of databases appear to the user as a single centralized database. This achieves the advantages of standardized access and search procedures while also ensuring economic efficiency and allowing the user to choose between logically distinct applications of competing variations of the same service.

DATABASE STRUCTURE, SEARCH PROCEDURES, AND STORAGE

There are three main types of database structure for teletext and videotex systems (Date, 1981):

- hierarchical structures
- relational structures
- network structures

In the *hierarchical* structures, which are used in most existing teletext/videotex systems, pages of data are stored in a simple tree format, each page being linked to all others in a strictly hierarchical arrangement (Figure 6.3).

In typical information retrieval implementations, the upper levels of the tree are used for indexing purposes to guide users who are interested in particular subject matter to the information they require. The number of selections at each branch in the tree is usually less than 10. Alternatively, users may move directly to the information page they require by keying in the page number (obtained from a subject directory or from any one of a number of local index pages). The third alternative procedure for locating information in a tree-structured database is keyword searching. Users type in a subject name; if it corresponds with a page or group of pages indexed in the database, an index page for the group is immediately displayed. Users will of course require a full alphanumeric keyboard to implement this option.

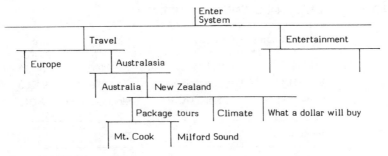

FIGURE 6.3   Hierarchical Database Structure.

Hierarchical database structures have been popular in early videotex system designs because of their simplicity for use by untrained users; at each branch point the user is presented with a simple choice of up to 10 alternatives. However, experience with the U.K. Prestel videotex system has suggested that as many of 50 percent of all users fail to negotiate four branch points without making an error. Further, recent experience with tree searches has suggested that users quickly find the simplistic step-by-step procedures too cumbersome and tend to prefer more direct access methods.

In *network database structures,* the data are represented by records and multiple links. Any data entry may have any number of links to other entries and is not constrained by a fixed hierarchical association. For example, the frame indicating United Airlines' coast-to-coast flights may have a simple one- or two-digit link to the schedules of other major carriers on the route. These could be accessed without having to return to the top of the tree and then search down a new branch. Although network structures provide greater flexibility in data handling, they suffer undue complexity in creating a cross-referencing mechanism. As such they are difficult to update and require complete knowledge of the various links. For videotex applications, network database structures fall in between the simplicity of hierarchical structures and the flexibility of the relational structures discussed below; they are therefore an intermediate stage in the evolution of videotex systems.

In *relational database structures,* data are stored in a set of interrelated tables. These tables are typically two-dimensional, in which the rows are the data records and the columns are the data fields. The records are related through fields having matching data. Table 6.1 illustrates two relational files that could possibly be part of a database used to provide a transaction service such as teleshopping or telebanking.

According to Schuster (1981), database files in relational structures have three characteristics that distinguish them from hierarchical and network-based structures:

All records in a relational file have, from the user's point of view, the same structure.

Relations between records are manifested by comparing common field values.

Users need not be concerned with how the data are physically accessed or stored.

TABLE 6.1   RELATIONAL DATABASE STRUCTURE

Customer name	Address	City	Zip
Smith J. R.	1624 First St. #201	San Francisco	94114
Brown H. L.	193 Jessop Ave.	Palo Alto	94303
Jones M. N.	41 Nancy Ave.	Mountain View	94040

Account #	Names	Balance
100046	Jones M. N.	$ 164.00
100047	Brown H. L.	$ 2.03
100048	Smith J. T.	$4,610.00

One key advantage of relational databases is that the files are logically interdependent but physically independent. This means that files can grow and that new files can be added without requiring structural changes to related files. Such a development is not easily accomplished in hierarchical or network designs.

Relative to other database structures, relational systems are a recent development. By 1981, about a dozen companies were offering relational database systems for mainframe computer applications, the most recent entry being IBM's Structured Query Language Data System (Hamilton and Manuel, 1981). By the mid-1980s, it is expected that the majority of databases will be of the relational type (Schuster, 1981). Videotex databases are almost certain to follow this trend.

The extent of storage required for teletext and videotex services depends on the particular application. Typical storage requirements are listed in Table 6.2, where we have assumed an average of 1,000 bytes of information per page, although depending on the number of characters and the extent of graphics, there may be between 500 and 2,000 bytes per page. We have assumed an average of 2 kbytes of message storage for each user and 8 kbytes for each available computer program/game. The major evolutionary development will be toward the remote user storing this information locally.

TABLE 6.2   DATABASE STORAGE REQUIREMENTS

Application	Units	Storage MBytes
Teletext (2 lines VBI)	100 pages	0.1
Teletext (full channel)	25,000 pages	25
Information retrieval (videotex)	100,000 pages	100
Transactions	2,000 pages	2
Messaging (mailbox)	10,000 users	20
Computing	500 programs/games	4

ACCESS PROCEDURES AND ACCESSIBILITY

One of the key characteristics that distinguishes teletext and videotex from other information services is their assumed ease of use by unskilled users. The procedures for accessing teletext and videotex databases must therefore be carefully designed in order to satisfy this "ease-of-use" criterion. Access procedures involve two main components:

connection to the videotex database or service (equivalent to "log-in" in computer timesharing applications)

location and/or modification of information in the database (database structure and search procedure)

For existing teletext and videotex systems, a number of different procedures are used for establishing a connection to the system database (Table 6.3).

As teletext and videotex develop into mass-market services, it is likely that a standardized connection procedure similar to those used by other mass media services will emerge; e.g., picking up a telephone handset establishes a connection to the local telephone exchange and turning the channel selector switch on a television set links the set to a television broadcast station.

Considering the range of teletext and videotex services that may emerge during the next two decades, the connection procedure is likely to reduce to two functional steps:

- service selection
- user identification

As will be discussed in Chapter 7, the number of communication links into homes or businesses is likely to increase. In selecting an information service, users will therefore be able to choose the most economical communication medium. For example, for "news headlines" a broadcast teletext service is likely to be selected; for

TABLE 6.3 EXISTING PROCEDURES FOR ACCESSING SYSTEM DATABASES

Teletext/Videotex system	Access procedure
Teletext	Select TV channel; press button for teletext service.
Prestel	Select TV channel; switch decoder on; dial Prestel computer (some decoders automatically dial Prestel); key in identification code.
The Source	Dial Telenet/Tymnet number; insert telephone receiver into audio coupler; type terminal identifier; type log-in characters; type identification code.
INDAX (Cable TV)	Turn on keypad; press "text" button. For transaction services, e.g., banking, enter password and identification number.
Viewtron	Turn on keypad; press "phone" button; dial Viewtron number on keypad; press "video" button on keypad; enter 4-digit ID number; enter 5-digit password.

teleshopping, a cable-based service may be selected; for electronic messaging, a telephone-based service may be selected.

Service options and communications links would be preselected by the user and programmed into the terminal; when a service option is selected (by pushing a key on a control box similar to today's automatic telephone dialer units), a connection is automatically established to the desired service using the preselected communications medium.

The second step in the connection procedure is for the user to provide some *identification* for security and/or billing purposes. Identification procedures may range from the relatively insecure password systems in common use today to magnetic cards to signature validation to voice recognition. All four methods are technologically feasible today; by the year 2000, it is likely that all four will be in common use.

One final aspect of the computer database that is of concern to the user is the waiting time for access to a page or frame of information. There are two fundamentally different situations: one-way broadcast and two-way cable or telephone services.

Consider first the one-way broadcast system. The average access time depends on:

- the number of pages per cycle (N)
- the amount of information per page (b_i = the number of bytes on the ith page)
- the number of VBI or TV lines out of 525 devoted to the teletext system
- the broadcast transmission speed or data rate (t bits/sec)
- the number of times a page is replicated in a cycle or omitted from cycles

The access time will also depend on the size of the buffer memory in the local terminal, the degree to which page selection may be "preprogrammed," and the extent to which successive pages are chained, i.e., when page j is accessed, the buffer stores a sequence of related pages $j + 1, j + 2$, and so on (Alternate Media Center, 1981).

If there is no buffer, then the speed of transmission becomes

$$\frac{kt}{525} \text{ bits/sec}$$

where k = number of VBI lines used in the teletext service.

Since there are 8 b_i bits of information on the ith page (1 byte being equal to 8 bits), the time to transmit (i.e., cycle) the whole database of N pages becomes

$$T = \sum_{i=1}^{N} \frac{4,200\, b_i}{kt}$$

For example, if $N = 100$ pages in the database, $b_i = 1000$ bytes/page, $k = 4$ VBI lines, and $t = 4.6$ Mbits/sec, then the total time to cycle the database is 22.8 seconds (plus any system overhead time), and the average waiting time is 11.4 seconds. The trade-off between information per page, transmission speed, number of vertical blanking intervals used, and number of pages in the database can easily be demonstrated (Table 6.4).

TABLE 6.4 TRADE-OFFS BETWEEN AVERAGE WAITING TIME AND TELETEXT SERVICE CHARACTERISTICS

Number of pages in database	Average information (text, graphics, codes) per page (bytes/page)	Transmission speed (Mbits/sec)	Number of VBI lines used	Total cycle time (sec)	Average waiting time per average frame (sec)
100	500	4.6	2	22.8	11.4
200	500	4.6	2	45.6	22.8
100	1000	4.6	2	45.6	22.8
200	1000	4.6	2	91.2	45.6
100	500	5.7	2	18.4	9.2
200	500	5.7	2	36.8	18.4
100	1000	5.7	2	36.8	18.4
200	1000	5.7	2	72.6	36.8
100	500	4.6	4	11.4	5.7
200	500	4.6	4	22.8	11.4
100	1000	4.6	4	22.8	11.4
200	1000	4.6	4	45.6	22.8
100	500	5.7	4	9.2	4.6
200	500	5.7	4	18.4	9.4
100	1000	5.7	4	18.4	9.4
200	1000	5.7	4	36.8	18.4

The choice of numbers in Table 6.4 reflects current systems. The British Oracle system offers 200 pages, whereas the WETA trial in Washington and the U.K. Ceefax system offer about 100 pages at any one time. The average information per page reflects the approximate balance between a low-resolution alphamosaic system and a picture description instruction alphageometric system with lots of graphics. The two transmission speeds correspond to the speed originally recommended by the FCC for use in the United States (5.7272 Mbits/sec) and that used by WETA in its Washington trial (4.58 Mbits/sec). The number of VBI lines used has been between 2 and 4 in most applications and is currently 4 in the above three systems.

By replicating pages within the cycle, chaining pages, or selectively introducing certain pages, the average waiting time per frame may be reduced. The size of the database may be increased substantially by allocating more lines to transmit teletext or by cycling different information at different times. The introduction of a buffer memory within the home terminal and the down-loading of information during non-viewing hours change the role of broadcast as a limited database information retrieval system. For example, utilizing the full 525 lines, with 1000 bytes per page on the slower transmission cycle, still allows the user to "receive" 576 pages or frames of information per second, or for an average waiting time of 10 (20) seconds a database of 11,520 (23,040) pages is supported.

Consider now two-way systems using either coaxial cable TV or a telephone network. Once access to the database is established through one of the ports in the relevant system computer, there is no theoretical limit to the size of the database that can be accessed almost instantaneously. There will, however, be a transmission time to get the frame of information to the user terminal. Its length will depend upon the amount of information stored on the frame and the downstream transmission speed, i.e., 8 b_i/t. For example, in a telephone-based service at 300 bits/sec, the time to display a 1,000 byte page is 33.3 sec. The trade-off between transmission speed, page size, and display time is

Transmission speed (bits/sec)	Page size (bytes)	Display time (sec)
300	500	16.7
1200	500	4.2
300	1000	33.3
1200	1000	8.3

In a two-way cable-based system with a downstream transmission speed of several megabits per second, the display time is essentially instantaneous.

The number of simultaneous users becomes an issue when considering widespread penetration. For two-way cable, the simultaneous access problem is addressed by multiplexing data messages on the upstream bandwidth. In the telephone system there are three alternatives. The first is to route individual calls through the local branch office directly to computer ports on a central computer. This has been shown by Bloom (1980) to increase the load at the switch and to disproportionately affect the switch control element (i.e., the number of calls to be handled). The second option is a gateway design in which the telephone calls are routed via a packet-switched network to the respective information provider computer. This is the design being used in the Bildschirmtext system, in which regional computers store the most frequently accessed pages from the general centralized database; remote databases are accessed via the packet-switched network. The third option is to "packetize" or multiplex the incoming calls at the local loop office. AT&T's proposed advanced communication system (ACS) is one such possibility.

CONCLUSION

The key factors involved in using videotex and teletext databases are ease of access, search procedures, and average waiting time. Complicated access procedures and protocols will definitely retard the acceptance of videotex systems. Given the difficulties experienced by users with tree-structured indexes, the issue of an appropriate search procedure must be addressed. Unfortunately, no one has really come up with a simple search procedure that is both powerful and easy to use. Keyword searches are steps in this direction.

The average waiting time, at least in the case of videotex systems, is a function of the investment in the communications network decoder. It is most important in telephone-based systems since high-speed modems are now relatively expensive. Experience with interactive computer systems has shown that users become impatient if they have to wait too long for a response from the system. In this regard, Miller (1968) has postulated a set of ''maximum'' response times for various categories of tasks. We can, in fact, draw analogies between Miller's tasks and teletext and videotex applications to postulate some comparable response times. These will be introduced in Chapter 9 when we finally draw together teletext and videotex applications with the technology components.

CHAPTER 7

TECHNOLOGY FORECAST: COMMUNICATION NETWORKS

The purpose of this chapter is to identify the communication networks that can potentially be used for teletext and videotex services and to forecast their development during the next two decades. Communication networks that deliver services to the home typically have two main components:

1 Local distribution networks, which provide a communication link from the computer database or videotex center into each user's home, e.g., telephone local loops and television broadcast transmissions

2 Long-haul networks, which interconnect local communication services with remote locations, e.g., telephone long-distance networks, national television networks, and packet-switched networks

The local distribution and long-haul network components of teletext and videotex systems may in fact all be part of the same underlying communication network. In the British Prestel system, for example, all the elements are part of the telephone network. Alternatively, each element may be provided by a different underlying network, e.g., local distribution may be provided by a cable television network and long-haul by a satellite common carrier. Our major focus is on the local distribution networks—the links to the user.

The local distribution network has a number of attributes relevant to the user. These are:

- penetration
- transmission speed
- interface unit cost
- communications link cost

Each of these attributes affects the nature, scope, or cost of a teletext or videotex service. Penetration—the percentage of all households connected to a given network—obviously influences the economic potential of a service. The transmission speed of the communication link has a major impact on the quality of the display, i.e., text, graphics, television-quality pictures, and combinations of these. In considering the cost of communication network services, it is helpful to distinguish between the cost of the interface device and the cost of the communications link. In traditional telecommunications services such as the telephone, it has been common to include both these costs with the service cost and charge users a uniform flat rate rental. However, as a result of increased deregulation of the telecommunications industry during the last decade, users may now purchase their own telephone instruments or data modems (modulator-demodulator) from independent suppliers. This has led to a separation of the terminal equipment costs from those of the communications network.

For teletext/videotex systems, further cost separations are required—many suppliers might potentially contribute to the provision of a system. For example, a local information provider (newspaper) might have a national information service provider (CompuServe) assemble the services for distribution locally (cable) or to remote areas (packet-switched networks). The local distribution network may even be provided by two separate organizations. In hybrid systems, for example, users may request information using the dial-up telephone network, but the information may be transmitted to the user over a cable television channel.

The interface device provides a physical and electrical connection between the user's equipment and the communications link. It also carries out any signal conversion necessary to ensure that information generated by the user's equipment is compatible with the communications link and vice versa. All teletext and videotex systems require a decoder to translate incoming digital signals into a video display format and, in addition, telephone network-based videotex systems require a modem to convert analog signals transmitted over the network into a digital format. Modems and decoders are either leased from the communications network provider or purchased outright from the manufacturer.

The forecasts presented in this chapter focus on known and proven communication technologies. Historically, the implementation of new communication technologies has proceeded over a number of decades, e.g., the telephone was invented in 1876, but it was not until 1952 that 66 percent of all U.S. households had one. The development of television, however, was more rapid; commercial television was introduced in the late 1940s and by 1955, 66 percent of all U.S. households had a TV set. Likewise, cable television systems were first installed in 1949; by 1982, 32 percent of all homes were receiving cable TV service (Table 5.16).

There are many factors that have influenced the growth rates of these communication technologies—technical, economic, regulatory, consumer demand, etc. It is difficult to ascertain the exact influence of each, but the lesson is clear: the implementation of a new communication technology takes many years to achieve widespread penetration. This is especially true for point-to-point technologies using physical communication links (e.g., paired wire, coaxial cable, optical fiber cable) because of very high capital and manpower requirements. Thus, if a communication technology has not yet been invented, it is unlikely to achieve widespread implementation this century. For a 20-year forecast period, therefore, it is reasonable to focus on existing communication technologies.

COMMUNICATION TECHNOLOGIES

The following communication network technologies are considered to be the most likely alternatives for local distribution of teletext and videotex services during the next 20 years:

One-way technologies	Two-way technologies
Broadcast television	Cable television
Multipoint distribution system (microwave)	Switched telephone network
Direct broadcast satellite	Packet-switched network
FM radio	
Cable television	

The alternative network arrangements supported by these technologies are illustrated in Figure 7.1. Many of the one-way technologies can, however, be used to provide hybrid two-way (videotex) services. In fact, we believe these hybrids offer an attractive alternative because each network component can be used to its full advantage. For example, broadcast technologies such as a direct broadcast satellite that provide wideband one-way transmission can potentially be used in conjunction with a narrowband point-to-point communications link such as a telephone to provide a two-way videotex service. Table 7.1 summarizes current attributes of the various communication technologies available in the local distribution network.

BROADCAST TELEVISION

The broadcast television network provides a means of communicating information on a wide-scale basis. In broadcast television teletext systems, digital data is typically transmitted during the vertical blanking intervals of the television signal. A special decoder is used to extract the data and display the information on a television receiver. The decoder may be either built into the receiver or connected to the antenna input leads.

In the United Kingdom, where there were over 300,000 teletext-equipped TV sets at the end of 1981, set-top teletext decoders ranged in price from $90 to $180 (U.S.). U.S. manufacturers expect to be able to market decoders for similar prices, and even lower prices can be expected with volume production and the continuing decreases in the cost of microelectronics.

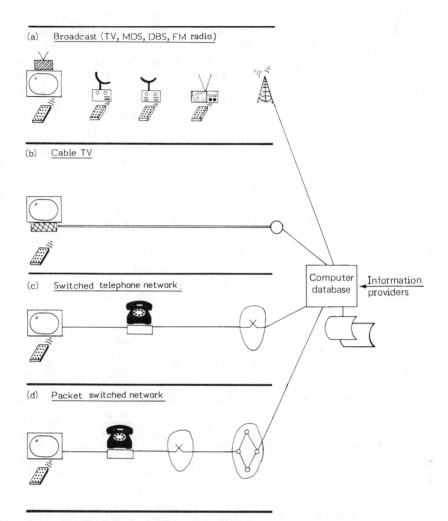

FIGURE 7.1 Communication Networks for Teletext and Videotex.

Most existing teletext systems use between two and four lines of the vertical blanking interval, providing an effective data transmission rate of 13 kbits/sec. The capacity of teletext systems can be increased by using more lines within the field blanking interval, e.g., the U.K. teletext system provides for up to 16 lines (Tanton, 1979), by cycling pages at certain times only or in certain sequences, and by "chaining" successive pages. (This last technique requires the user to have some storage capacity in the local terminal.)

By allocating all lines of the television broadcast to teletext use, i.e., a full television channel, the number of pages of information in the system could be increased to over 25,000 without increasing the user access time (see Chapter 6). Alternatively, much larger data sources (millions of pages) could be accommodated using cycle

TABLE 7.1 COMPARISON OF COMMUNICATION TECHNOLOGIES BY ATTRIBUTES

Network	1982 Penetration (% households)	Communications link cost (1982 $/month)	Transmission speed (bits/sec)	Interface unit cost ($1982) Modem	Decoder
Broadcast television (2 VBI lines)	99%	Nil	13,000		$200
Multipoint distribution system	<1%	Variable	8M		$300
Direct broadcast satellite	<1%	—	(1)		$4,000
FM radio	95%	Nil	9,600		$400
Cable television (full-channel)	28% (1-way) <1% (2-way)	$8–30	8M		$300[2]
Telephone public switched network	98%	$0–100[3]	300 1,200	$ 150 $ 700	$400–1,000 $400–1,000
Leased circuits		$200–2,000[4]	2,400 4,800 9,600	$1,600 $2,000 $3,000	$400–1,000 $400–1,000 $400–1,000

[1]Channel bandwidth of 27 MHz defined at 1977 World Administrative Radio Conference for Region 1 (Europe and Africa) and 3 (Asia); to be defined for Region 2 (North and South America) at 1983 Regional Administrative Radio Conference.

[2]Decoder is very often provided free with an added monthly cable service charge such as with INDAX ($4-6).

[3]Cost depends on usage and tariff structure; for calls within the flat rate area, users would not pay any additional charges. For toll calls, a typical cost would be $100/month (assuming average monthly usage of 6 hours with 25% day, 50% evening and 25% night rates). For access to a packet switched network, such as TYMNET, there is an additional cost of approximately $30/month (assuming average session length of 12 minutes, average monthly usage of 6 hours, 8,000 characters per session).

[4]Cost depends mainly on the distance of the leased circuit (typically, a per-mile charge).

times of, say, one hour if intelligent terminals were used to identify the pages required and store them for subsequent access. However, the number of channels that can be used for normal over-the-air television broadcasts is limited by the availability of frequency spectrum in the very high frequency (VHF) and ultrahigh frequency (UHF) bands. Full-channel teletext systems are more likely to evolve either on coaxial cable systems, where the reuse of frequency spectrum is not constrained by the need for geographical separation; on low-power television stations that have recently been authorized by the FCC; or at higher radio frequencies, where the spectrum is less congested, e.g., direct broadcast satellite systems in the 14/12 GHz band (see later discussion).

Twenty-Year Forecast

After two decades of rapid expansion, conventional broadcast television systems have reached a plateau. Virtually every home has a television receiver (most in color), and most are able to receive three national commercial network transmissions as well as a number of local transmissions. Although an increasing number of households are subscribing to cable television systems to obtain interference-free reception as well as an expanded selection of programs, it is likely that broadcast television as it exists in 1980 will still exist in 20 years' time. There will, however, be an increasing emphasis on diversity in programming. The growing number of pay TV subscribers has already demonstrated there is a market demand for this diversity.[1] Over-the-air subscription television (STV) service is a possible medium for teletext services. Low-power television (LPTV) broadcasting is another.[2] About 450 stations are expected to be on the air by the mid-1980s (Baker, 1981), although the FCC may eventually license as many as 4,000 new stations. Each of these stations, which can be built for less than $100,000, and target special audiences could potentially provide a teletext service. They may be owned and operated by organizations not currently in broadcast television and could radically alter the current structure of broadcast television over 20 years.

Teletext decoders will decline rapidly in price as they are integrated into the television set and the price of computer and memory chips declines. By the end of the 1990s, Zenith estimates the current price should be no more than the cost of the keypad.

There have also been two recent developments in television technology, both of which are likely to filter into the mass consumer market during the next two decades. One is the development of digital video systems; the other is high-definition television (HDTV). Digital video equipment has been demonstrated to produce improved picture and sound quality in a studio environment; however, other characteristics such as editing are not yet as well developed as in analog equipment. Furthermore, digital video signals require about twice the bandwidth of analog signals, i.e., about 12 MHz. Thus even though very large storage capacities are required to deal with digital video signals, the rapidly declining costs of computer memory will eventually allow the technology to become economically feasible. When this occurs, digital video will permit greatly enhanced teletext/videotex systems. Full-motion video will replace existing slow scan video and graphic/text displays.

High-definition television will also have an important influence on future television systems. However, it is unlikely to influence teletext/videotex developments to the same extent as digital video. HDTV has been developed by the Japanese and some experimental equipment has already been demonstrated, e.g., receivers and cameras (Price, 1981). This equipment uses 1,125 scanning lines per picture frame (compared to 525 lines in existing U.S. television), has an aspect ratio of 8 × 5 (compared to 4 × 3 in existing TV), and requires a bandwidth of some 30 MHz (compared to 6 MHz for existing TV). The existing VHF band would be unable to cope with such a

---

[1]During 1980 it was estimated that pay television would generate nearly $800 million in revenue (*Broadcasting*, 1980*b*). This includes subscription television (over-the-air and cable) and multipoint distribution services (MDS).

[2]A 1,000-watt television transmission has a broadcast range of 10 to 15 miles; a typical full-power 50,000-watt transmission has a 60-mile range. By March 1982 when the FCC formally authorized a low-power television broadcasting service, they had received over 6,500 applications for licenses.

large bandwidth service, but direct broadcast satellite systems could. HDTV is likely to evolve through specialized applications, e.g., the Japanese are considering using HDTV for transmitting movies to remote theaters as a substitute for celluloid film.

MULTIPOINT DISTRIBUTION SYSTEMS (MDS)

An MDS system consists of a fixed broadcast station transmitting omnidirectionally in the 2 GHz (microwave) frequency band to fixed receivers. The range of such systems is generally limited to about 25 miles. Information providers lease channels from an MDS common carrier at tariffed rates. The most common users of MDS are pay television systems (e.g., Home Box Office). MDS pay television requires a dish antenna and a signal converter for apartment or home reception. In addition, each subscriber rents a decoder to unscramble the television signal; typical monthly rentals are in the range of $15 to $20.

MDS facilities are suitable for either analog or digital transmissions, and potentially could be used for teletext/videotex services. In the teletext mode, either the vertical blanking interval could be used, as in other broadcast television systems, or a complete channel could be dedicated for teletext use (providing data transmission rates of up to 8 Mbits/sec). In the videotex mode, the telephone system could be used as the user-to-network communications, and MDS could be used to deliver the required information to the user.

MDS teletext decoders are currently slightly more expensive than VHF/UHF decoders because of the need to convert the received microwave frequencies to VHF for compatibility with the home television receiver. The 1982 cost is approximately $300 (cf. $100 to $200 for VHF/UHF decoders). This cost is expected to decline, and therefore the price differential between teletext decoders for different communications media will become relatively insignificant.

By the beginning of 1980, almost 50 cities were being served by MDS, another 93 systems were being constructed and over 500 construction permits for new systems were pending before the FCC (Mosher, 1980). In early 1982 Microband Corporation of America (a subsidiary of Tymshare) announced plans for "Urbanet," a multichannel MDS service that can be used for teletext, pay TV and data transmission services, and videotex in conjunction with a telephone system.

Twenty-Year Forecast

MDS pay television systems have emerged primarily to provide service to areas not served by cable. As the penetration of cable television systems grows, the need for MDS as a television distribution medium will diminish and MDS is likely to become an important distribution for high speed business data communication (including business videotex). Despite the recent shelving by Xerox of its planned XTEN data communications network using MDS links, during the late 1980s and 1990s, networks similar to that proposed by Xerox will re-emerge. Current FCC rules allow for two MDS channels in each of the 50 largest markets and one channel in all smaller markets. However, the FCC is considering reassigning part of the spectrum now used for educational services (ITFS) to provide up to 10 MDS channels per market. The Microband system, which proposes to offer up to 14 channels in the top 50 U.S. markets, suggests that MDS could become a major medium for distribution of teletext and videotex services.

DIRECT BROADCAST SATELLITES (DBS)

Satellite technology is well established as a reliable and low-cost (per circuit-mile) point-to-point communications medium. In 1980, 60 percent of all intercontinental telecommunications traffic was transmitted via satellite circuits (LeMasters, 1980). During recent years many countries have also established their own domestic satellite communications systems. By 1980, 12 domestic satellites were serving North America and another 14 are expected to be in orbit by 1984.

A major user of domestic satellite communications in the United States is the cable TV industry. Over 55 percent of the 4,100 existing cable TV systems are equipped with satellite earth terminals (LeMasters, 1980).

Signals transmitted from satellites are widely dispersed by the time they reach earth, and home experimenters quickly discovered they could build relatively inexpensive home receivers to pirate the cable TV transmissions. Some experimenters claim to have built systems for under $1,000 (Cooper, 1980). Since the deregulation by the Federal Communications Commission of receive-only satellite earth stations in 1976, many firms have emerged marketing home terminals. Prices currently range from $4,000 to almost $40,000.[3] These systems require large dish antennas, typically between 10 and 15 feet in diameter. Because of this and the high equipment costs, these satellite receivers are unlikely to be used on a wide-scale basis. One estimate puts the size of this market at between 50,000 and 100,000 (Hopengarten, 1980).

However, direct broadcast satellite systems specifically designed for home reception are much more likely to attain wide-scale penetration. By increasing the power of the transmitted signal, much smaller and less expensive receivers are required. It has been estimated that in order to cover a time zone of the United States with sufficient radio energy for reception by 3-foot dishes, between 170 watts and 300 watts of transponder[4] power are needed (*Broadcasting*, 1980a). Experimental DBS systems have already demonstrated the technical feasibility of these high-power transponders in the 14/12 GHz frequency range (Ku band), e.g., the Japanese experimental broadcasting satellite (BSE) had a 100-watt transponder; the Canadian CTS[5]/Hermes satellite had a 200-watt transponder. France and Germany are each planning to launch a ''preoperational''[6] DBS satellite with 260-watt transponders in 1983.

---

[3]Charlie and Company, San Jose, Calif.: Model ESSI—$3,995; Neiman-Marcus, Dallas, Texas: Scientific-Atlanta—$36,500.

[4]Transponder: A receiver-transmitter combination that retransmits the received signal greatly amplified at a different frequency.

[5]CTS: communications technology satellite.

[6]The ''operational'' satellites are expected to be launched some two years later.

In the United States, the Satellite Television Corporation (a subsidiary of Comsat) is proposing to establish a three-channel satellite subscription television (SSTV) service by the mid-1980s. The first satellite launched will serve the Eastern time zone of the continental United States; three additional satellites will subsequently be launched to serve the other time zones. Subscribers to the service will need to purchase a 2.5-foot-diameter dish antenna and associated frequency translation equipment as well as lease a decoder to unscramble the transmitted signals (approximately $10 per month). In addition, there will be a monthly program service charge of $14 to $18, similar to other pay TV services (*Telecommunications Reports*, 1980). A major electronics manufacturer has quoted prices for home receivers with a one-foot-diameter antenna for this proposed service in the $200 to $250 range, in manufactured quantities of 10 million (*Broadcasting*, 1980c).

A key element in the development of low-cost home receivers has been the design of the down-converter earth terminal. Most satellite earth stations in operation today use a low-noise amplifier (LNA) at the focal point of the dish or attached to the back of the dish. The high cost of amplifiers (typically $1,000 to $6,000 even for simple home terminals) has been a major deterrent to wide-scale implementation. The down-converter terminals overcome this problem by converting the received 12 GHz signal to a lower frequency (e.g., 700 MHz), which can then be fed over standard coaxial cable to an FM-AM converter/decoder unit associated with the television receiver (Figure 7.2). The cost of mass-produced down-converter terminals is currently an order of magnitude below that for LNA terminals. However, with the continuing decline in the cost of semiconductor circuitry, the cost of LNA terminals may eventually become competitive.

Twenty-Year Forecast

Although DBS technology has been successfully demonstrated, it will be at least another 5 to 10 years before service becomes available on a wide-scale basis in the

FIGURE 7.2   Components of Home Satellite Earth Terminals.

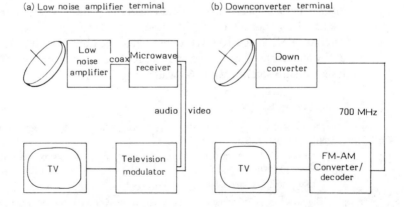

(a) Low noise amplifier terminal   (b) Downconverter terminal

United States. Even then DBS, as a pay TV service, will compete with cable TV, UHF subscription TV, and MDS pay TV services.

DBS systems, however, do have an enormous potential capacity in terms of bandwidth. The bandwidth available in the 14/12 GHz band is adequate to eventually allow 50 to 100 television channels to be broadcast to the home (Taylor, 1980*a*). Given this capacity, it will be possible to allocate one or more complete channels for new services such as teletext. Alternatively, the increased bandwidth availability provides an opportunity for the implementation of new high-definition wide-screen television systems with stereophonic sound.

There is even greater potential at the higher satellite communication frequencies—17.7 to 21.2 GHz (Kc band) and 27.5 to 31.0 GHz (K band). Communications at these frequencies are still mainly experimental, but within a 20-year time frame they are likely to enable an expansion of communications services into the home.

FM RADIO

FM (frequency modulated[7]) signals are used for radio broadcasts as well as for the sound component of television broadcasts. Since 1955, the FCC has permitted FM broadcasters to provide supplemental services in unused frequency spectrum of the FM channel allocations. The most common usage of these ''subchannels'' (or ''subcarriers'') on FM radio has been for background music services to businesses and department stores, e.g., Muzak. These subchannels could also potentially be used for broadcast teletext transmissions. The subchannel bandwidth is adequate to allow a data transmission speed of up to 9,600 bits/sec, which comes close to that achievable on two lines of the vertical blanking interval in broadcast television transmissions (13,000 bits/sec). By the beginning of 1981, there were almost 3,300 commercial FM stations in operation (cf. 4,600 AM stations), and the number is expected to grow. Most FM radio broadcasts transmit stereophonic sound, using the main carrier frequency for one channel and one subcarrier for the other. The second subcarrier[8] is already used by some FM stations for Muzak distribution or similar application. However, the subcarriers associated with television sound transmissions remain largely unexploited.

Decoders for extracting teletext information from the FM subcarrier and displaying it on a television screen would cost about $400, using 1982 technology. However, this cost could be expected to decline with volume production and reducing costs of microelectronics. In addition, Dataspeed, Inc. proposes a teletextlike service in which a portable, speakerless FM receiver visually displays current prices on up to 50 stocks selected by the user.

---

[7]Frequency modulation: One of three ways of modifying a sine wave (carrier) signal to make it ''carry'' information. The carrier has its frequency modified in accordance with the information to be transmitted. The other two methods of modulating a carrier are amplitude modulation (AM) and phase modulation (PM).

[8]Only two are permitted by the FCC.

Twenty-Year Forecast

FM broadcast technology could become an alternative broadcast distribution medium for teletext or hybrid videotex services even though no trials currently use this transmission medium.

CABLE TELEVISION

Cable television systems were initially established as a means of distributing television broadcasts to subscribers in weak signal areas. Typically, a receiving antenna would be located in a strong signal area and a coaxial cable feed used to distribute the signal to individual subscribers.

Unlike broadcast television channels, which are limited by the availability of frequency spectrum, there is no theoretical limit to the number of channels that can be distributed on a coaxial cable system. The capacity of an individual coaxial cable is largely determined by the characteristics of the cable amplifiers and the spacing between amplifiers along the cable. Current technology has achieved a bandwidth of 400 MHz, which provides for over 50 television channels. The choice of amplifier spacing is essentially an economic trade-off between the cost of amplifiers and the cost of an additional coaxial cable. The decreasing cost of microelectronics is likely to favor the adoption of even closer amplifier spacings in future cable systems, providing further increases in the number of channels available (Figure 7.3). The availability of such large numbers of television channels makes it possible to dedicate complete channels for teletext/videotex services. Each 6 MHz channel is capable of carrying between 5 Mbits/sec and 10 Mbits/sec of data (Frank, 1981). Furthermore, cable television technology provides an excellent transmission quality (bit error rate better than 10^{-8}). The major incremental cost for adding data to a conventional two-way cable television system is the cost of the interface device ($300 in 1982). However, like the cost of telephone network data modems, the cost of cable television adapters is also likely to decline substantially during the next decade.

FIGURE 7.3 Forecast Growth in Capacity of Cable TV Systems.

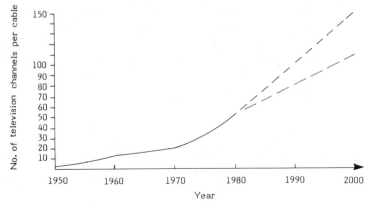

After 1990, it is likely that optical fiber cable systems will supersede coaxial cable technology. Optical fiber systems will be able to provide much higher bandwidths than coaxial cable, yet at the same time avoid the need for reduced repeater spacings. Even with today's technology, more than twice the repeater spacing is possible with optical fiber systems (Bigi et al; 1979).

By 1982, 28 percent of the 80 million households in the United States subscribed to a cable television service. The explosive growth in cable television during the last decade (Figure 7.4) has been spurred by three key regulatory developments (Mosher, 1980). First, in 1972 the FCC repealed a 1968 regulation that had effectively barred cable systems from nearly all of the 100 largest cities. Second, in 1976 the FCC deregulated receive-only satellite earth stations, thereby expanding the range of programs available to cable viewers. Of the 4,100 cable TV systems in operation at the beginning of 1980, half were linked to a satellite earth station. And third, in 1977, the courts overturned FCC rules that had restricted pay TV programming of first-run movies and sports that had been under contract to broadcast television within a five-year period. By 1980, some 36 percent of all U.S. homes were able to be served by a cable TV system (Hopengarten, 1980), i.e., a feeder cable passed 36 percent of homes. This figure is expected to rise to 66 percent as early as 1985 (Taylor, 1979*a* and *b*).

Some 98 percent of the 4,100 cable TV systems in operation in 1980 were designed for one-way operation, i.e., the distribution of television signals *from* a central point *to* the network subscribers. However, the pioneering Warner-Amex QUBE system in Columbus, Ohio, has demonstrated the potential of coaxial cable systems for two-way communications. The simplest two-way cable configuration involves the use of two separate coaxial cables, one for each direction of transmission. Alternatively, a single cable can be used with different frequencies being employed for the two directions of transmission (Figure 7.5). This alternative is technically more complex but is somewhat less expensive than the two-cable option. Despite some uncertainty in the cable TV industry regarding the optimum timing for

FIGURE 7.4   Forecast Growth in Cable TV.

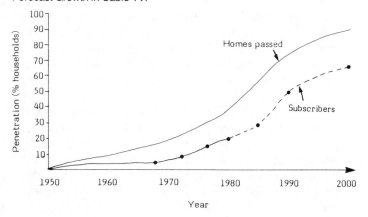

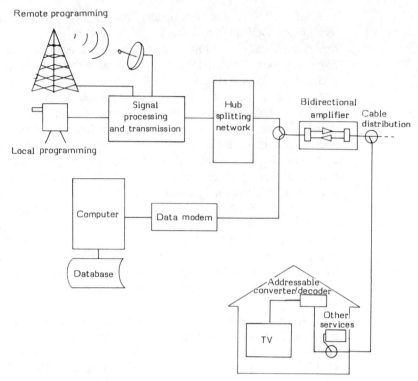

FIGURE 7.5 Two-Way Interactive Cable TV System.

introducing two-way cable TV services, there is general agreement that two-way cable will become increasingly important in the future (Figure 7.6). The FCC has recognized this trend and now requires that all new coaxial cable systems be at least convertible to two-way operation. Thus, the stage has been set for the evolution of a second *two-way* communications link into the home (the first being the telephone, of course). It will be difficult and expensive to upgrade most existing one-way systems. Many will require new cable. Nevertheless, the high growth rate forecast for new cable TV systems means that widespread penetration of two-way systems is feasible within a 20-year time frame. Furthermore, all existing franchises will be due for renewal during this time frame, thus providing an opportunity for upgrading. The cable TV network therefore becomes an alternative medium for the development of interactive videotex systems.

A development associated with the move to two-way cable technology has been the addressable converter. Television signals are scrambled at the cable TV control center (the "head end") and a tagging system is used to identify which home decoders are authorized to unscramble the signals. Address data presets the decoder in the home to respond to any tagged channel that matches the particular authorization level. When a subscriber requests a specific program or level of programs, the information is entered into the system computer. The appropriate authorization signal is

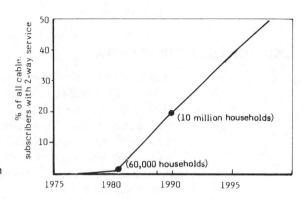

FIGURE 7.6 Forecast Growth
in Two-Way Cable.

sent to the central controller from the computer, which in turn sends the information to the subscriber's terminal. Addressable converters were originally developed to prevent basic cable subscribers from receiving free pay TV programs. The converters, however, also provide for selective access to teletext/videotex services as well as a means for billing individual subscribers. The 1981 cost of addressable converters is in the range of $150 to $200.[9] As production quantities increase and the cost of microelectronics declines, the cost of these converters is likely to decline.

Another important emerging application of coaxial cable communication networks is in short-haul private data communication networks, a variation on the basic cable theme. A coaxial cable is used as the bearer; users tap into the cable as required and broadcast data packets along the cable. Typical packet cables are no more than 0.5 mile in length, providing communications within a building or between buildings in the same vicinity. Even though coaxial cables allow very high transmission speeds (up to 400 Mbits/sec), the cost of the user interface rises with the transmission speed. Practical implementations to date have employed much more modest speeds, e.g., Ethernet operates at a data transmission speed of 3 Mbits/sec (Metcalfe and Boggs, 1976). Packet cable networks are likely to become an important communication medium for intracompany teletext/videotex systems, similar to the private branch exchange (PBX) in the telephone network.

Twenty-Year Forecast

Cable television is currently experiencing rapid growth. During the next two decades, cable television distribution networks will become a major medium for distributing teletext and videotex services. The availability of a large bandwidth communication

[9]Scientific-Atlanta is supplying Ameri-Cable (Miami, Florida) with 19,300 addressable converters during 1981 at a cost of about $145 each. TOCOM (Dallas, Texas) expects to be producing 10,000 55-Plus (55 video channels + 55 teletext channels) addressable converters per month at a cost of about $200 by the end of 1981 (*Cablevision*, 1980).

link on today's coaxial cable system (and tomorrow's fiber optic systems) makes possible the eventual inclusion of full-motion digital video signals in teletext and videotex systems.

SWITCHED TELEPHONE NETWORK

The telephone subscriber's local distribution network, connecting each individual subscriber's telephone to a local exchange (a central office), is known as a local loop. The technology used almost universally in the telephone subscriber's local loop is paired copper cable. This technology has proved to be remarkably stable for almost a century. Copper cables were first used to replace open wire lines around the turn of the century and will most likely still be the dominant technology in the subscriber's local loop at the end of this century. The main reason why paired cable technology has proved to be so stable is that the local distribution network represents a substantial proportion (15 percent[10]) of the telephone company's capital investment. It has been estimated that the total book value of local loops in North America is between $10 and $20 billion (Martin, 1976). As long as paired cables continue to provide adequate service, there is little motivation to invest in some other technology. Furthermore, the penetration of the telephone network has reached saturation and hence there is no potential for rapid growth to stimulate the implementation of some new technology.

Even though the average telephone subscriber uses the telephone for no more than 20 minutes per day (Pacific Telephone, 1981), the availability of the network for speech communications is clearly reduced if the local loop is also used for videotex services; i.e., users can not make or receive telephone calls while videotex information is being accessed. However, it is technically possible to multiplex speech and data onto the same physical cable pair, thereby allowing videotex services to be provided without affecting the availability of the loop for telephone service. Thus, the use of multiplexing techniques in subscribers' local loops can enhance the availability of the local loop for telephone-network-based videotex services.

Another enhancement to the telephone network local loop is the mobile telephone. These services have been available for many years but on a very limited scale. In New York, for example, only about two dozen people can use mobile telephones simultaneously. In 1980, the Bell System served about 50,000 mobile telephone customers; a further 20,000 were on waiting lists (*San Francisco Chronicle*, 1981). Current mobile telephone technology uses centrally located medium-power radio transmitters, and the number of customers that can be served simultaneously is constrained by the limited availability of frequency spectrum. In April of 1981, the FCC allocated 40 MHz of spectrum for the implementation of a new, expanded mobile telephone service called cellular radio. Cellular radio systems require a given geographic area to be subdivided into a number of "cells," each served by its own low-power transmitter. As mobile units move from one cell to the next the call is automatically rerouted via the new cell's transmitter. This allows a given channel to be used several times simultaneously in an area, thereby increasing many times the

---

[10]For the Bell System the cost breakdown is: switching, 45%; subscribers' instruments, 23%; trunk circuits, 17%; local loops, 15% (Martin, 1976).

number of possible simultaneous conversations. Initial applications of cellular radio systems will focus on voice communications, but the system could also potentially be used for mobile videotex services.

While the local loop is relatively stable, the switching component of the telephone network is experiencing considerable change. Telephone exchange technology has evolved through a number of generations of analog switching—manual, electromechanical (step-by-step, crossbar, reed), and electronic. During the 1970s, digital switching systems began to emerge, and toward the end of the decade, these became competitive with equivalent analog systems. During the next 20 years, digital systems are expected to become considerably less expensive.

Initial implementations of digital switching have been in the long-distance toll network. The first Bell System No. 4 ESS (electronic switching system) was put into service in Chicago in 1976. By 1990, nearly all of the Bell System toll switching network will have been converted to digital operation (Vigilante, 1980). Wide-scale implementation of digital technology in the Bell System's local office network is expected to start during the mid-1980s.[11] However, since *two* communications channels are required for a digital circuit (one channel for send, the other for receive), the wide-scale extension of digital links into subscribers' premises is likely to proceed only in concert with a major redevelopment of local loop technology. Existing cable technology in general provides only a single cable pair to each subscriber. The cost of the provision of an additional cable pair into every house would be prohibitive in terms of both capital and installation effort. While subscriber carrier systems can increase the capacity of existing cable plants, the number of systems that can be installed on any particular cable is also limited with today's technology (because of interference between systems, e.g., cross talk). The most likely development is that the digital link to the home will evolve in concert with the upgrading of subscribers' local loops with optical fiber cables.

But because of the huge investment in the existing telephone network (approximately $169 billion[12] at the end of 1980), it is expected that the conversion toward a total digital network will proceed slowly during the next 20 years. Thus, at least for a large portion of the next 20 years, the switched telephone network will present an analog interface to the videotex user. Modems will therefore continue to be required to interface digital user equipment with the telephone network.

Most modems in service today in the public switched telephone network operate at 300 bits/sec, although many users are increasingly opting for 1,200 bits/sec. Modems for the public switched network are also available at 2,400 bits/sec, 4,800 bits/sec, and 9,600 bits/sec, but the high cost of these units has tended to restrain demand. The high cost arises largely because the modem must dynamically condition[13] the circuit

---

[11]Some independent telephone companies in the United States have already installed digital local exchanges. On a worldwide basis, by the end of 1980 some 15.7 million lines of digital exchange equipment had been installed (the number of telephones in the world is 400 million) (*Telephony,* 1981*a*).

[12]Bell plus independents (*Telephony,* 1981*b*).

[13]"Conditioning" is a procedure used to make transmission impairments (i.e., signal distortion) of a circuit lie within certain limits that are specified in a tariff. Conditioning is frequently used on telephone lines leased for data transmission to improve the possible transmission speed while maintaining an acceptable error rate.

to compensate for differing line conditions encountered on a typical switched telephone connection. Most often, however, the higher transmission speeds are used on manually conditioned lines (leased circuits). This reduces the modem cost but increases the cost of the communications link for low-volume users. In addition, because of the limited availability of additional circuits in the telephone local distribution network, it is unlikely that leased circuits will be used on a wide-scale basis for the local distribution of videotex services.

In addition to a modem, the videotex user also requires a decoder to display the information on his or her terminal unit. The decoder can be integrated within the terminal unit or be a stand-alone device much in the same way as a modem. The complexity of the decoder unit depends on the data coding standard (e.g., Antiope, ASCII, AT&T Presentation Level Protocol (PLP), Prestel, or Telidon) and the type of terminal unit (TV, monitor, or CRT terminal). The cost of the decoder depends on these factors as well as the number of units sold.

For example, many existing computer terminals have a built-in ASCII decoder and connect to a modem via a RS232-C connector. As such the cost of the decoder is simply part of the terminal unit cost. While this situation may become widespread later for full videotex protocols, most vendors sell decoders separately for connection to a monitor. Prices for these decoders range from $400 for a Prestel decoder to $1,000 for a Telidon decoder. Since these prices are for producing relatively small numbers of units, the costs will decline markedly with mass production.

The cost of the communications link in the switched telephone network is more difficult to forecast. When buying a communication service, the user is in fact obtaining a number of services, e.g., transmission, speed of transmission, availability of access, and switching functions. The inherent costs of technical implementation often bear very little relation to the tariff structure. Expenses for administration, marketing, and initial development frequently become major cost factors. Marketing considerations tend to dominate the price structure.

In the switched telephone network, tariff structures often allow unlimited free local calling (flat rate tariffs in which charges are included in the monthly line rental of between $6 and $15) but charge on a time-usage basis for long-distance calls. Long-distance (over about 5 miles) charges are typically based on the

- geographic distance of communications link
- time duration of call
- time of day
- degree of operator assistance required

In 1982, long-distance charges for direct-dialed calls during business hours[14] varied from about $0.15/minute for short-haul calls (less than 50 miles) to $0.38/minute for long-haul calls (over 400 miles).[15] Thus the communications usage charge for a videotex service would depend primarily on the distance between the user

[14]Business hours: 8 a.m. to 5 p.m. (Monday through Friday); evening rates are discounted by 35 percent and night and weekend rates by 65 percent.

[15]Charges for second and subsequent minutes of a call; the charge for the first minute of a call is typically some 50 percent higher to allow for the call set-up time.

and the videotex computer and the time of day the service is used. Given a widespread implementation of videotex, it is likely that most users will be able to access a videotex computer (or system concentrator) with only a local call. If such calls remain uncharged, the telephone network provides an attractive (cost-wise) communication medium to the user. However, it is unlikely that videotex "calls" would remain uncharged. There is already mounting pressure within the telephone companies and regulatory commissions to introduce usage-sensitive pricing for local calls (Mitchell, 1978; Gelder, 1980). Local telephone calling rates per subscriber have been increasing by an average of 2 percent per annum during the last decade, and with flat rate tariffs, these increases in local calling have added to telephone company costs but not to their revenues. The AT&T divestiture decision of early 1982 will, if approved, contribute to a substantial increase in the costs of local calling. Widespread implementation of videotex would add further to these costs. If advantage is to be taken of the unused capacity of the local telephone network, tariffs must be suitably structured to discourage videotex use during the telephone network "busy hour."[16]

Twenty-Year Forecast

The theoretical limit on the capacity of a channel in the switched telephone network is between 20 and 30 kbits/sec (depending on the degree of optimism shown in the choice of channel bandwidth and signal-to-noise ratio). The practical limit, however, is currently 9,600 bits/sec. It is likely that this will remain the practical limit for the switched telephone network during the next decade, although higher speeds may be realized within a 20-year time frame. However, because of the rapidly declining cost of semiconductors, in particular large-scale integrated circuits, the cost of these high-speed modems could be expected to gradually approach that of the lower-speed modems. Already there are indications of this trend, e.g., 1,200 bits/sec modems on a card currently cost about $200 in OEM (original equipment manufacturer) quantities,[17] which is very close to the final cost of 300 bits/sec modems ($150). By 1990, a 9,600 bits/sec modem will cost between $500 and $1,000; by the year 2000 it will cost less than $100. During the next decade 300 bits/sec and 1,200 bits/sec modems will both fall to under $100; during the 1990s these modems will generally be incorporated within the terminal equipment at no significant additional cost (i.e., less than $20). Multiplexers for a "data + voice" service will similarly decline to under $100 during the next 20 years.

The costs of videotex decoders will also decline rapidly. For example, the Norpak Mark III Telidon decoder, which now retails for $1,000, will upgrade to include the AT&T PLP standard and has a targeted OEM price of $150 in mid-1983 (*Computerworld*, 1981b). Percentage price changes in decoders for other videotex standards are expected to be the same. Thus we can reasonably expect that during the late

---

[16]The "busy hour" is a telecommunications engineering term that refers to the four consecutive 15-minute periods during a business day when telephone traffic is the highest. Typically, the busy hour occurs sometime between 9 a.m. and 11 a.m.

[17]OEM quantities: as supplied by the modem manufacturer to the communications service provider or to the retailer.

1990s, the videotex decoders will become like ASCII decoders and be incorporated within terminal equipment at a cost of less than $50.

For full-motion video and high-quality still pictures, existing transmission speeds are still inadequate. During the next 20 years, however, image compression techniques will likely be developed that permit video signals to be transmitted over the switched telephone network without incurring excessively long delays. In concert with these developments, the existing analog telephone network will gradually be upgraded to a fully digital network (providing telephone subscribers with a 56-kbits/sec local loop). Toward the end of the century, optical fiber local loops will begin to replace the existing copper cable loops, providing capacity for voice, data, and video signals.

Beyond the local loop, the existing analog telephone network is less suitable for videotex services. Videotex traffic is characteristically bursty, and therefore for a substantial proportion of time a telephone circuit carrying videotex traffic is idle. This inefficiency can be avoided by using communication networks that are more suited to carrying bursty traffic, e.g., packet-switched networks (see the next section). Thus, telephone-network-based videotex applications are most likely to evolve in conjunction with other networks rather than use the telephone network exclusively.

PACKET-SWITCHED NETWORKS

Packet-switched networks have evolved during the last decade in response to a need for more accurate and less expensive data communication services, in particular for transaction-type services. Computer-generated or terminal-generated data is divided into blocks (typically no more than 1,024 bits), addressing and control information is added, and the resulting data "packet" is transmitted through a store-and-forward[18] communications network. Assembly of data into packets allows a much more efficient utilization of channel capacity, thus reducing the communication cost.

Connections from a user terminal to a packet network are typically provided by leased or dial-up telephone lines. Connections between switching nodes use high-capacity terrestrial or satellite communication channels. Transmission speeds in typical packet networks vary from 300 bits/sec to 9,600 bits/sec in the subscriber access link and from 56 kbits/sec to 1.5 Mbits/sec in the internode links; e.g., AT&T's proposed Advanced Communication System (ACS) and TELENET use 56 kbits/sec backbone links (Bolt, Beranek and Newman, Inc., 1979). An upgrading of TELENET links to 1.5 Mbits/sec is currently in progress (Wessler, 1980).

Despite competitive tariffs and widespread availability, public packet-switched networks still attract only a very small proportion of data communications traffic. In 1979, TYMNET's[19] revenues were $24 million compared to $4 billion for all other data communication services in the United States (*Datamation*, 1980). Part of the reason for this large difference is that packet-switched networks have been established only relatively

---

[18]Store-and-forward: packets are stored at each node (switching exchange) on the transmission path before retransmission to the next node.

[19]TYMNET is the largest public packet-switched network in the United States.

recently (TYMNET in 1969 and TELENET in 1975), and thus most users already have an investment in existing circuit-switched data communication services. Another reason for the difference is the nature of data traffic. Most data communicated by electronic means is of the high-volume type, for which the economic advantage of packet switching over circuit switching (or leased circuits) becomes less significant.

Packet-switched networks are currently used mainly for long-haul communications, where the high cost of the communications link has encouraged the development of a more efficient method of transmission. High-volume business users typically lease a direct line between their terminal equipment and the packet-switched network, using special interface equipment to assemble (and disassemble) data into packets. However, the wide-scale implementation of any new service based on packet-switched networks as the primary communications medium is still likely to rely on the analog-switched telephone network to provide access to the packet network, at least for a large proportion of the next two decades. Hence, users will be faced with the same interface costs (i.e., modem costs) as for the switched telephone network. The main difference will be in the communications link cost. The access charge on dial-up ports is based on an hourly rate (typically between $3 and $6 per hour), whereas that for the private port is a fixed monthly rate (typically between $200 and $1,500). In addition, private port access has an associated leased line charge (to connect the subscriber to the packet network). For both types of access, there is also a transmission charge based on the number of bits or characters transmitted; e.g., TYMNET transmission charges are $0.04 per 1,000 characters and TELENET transmission charges are $1.00 per 1,000 packets (up to 128 characters per packet). Private port access becomes a more attractive economic alternative when usage exceeds about 150 hours per month. Although data on teletext/videotex usage patterns is rather limited, it is unlikely that monthly individual usage will come anywhere near 150 hours per month. Plummer (1980) reports that average usage of the "Source" is 5.9 hours per month. Thus, as in the switched telephone network, dial-up access to packet networks is likely to be the more relevant alternative for evolving teletext/videotex systems.

Twenty-Year Forecast

During the next two decades, the volume of transactional-type data traffic is expected to grow. Many of the new communication services associated with videotex systems will generate data traffic of short duration and small volume. Packet-switched telephone networks can therefore be expected to capture an increasing proportion of the data communications market.

Packet radio systems are also potential carriers. The radio frequencies used in packet radio systems (typically UHF) limit the range of transmissions to essentially line of sight. However, with a network of transmitters and repeaters, good coverage of a local area can be achieved. The attractive characteristic of packet radio systems is the potential for portable terminals. The Advanced Research Project Agency (ARPA) of the Department of Defense has developed a portable 10-watt packet radio terminal measuring about a cubic foot and weighing 40 pounds. The cost of the ARPA unit is

$65,000, but the consensus of potential manufacturers is that the cost will be as low as $5,000 by 1985 (*Data Communications*, 1980). In the telephone network, there is a substantial demand for portable voice communications—either as an extension of the public switched network (e.g., car telephones) or in private radio networks (mobile radio). A similar demand might be expected for portable teletext/videotex systems. Already GTE-Telenet has announced plans for extending TELENET with packet radio systems (Wessler, 1980).

CONCLUSION

It has been shown that there are numerous possible communication channels for teletext and videotex services.

For teletext there are at least six distinct options for one-way transmission of signals:

- Limited channel   vertical blanking interval broadcast or cable TV
- Full channel   full-channel broadcast TV, e.g., low-power television
                     multipoint distribution systems
                     direct broadcast satellite
                     FM radio
                     full-channel cable (utilizing one-way facility)

And for videotex the various options may be grouped under three broad classes of networks, namely, telephone, cable, and broadcast hybrid. For telephone, there is the local loop and a packet-switched network, for cable there are two-way cable networks and packet cable, and broadcast hybrid options include MDS/telephone, FM/telephone, satellite/telephone, and cable/telephone. These network options are summarized in Table 7.2.

The U.S. business and consumer market is sufficiently large and diverse to allow some or all of these options to exist. The extent to which the networks are utilized will depend upon a trade-off among the general performance attributes of penetration,

TABLE 7.2 COMMUNICATION NETWORK OPTIONS FOR TELETEXT AND VIDEOTEX

Teletext		Videotex		
Limited channel	**Full channel**	**Telephone**	**Cable**	**Broadcast hybrid**
VBI-broadcast TV, Cable TV	MDS DBS FM radio Low-power TV Cable TV (one-way)	Local loop Packet-switched network	Two-way cable Packet cable	MDS/telephone* DBS/telephone* Cable/telephone FM radio/ telephone

*or cable

TABLE 7.3 PENETRATION* OF COMMUNICATIONS NETWORKS
FOR TELETEXT AND VIDEOTEX SERVICES:
LOCAL DISTRIBUTION (1982–2000)

	1982	1990	2000
Broadcast TV	99%	99%	99%
Multipoint distribution system	<1	5	10
Direct broadcast satellite	<1	5	30
FM radio	95	95	95
Cable TV	28	50	66
Telephone network (analog)	98	98	98
Telephone network (digital)	0	20	60
Packet-switched network	0	5	40

N.A. = not available

*Penetration = communications link available into the
home plus service option taken up

transmission speed, interface unit cost, and communication link cost. The penetration
of the various alternative networks is shown in Table 7.3, FM radio being an unknown potential at this time. Some further insight is obtained by considering a forecast of the cost of the interface devices for various transmission speeds for each of the
alternative networks (Table 7.4).

TABLE 7.4 TELETEXT/VIDEOTEX INTERFACE DEVICES: COST FORECAST

	Transmission speed (bits/sec)	Interface costs ($1982)		
		1982	1990	2000
Television or coaxial cable (2 or more VBI lines)	13,000	200	20	10
Multipoint distribution system	8M	300	100	40
Direct broadcast satellite	N.A.	4,000	200	50
FM radio	9,600	400	100	40
Coaxial cable (full channel)	8M	300	100	30
Telephone modems	300	150	50	10
	1,200	700	100	20
	2,400	1,600	250	30
	4,800	2,000	400	50
	9,600	3,000	800	100
Telephone decoders		400–1,000	100	20

N.A. = not available

CHAPTER 8

TECHNOLOGY FORECAST: USER TERMINALS

Just as there is no single network to bring teletext/videotex into the home, there is also no single piece of technology that constitutes a teletext or videotex terminal. Hence, we cannot simply forecast that "the" terminal will get cheaper, bigger, or faster. We must first look at what *might* constitute a user teletext/videotex terminal, and then describe the expected changes in the cost of these components.

User terminals for teletext/videotex services include up to four main functions, namely, displaying information, inputting data, processing data, and storing data. Simple teletext systems, such as closed captioning, display information on a television receiver. At the other end of the spectrum, there are intelligent computer terminals with full alphanumeric keyboards, disc storage, dedicated color-television monitors, voice synthesis, and even voice recognition units to support the full range of functions.

This chapter examines the technological alternatives for each of the above four terminal functions in relation to a set of technological performance attributes and forecasts likely cost trends. As with the communication networks forecasts in the previous chapter, a 20-year time scale is considered for terminal developments. Historically, terminal equipment has developed more rapidly than communication networks, and therefore we can be less confident that some as yet uninvented device will not achieve widespread penetration within the next 20 years. However, there is still a considerable time lag between invention and mass production; manufacturers like to be sure of a market before investing in large-scale production. In addition, the development of a product is often constrained by the technology used for implementation, e.g., the development of higher and higher density semiconductor chips has been closely related to very large scale integration (VLSI) techniques. Of even greater importance, is the time lag inherent in the widespread acceptance of new devices and services.

Thus, as with the communication technologies discussed in the previous chapter, only currently *known* terminal technologies are likely to achieve wide-scale penetration during the next two decades.

Cost is a primary determinant in the level of penetration of a new technology and therefore is one of the most useful attributes for forecasting. However, because many of the teletext/videotex component technologies are still relatively new, historical cost data tends to be rather limited. Nevertheless, that which is available does provide a base for the forecasts included in this chapter. In addition, the presentation of technological performance attributes for various alternatives offers a set of noncost criteria for use in comparative analysis. In general, the forecast data has been derived from industry forecasts, prepared primarily for use by manufacturers. Since computer and communication equipment manufacturers hold the key to mass-market penetration of the component technologies, it is considered that their forecasts are the most relevant.

DISPLAYING INFORMATION

The possible attributes of the display unit reflect the attributes of the media used by teletext/videotex. Considering each of these media, we may expand on the list of attributes associated with current teletext/videotex applications (Table 4.3) as follows:

- transmission speed
- text format
- graphics
- frame video
- hard copy
- motion video
- animation
- color
- speech
- music
- cost

While early applications of teletext and videotex have used the existing home television set, other types of display devices may be used to augment or even replace the TV set for selected applications. Looking ahead 20 years, the probable display units capable of supporting the above attributes are:

- television set
- computer terminal:
    (1) paper-based
    (2) video
- printer
- printer/plotter
- speech synthesizer

The existing attributes of these display devices are summarized in Table 8.1

Television Set

The home TV set is an ideal teletext/videotex medium because of its universal penetration and media flexibility. It can display text, graphics, voice and still frame as well as full-motion video, thus supporting a complete range of audio and video communications. While nominally capable of receiving data at speeds up to 8 Mbits/sec,

TABLE 8.1   ATTRIBUTES OF TELETEXT/VIDEOTEX DISPLAY DEVICE

Attribute	TV	Computer terminals				Printer/ plotter	Speech synthesizer
		Paper-based	CRT	Graphics	Printer		
Speed (bits/sec)	5.7 M	300–1200	300–19.2 K	1.2–19.2 K	300–1,500	300–1,500	2.4–9.6 K
Text format	24 rows 40 col.	80–132 col.	24 rows 80–132 col.	24 rows 80–132 col.	15–132 col.	80–132 col.	No
Graphics	240 lines 320 pixels/ line	Line-printer	Line-printer	250–500 lines 300–1,000 pixels/line	Line-printer	900 pixels/ line	No
Frame video	Yes	No	No	Yes	No	Yes	No
Hard copy	No	Yes	No	No	No	Yes	No
Motion video	Yes	No	No	Yes	No	No	No
Animation	Yes	No	Yes	Yes	No	No	No
Color	Yes	No	No	Yes	No	Yes	No
Speech	Yes	No	No	No	No	No	Yes
Music	Yes	No	No	No	No	No	No
1982 cost	$100–500	$1,200–2,000	$500–1,200	$1,000–5,000	$200–1,500	$800–1,500 (b & w) $2,000–12,000 (color)	$200–400

only dedicated full-channel applications will make use of the wide bandwidth available. Existing teletext applications receive data at much lower speeds (13 kbits/sec for systems using 2 VBI lines) and telephone network-based videotex systems receive data at still lower speeds (300 to 1,200 bits/sec). The display of text on television sets is generally limited to at most 40 columns and 24 rows (960 characters per screen page). The main constraints on the quality of text that can be displayed on a television screen is television tube resolution. The graphics resolution achievable on existing 525-line television receivers is approximately 300 pixels per scanning line. During the next 20 years, as high-definition, large-screen television sets become available, it will be possible to increase the number of characters displayed by a factor of up to 4. However, such displays will be incompatible with the 40 × 24 and 40 × 20 proposed standards, and therefore high-definition displays are likely to develop as a component of enhanced teletext/videotex services.

The cost of a color television set has remained remarkably stable in real terms during the last decade at about $500 (U.S. Bureau of the Census, 1980). During this period the costs of components in a television set have declined steeply, and similar decreases can be confidently predicted during the next two decades. However, costs of assembly have continued to rise, and new facilities have been added to encourage consumers to buy the higher-priced sets, e.g., those with remote tuning devices. In 1980, domestic production of color TVs reached a level of 10.0 million units and imported sets totaled 1.2 million. Thus, despite reaching national saturation in the penetration of TV sets (over 99 percent of all households by 1980), some 10 percent of all households purchase a new color set annually. No doubt, the television manufacturing industry will be anxious to maintain this level of production, and hence there will be continuing pressure to market sets with additional facilities, e.g., teletext and videotex decoders.

Twenty-Year Forecast

During the next ten years, the average retail price of a color television set will remain at about $500, but new facilities will progressively be included in the standard TV set, e.g., teletext/videotex decoders, built-in telephones, and data modems. A 10-year projection of sales of TV sets with videotex decoders made by International Resource Development, Inc., of Norwalk, Connecticut (Mokhoff, 1980), forecasts 12 million sets installed by 1990 at this average price. Between 1990 and the year 2000 the penetration of these multifacility television sets will expand to reach the majority of households. In addition, new developments such as high-definition television, large-screen and flat screen display (Chase, 1981), and digital TV receivers will start to emerge, offering an enhanced video display in the home. Penetration of these enhanced video displays is likely to remain quite small (< 10 percent) until after the year 2000, and therefore the price of these receivers will be at least two to three times that of the basic multifacility sets.

Computer Terminals

Computer terminal displays may be either paper-based or video-based. Although both types of display can be used in teletext/videotex systems, the computer terminals available today are likely to achieve wider penetration in the business sector than the mass consumer market because of their high price, complexity, and limited display potential (as indicated in Table 8.1). According to International Resource Development, the 2 million white-collar workers who routinely use computer terminals today are only a small fraction of the potential market (Bulkeley, 1981). IRD expects this business market to grow at least sixfold by 1990.

Paper-based terminals are only able to print alphanumeric characters in a uniform text format of 80 or 132 columns and only very limited "line-printer" type graphics are possible. Most existing paper-based terminals operate at 300 or 1,200 bits/sec. While such paper-based terminals may be quite satisfactory for electronic messaging services and some basic information retrieval activities, they fall well short of utilizing the full potential of teletext/videotex sytems. A further deterrent to widespread implementation is cost. Typical paper-based terminals include an input keyboard as well as the printing mechanism (and often local processing power and memory as well); in 1982, costs ranged from around $1,200 to $2,000.

Video terminals are based on a cathode ray tube (CRT), much like a television set is. They are capable of operating much faster than paper-based terminals, typically up to 9.6 kbits/sec, and more recent models are able to operate at 19.2 kbits/sec. Applications that use the switched telephone network, however, are generally limited to 1,200 bits/sec (the cost of higher-speed modems being a major constraining factor). Basic CRT terminals usually provide a text format of 20 to 24 rows and 80 or 132 columns, and, as for paper-based terminals, only line-printer type graphics are possible. Costs in 1982 ranged from $500 to $1,200 (including the keyboard). A typical CRT terminal is illustrated in Figure 8.1.

Graphics terminals, like basic CRT terminals, also have a video display. The main distinction is in the level of resolution. While CRT terminals have only the capability of line-printer graphics, graphics terminals may have a resolution of up to 1,000 pixels per scanning line. Although text formatting on graphics terminals is normally the same as on CRT terminals (24 rows × 80 columns), the increased resolution permits software emulation of any character set. In order to function effectively, operating speeds typically range between 1,200 bits/sec and 19.2 kbits/sec. While most graphics terminals in the past have been black (or green) and white, newer, more expensive models have included color. Typical prices range from $1,000 to $8,000 but can be as high as $50,000, depending on the resolution, color, and other microprocessor-controlled options.

Twenty-Year Forecast

All three types of computer terminal (paper based, basic CRT, and graphic CRT) have experienced substantial price reductions during the last decade, e.g., graphics terminals priced between $60,000 and $200,000 in 1970 were available from the

FIGURE 8.1   Televideo 912 CRT Terminal. (*Source: Televidio Systems, Inc.*)

same manufacturers for less than $2,000 in 1979 (*Computer Business News*, 1979). During the next two decades, further price reductions can be expected as the terminal market continues to expand. A market study by Venture Development Corporation in Wellesley, Massachusetts, forecasted that the graphics terminal industry will grow from today's $2 billion market into a $15 billion market by the end of the decade (Allen, 1980). Main applications are expected to be in the industrial and business sectors, although when prices become low enough, graphic terminals could become the major display component for enhanced teletext/videotex services. During the next two years, new low-priced basic CRT terminals aimed at the business sector are to be introduced. Northern Telecom, Ltd., working with Bell Canada, has developed a telephone-terminal combination priced between $1,000 and $2,000. Texas Instruments is planning to introduce a CRT terminal priced at $995 (Bulkeley, 1981). In mass production, these prices will be driven even lower (Figure 8.2).

Printers and Plotters

There are a number of different types of electronic paper-based printers—daisy wheel, drum and band-type line printers, thermal, impact matrix, electrostatic, laser, and magnetic. However, these different types can be classified into two broad categories: impact and nonimpact.

Impact printers include the common daisy-wheel character printers, the high-speed

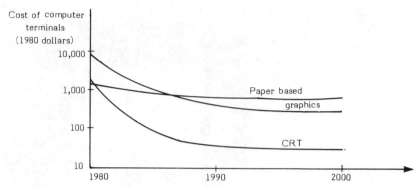

FIGURE 8.2   Twenty-Year Forecast: Computer Terminals.

computer line printers, and the low-priced serial impact matrix printers. Nonimpact printers include mainly thermal and electrostatic machines, although laser and magnetic printers are expected to become increasingly available during the next decade. The market for nonimpact printers is expected to grow at over 20 percent per annum and will be around 40 percent of total printer shipments by 1985 (*Computer Business News*, 1981).

For general teletext/videotex applications, low-cost printers are likely to be the most relevant (i.e., less than $1,000). These printers, which rely on impact or thermal rather than daisy-wheel, drum, or band-type technologies, are available for as low as $350 (Table 8.2).

In addition to these black-on-white printers, color printers are currently available at prices between $2,000 and $11,000 (Table 8.3).

TABLE 8.2   COMPARISON OF SOME LOW-PRICED PRINTERS

Model	Manufacturer	No. of columns	Print method	Graphics	Speed (char/sec)	Cost ($ 1982)
Sprinter 40	Alphacom	40	Thermal dot matrix	Yes	160	$350
Sprinter 80	Alphacom	80	Thermal dot matrix	Yes	160	595
GP-80M	Seikosha	80	Dot matrix	Yes	30	399
IMP-2	Axiom	132	Impact wire matrix	Yes	80	644
MX-70	Epson	80	Impact dot matrix	Yes	80	450
Microline	Okidata	132	Impact wire matrix	Yes	400	399
800B	Base 2 Inc	132	Impact dot matrix	Yes	960	699

TABLE 8.3   SOME COLOR PRINTERS

Model	Manufacturer	Technology	Colors	Print speed	Cost ($ 1982)
Prism printer	Integral data	Dot matrix	4	110 char/sec	$ 2,000
3287	IBM	Dot matrix	4	120 char/sec single color	6,800
IS8001	Printacolor	Ink jet	3	2 min/page	5,700
4100	Ramtek	Dot matrix	4	3 min/page	12,000
Colorplot 100	Trilog	Dot matrix	3	150 char/sec	11,800
Omnicolor	Omnico	Dot matrix	3	40 char/sec	9,000

Sources: Scannell, 1981; Omnico brochure.

Twenty-Year Forecast

In 1980, impact printers captured over 75 percent of the low-cost printer market. However, nonimpact printers are expected to gain an increasing share during the next decade. Strategic Business Services, Inc. (SBS), of San Jose, California, has forecasted that the market share will grow from the existing 24 percent to 37 percent by 1985. Impact devices are limited by inertia in very high-speed printing applications; but in nonimpact printers, there are fewer mechanical components and thus inertia does not create the same problem (*Computerworld*, 1981*a*).

As is evident from Table 8.2, the cost of printers has already become very low. Creative Strategies International (CSI) has forecast further decreases, with prices ''bottoming out'' between 1982 and 1983 (*Computer Business News*, 1980*b*). Developments during the remainder of the decade and in the 1990s will focus on improving the quality of the printer output. For example, impact dot matrix methods are not yet able to achieve a print quality comparable to that of a daisy-wheel or golf-ball machine, but they are expected to. In 1982, Tokyo Juki Industrial Company plans to modify its recently announced model 3300 golf-ball electric typewriter to create a machine costing less than $600 that can be wired to personal computers for letter-quality text editing.

Other developments during the next two decades are likely to concentrate on reducing the cost of color printers. Effort is also likely to be directed toward reducing the cost of devices for printing video images from television screens. Tudorcape in the United Kingdom has developed a device called the Vidiprinter that is capable of recording and developing 960 images per hour; the equipment costs between $25,000 and $35,000. Figure 8.3 projects the cost of printers/plotters suitable for widespread use as a teletext/videotex peripheral during the next 20 years.

Speech Synthesizers

Speech synthesizers, which convert digital data into audio signals, are already commercially available and make possible the addition of an audio output for

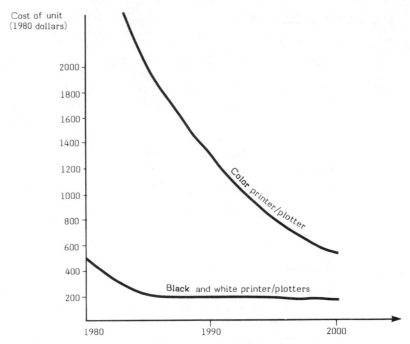

FIGURE 8.3   Twenty-Year Forecast: Printer/Plotters.

teletext/videotex displays. A number of techniques for synthesizing speech have been developed. In general, these fall into two classes—phoneme and time domain (Anderson, 1981). Phoneme (a distinct word sound) synthesizers are essentially electrical analog models of the human vocal tract; words are constructed by adding together the component sounds, which are stored in computer memory, e.g., "seven" could be constructed by adding the "s" sound to an "eh-vun" sound. These synthesizers are able to produce speech from very low data transmission speeds (600 bits/sec or even less) and are most suited to applications where bandwidth or memory is at a premium. For a given language, these synthesizers have a virtually unlimited vocabulary. The quality of speech produced by phoneme synthesizers is quite intelligible but rather mechanical sounding. Examples of products available in 1982 are the Street Electronics Corporation (SEC) "Echo" unit ($225) and Votrax's "Type-'N-Talk" ($375). The SEC device is intended for use with user-developed software, whereas the Votrax unit can also be used to convert ASCII code directly into synthesized speech.

Time domain techniques (e.g., delta modulation, pulse code modulation) can also be used for digitizing speech. These techniques simply record speech-signal parameters as a function of time; mass-storage devices (e.g., disc drives) and higher transmission speeds (e.g., 9,600 bits/sec for delta modulation and 64 kbits/sec for pulse code modulation) are required. These systems have a limited vocabulary but are able to produce sounds that are almost indistinguishable from normal human speech.

In typical applications, digitized sentences are stored in a host computer and down-loaded to the synthesizer as required, e.g., Centigram's LISA (logically integrated speech annunciator) can process digital speech at 4,800 bits/sec and is able to store up to one minute of speech at a time (using 16 kbytes of storage). As a stand-alone unit LISA costs $3,450, but in a single board OEM configuration the cost is only $1,800 (Bassak, 1981).

Twenty-Year Forecast

Because of lower costs and the ability to operate at very low transmission speeds, phoneme synthesizers are more likely to emerge initially as an audio output for teletext/videotex systems. Even though these synthesizers have a mechanical-sounding voice response, they are likely to be accepted initially by users because of many years of conditioning by science fiction movies, i.e., they sound like people expect computers to sound. The cost of phoneme synthesizers is closely tied to the cost of semiconductor chips. As increasingly sophisticated chips are developed with larger and larger memories at decreasing costs, the cost of speech synthesizers will decline accordingly (Figure 8.4).

SRI International has estimated that the size of the market for speech input/output equipment (speech synthesizers *and* speech recognition units) in 1980 was no more than $20 million. However, SRI International forecasted that this market will grow to between $1.2 billion and $1.8 billion by 1988 (Allen, 1981*a*). Applications in consumer products are already beginning to emerge, e.g., Texas Instruments's "Speak and Spell" learning aid, Matsushita's Quasar microwave oven, and Sharp's C56500 desk-top calculator.

Beyond 1990, the price differential between different synthesis techniques is likely

FIGURE 8.4 Twenty-Year Forecast: Speech Synthesizers.

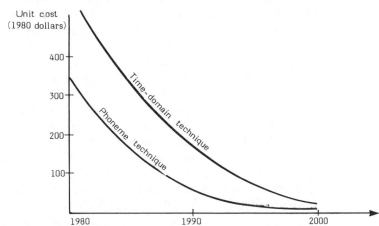

to become relatively insignificant. Furthermore, because complete synthesis units will be available for less than $50, they will become an integral component of home television sets.

INPUTTING DATA

While the user input channel currently has a much lower transmission rate than the display channel, for example, 75 bits/sec compared to 1,200 bits/sec on Prestel, the inequality may disappear in future teletext/videotex systems, e.g., cable-based systems. As a result, the range of user input devices in the future might encompass many of the same communication media as the display devices. A list of the key attributes of input devices includes:

- textual input (numeric/alphanumeric)
- graphics input (pointing/drawing)
- aural input (voice)
- video input (still frame/full motion)
- sensory input (e.g., fire/smoke detectors)

In a manner analogous to that for display units we can construct a list of technological options that accommodate the attributes of future input devices. Such devices include:

- touch-tone telephone
- numeric keypad
- alphanumeric keyboard
- joystick
- light pen

- touch-sensitive display
- tablet/pen
- speech recognition devices
- digitizing still-frame video camera
- sensor

Table 8.4 presents a comparison of the present attributes of these user input devices.

The *touch-tone telephone* supports both voice and numeric input and is the cheapest of the input options. Depending on the type of phone jacks in the home, the initial installation charge is usually a maximum of $30, a cost that may rise significantly given the AT&T divestiture decision. The additional annual cost of touch-tone service compared to the cost of standard dial services is in the range of $15 to $20.

Numeric keypads are interfaced into either the user's decoder or display unit depending on whether the service is teletext or videotex. Add-on keypads are relatively inexpensive, about $20 to $50 in large volumes.

Alphanumeric keyboards are currently available at prices in the range of $50 to $300. At the upper end of the scale are the more conventional reed switch, capacitance, and Hall effect units; newer membrane-type keyboards tend to be cheaper (Figure 8.5). Because the more expensive keyboards are electromechanical, it is not likely that there will be any large-scale decrease in prices, although semiconductors may be used to monitor and improve key-switch movement. Keyboard sales are expected to double from their 1980 value of $100 million by 1985 (IRD, 1980).

Graphic input devices require some intelligent interconnection with the display device. This may result from a direct integration with the display unit (e.g., the

TABLE 8.4   ATTRIBUTES OF USER INPUT DEVICES

Attributes	Text			Graphics				Voice	Video	Sensory
	Touch-tone tele-phone	Numeric keypad	Key-board	Joy-stick	Light pen	Tablet/pen	Touch-sensitive display	Speech recognition	Digitizing camera	Sensor
Numeric	×	×	×							
Alpha-numeric		×								
Pointing				×	×	×	×			
Drawing				×	×	×	×			
Voice	×						×	×		
Video									×	
Sensory										×
Cost ($ 1982)	$30	$20–$50	$50–$300	$10–$60	$75–$700	$800–$1,500	$1,000–$3,000	$900–$7,500	$600–$2,000	$1–$20

touch-sensitive display) or be through a microprocessor, memory, and input/output port. Because of the need to input alphanumeric information for most teletext/videotex applications, these devices are usually considered as secondary input devices. This may perhaps result from the fact that interactive graphic applications of teletext and videotex are still developmental and have not yet been adequately explored.

FIGURE 8.5   RCA Membrane Keyboard. (*Source: RCA Corporation.*)

The least expensive graphic input device is a *joystick*, or game paddles, as they are called in some home computer environments. Joysticks currently range in price from $10 to $60 per unit, depending on the degree of sophistication and the number of accessories. The price does not include the hardware/software needed to translate the X-Y output voltage to screen positions, although that is often provided at no extra cost in a home computer environment. Because of the low price and the electromechanical nature of the device, little price reduction is expected.

A *light pen* reads the X-Y coordinates of the CRT electronic beam and transmits this information to the computer processing unit, which is programmed to reform the display image in some predefined way, e.g., to draw a line from the selected X-Y coordinates to some other previously defined point. The current price of light pens depends on their resolution and additional computer hardware and software required for normal operations. Prices, which currently range from $75 (e.g., an ATARI CX70 light pen) to $700 (e.g., a RAMTEK 5911 graphic terminal light pen), are likely to decrease with volume production.

A *graphic tablet/pen* combination (Figure 8.6) is a more expensive graphic device, costing in the range of $800 to $1,500, again depending on resolution. Examples that span the resolution spectrum include a graphic tablet for an Apple II personal com-

FIGURE 8.6   Apple II Graphic Tablet/Pen. (*Source: Apple Computer.*)

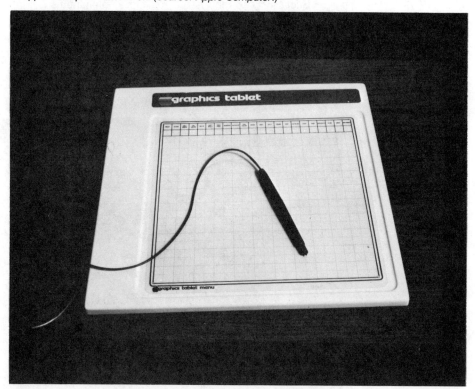

puter and a higher-resolution unit from Summagraphics. The graphics tablet is suitable not only for drawing but also for recording traced graphic information. The relatively high price for the unit leaves considerable room for volume decreases.

Touch-sensitive displays provide a very low resolution graphics input capability (e.g., 24 rows × 40 columns) by sensing the presence of an object (e.g., a finger) at the screen surface. Such units presently range in price from $1,000 to $3,000 (Shaw, 1980), and future price reductions can be expected as the cost of graphic terminals declines.

Speech recognition devices can substitute for keypad or keyboard input in teletext and videotex applications. Initially, a user enters a limited vocabulary of words and pronounces each 10 or so times. Then when in use, the device matches the words spoken by the user against the stored vocabulary. The recognition success rate is usually 95 percent or greater. Prices for speech recognition devices currently begin at $900 (for a Vet-2 voice-entry option for a Radio Shack TRS-80 home computer with a vocabulary of 40 words) and extend upward (Allen, 1981*a*). Because speech recognition is highly dependent on processing power and memory, the sophistication and vocabulary of these devices is likely to increase substantially without any increase in price.

A *digitizing camera* to capture still frame pictures currently costs from $600 to $2,000, depending on the level of resolution. The lower priced models with a low resolution interface easily with personal computer systems. However, video compression techniques to permit transmission over narrow band lines, e.g., the telephone network, can easily increase the cost by another $2,000. These input units are also very susceptible to price decreases in their component chips.

Solid-state *sensors* are increasingly being used to sense pressure, position, stress, temperature, magnetism, fluid level and flow, force, radiation, etc. (Allen, 1980). These sensors (or transducers) provide a low-cost interface between events in the analog world and semiconductor intelligence. Current prices range between $1 and $20. With volume production these prices will rapidly reduce to a few cents. By 1985 the market for all types of transducers is expected to be on the order of $1 billion (Allen, 1980).

A convenient device for accessing subscription teletext/videotex services from public terminals is likely to be the "smart card." Currently, smart cards are used to access automatic teller machines and provide a measure of security protection. In addition, these cards can greatly simplify the log-on procedures to databases.

Twenty-Year Forecast

Cost trends for most of the input devices discussed above are likely to "bottom out" rapidly and then remain constant for the remainder of the century (Table 8.5).

PROCESSING DATA

All teletext/videotex user terminals require some processing capability. In simple teletext applications, a relatively small amount of processing capability is required;

TABLE 8.5   TWENTY-YEAR FORECAST: INPUT DEVICES ($ 1982)

Input device	1980	1990	2000
Touch-tone telephone	$30	$30	$30
Numeric keypad	$20–$50	$20–$50	$20–$50
Alphanumeric keyboard	$50–$300	$30–$250	$30–$250
Joystick	$10–$60	$10–$60	$10–$60
Light pen	$75–$700	$40–$300	$25–$200
Tablet/light pen	$800–$1,500	$500–$1,000	$250–$500
Touch-sensitive display	$1,000–$3,000	$700–$1,000	$500–$800
Speech recognition devices	$900–$7,500	$500–$5,000	$100–$1,000
Digitizing camera	$600–$2,000	$200–$1,000	$100–$1,000
Silicon sensor	$1–$20	1¢–20¢	0.1¢–10¢

user requests for a particular page of information must be recognized, the page must be captured from the broadcast cycle, and the digital signals must be decoded to assemble a video picture frame. All these functions are performed within the decoder. At the other end of the spectrum the user's terminal may be an intelligent computer terminal with local processing capability in addition to providing timeshared access to remote computers. In videotex applications, intelligent terminals may be used as stand-alone computers or for enhancing the videotex service, e.g., as programmed information retrieval services.

In mass-market applications, processing capability is most likely to be preprogrammed, with user-created programs remaining in the domain of the computer hobbyist for at least the next two decades. Thus, in this discussion of user terminal technology we will leave aside the discussion of preprogrammed software, since it will be application dependent, and focus on the hardware aspects.

From a hardware perspective, the processing capability of a user's terminal is defined by the power of the central processing unit (CPU) and the extent of memory available within the terminal. These two attributes are discussed below.

CPU Power

The central processing unit carries out programmed instructions and provides intelligence for the teletext/videotex terminal unit. The power of the CPU is proportional to the number of bits it uses for its instruction set. The greater the number of bits, the more powerful the CPU. The increase in power arises from the smaller number of CPU cycles (and hence less time) needed to perform a particular function (such as adding two numbers) and the ability to access greater amounts of memory. Table 8.6 compares the current costs of successive generations of microprocessor chips.

With each new generation of microprocessor, the processing power available has expanded many times over. For example, the most recent addition to the microproces-

TABLE 8.6 MICROPROCESSOR
COSTS ($ 1982)

4 bit	$ 2
8 bit	$ 8
16 bit	$ 50
32 bit	$150

sor family is Intel's 32-bit "micromainframe" called the 432, which was announced in early 1981 (*The Economist*, 1981). It consists of three separate chips that together hold about 225,000 transistors. Jointly, they take up the area of a small postage stamp, but they are as powerful as a mainframe computer.

Twenty-Year Forecast

Even before the emergence of the microprocessor, which greatly simplified the requirements for electrical connections in and out of the processor chip, the overall cost of computations, including both logic and memory, has been falling during the last 25 years at an annual rate of about 25 percent compounded annually. This is expected to continue into the late 1980s. Since the scientific limits to further micro-miniaturization (which allows speed improvement and cost reductions) are well understood and the linear dimensions of silicon devices can be reduced by another factor of at least 3 to 6 (Branscomb, 1980), costs of existing microprocessor chips will continue to decline until the end of the 1980s, when they will tend to bottom out (Figure 8.7).

Memory

There are two types of memory technology commonly used in intelligent computer terminals: silicon chips, including random-access memory (RAM) and read-only memory (ROM), and magnetic bubbles. RAM chips have a read and write capability and provide volatile storage, i.e., information stored is lost when the power is switched off. ROM chips have a read-only capability but provide permanent storage (programmed by the manufacturer). Bubble memory, on the other hand, has a read and write capability and provides nonvolatile storage. Table 8.7 compares memory costs for different semiconductor technologies.

Developments in semiconductor microminiaturization have steadily increased the storage capacity of silicon chip memory devices. Dynamic random-access memories with a storage capacity of 64 kbits are becoming generally available from a number of manufacturers. In addition, the Japanese have already displayed 256-kbit devices and are actively developing 1-Mbit chips. This trend toward increasingly higher packing densities is expected to continue.

Magnetic bubble memory chips provide a read/write capability and are nonvolatile. They can also withstand extremes of temperature, vibration, dust, and humidity better

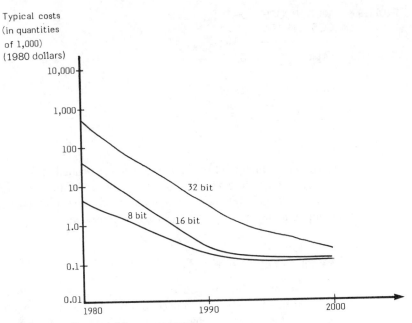

Typical costs
(in quantities
of 1,000)
(1980 dollars)

FIGURE 8.7   Twenty-Year Forecast: Microprocessors.

than other memory technologies. In 1980, 1-Mbit magnetic bubble memories became available, and the development of the next generation (4-Mbit) is in progress (Waller, 1980). Compared to other storage options, including semiconductor memory, bubble memory is still considerably more expensive per bit, and costs in the United States are not falling as predicted (*Business Week*, 1981*c*). If high-density 1-Mbit and 4-Mbit devices ever move into production quantities, the price per bit will decline, and bubble memory will become price competitive with other memory technologies as well as with external storage media, in particular fixed head disc devices (Allen, 1981*c*). Venture Development Corporation has forecast that fixed head disc drives will be displaced by bubbles when the price drops to about 15 millicents per bit (VDC, 1981).

TABLE 8.7   SILICON CHIP MEMORY COSTS
(¢ 1982)

RAM/ROM (bipolar)	0.2 ¢/bit
RAM/ROM (MOS)*	0.05 ¢/bit
RAM/ROM (CCD)†	0.03 ¢/bit
Bubble	0.1 ¢/bit

* MOS: metal oxide semiconductor.
†CCD: charged coupled device.

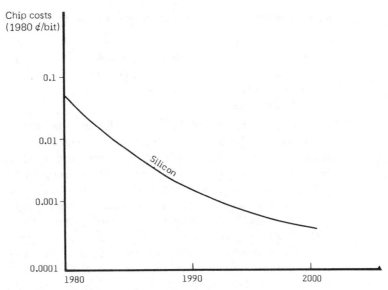

FIGURE 8.8   Declining Costs of Semiconductor Memory.

Twenty-Year Forecast

During the last decade, the cost of semiconductor computer memory has been reducing at a rate of 28 percent per annum, compounded annually (Branscomb, 1980). Given the combined effects of improved technology, which allows increasingly greater component packing densities, and a rapidly expanding market, which permits economies of scale, this continuing decline in semiconductor memory costs is expected to continue (Figure 8.8).

STORING DATA

In addition to the semiconductor memory associated with a terminal's processing capability, storage capacity may be desirable or necessary for some teletext/videotex applications. Information to be stored may be analog or digital, and storage facilities capable of supporting the range of communication links likely to be used in future teletext/videotex systems must be considered. Terminal storage attributes include:

- digital storage capacity
- audio storage capacity
- video storage capacity
- access time

- transfer rate
- replaceable or fixed storage medium
- read or read-write capability
- cost

In considering the technological options to support these attributes, we examine the devices at the lower end of the performance-price scale, since they are the most likely to emerge in the user's environment as teletext/videotex system components. These devices are:

- floppy disc
- hard disc
- audio cassette
- digital cassette
- cartridge tape
- videodisc
- videocassette

The current attributes of these storage technologies are shown in Table 8.8. We have chosen to consider the type of storage provided by each option in terms of its digital, audio, and video components. Of course, audio and video information can be digitized or digital information can be stored in analog form (e.g., audio cassettes), creating an interchangeability among the media capacities. The unit costs in Table 8.8 are typical retail prices of the various playback/record devices. The media costs are those of the actual storage medium, e.g., cassette tape or disc.

The flexible mylar 5-inch minifloppy and 8-inch *floppy discs*, which resemble small phonograph records, currently provide from 100 kbytes to 1 Mbyte of direct-access storage depending on the size, the storage density, and the number of surfaces used. The current disc drive cost of $500 to $2,500, which depends on storage capacity and OEM quantities, is unlikely to alter considerably in the future since these units are mechanical devices. Rather, the storage capacity is expected to increase as new writing techniques and greater floppy disc densities are implemented.

Hard discs (or rigid discs) provide an alternative high-capacity direct-access storage device. Compared to floppy discs, hard discs provide more storage capability, faster access times, and higher data transfer rates. In 1979, virtually all hard disc devices used a 14-inch disc; however, by 1984 it is expected that 8-inch hard disc drives will capture 50 percent of the market (Venture Development Corp., 1980). Most of these will be of the fixed type, i.e., use sealed, lubricated discs that cannot be removed (commonly referred to as "Winchesters"), although removable 8-inch cartridge discs are under development. The current price of fixed, hard disc drives ranges between $1,000 and $4,000 depending on OEM quantities and disc size. During 1981 an even cheaper Winchester drive using a 5¼-inch disc became available from a number of manufacturers. According to Roman Associates International, the total market for Winchester disc drives is expected to increase from $3.7 billion in 1981 to $18.3 billion by 1985 (Allen, 1981*d*).

Hard disc drives, like floppy disc drives, are mechanical devices and therefore further substantial price decreases are unlikely. However, the bit capacity per unit area of disc is expected to continue to increase (*Computerworld*, 1980). Venture Development Corporation (VDC) predicts that hard discs will gradually replace floppy discs for many applications (VDC, 1980).

For teletext and videotex applications the advantages of hard discs compared to floppy discs (faster access time and transfer rate and higher storage capacities) are not as important as in conventional computer-based applications. Nevertheless, as the penetration of hard discs in minicomputer-based systems increases, their availability for other applications such as teletext/videotex also increases.

TABLE 8.8 ATTRIBUTES OF STORAGE MEDIA

Technological option	Floppy disc	Hard disc	Audio cassette	Digital cassette	Cartridge tape	Video disc	Video cassette
Digital storage (Mbytes)	0.1–1	3–20	0.1	0.5	2.5	—	—
Audio storage (hours)	—	—	1	—	—	—	2
Video storage (frames)	—	—	—	—	—	108 K	216 K
Access time	250 ms	20–80 ms	40''/sec	300''/sec	30–90''/sec	2.5 sec	4.5 min
Transfer rate (Kbytes/sec)	60	5,000	75	2–8	70	30*	30*
Replaceable fixed (R/F)	R	F	R	R	R	R	R
Read/read-write (R/R–W)	R–W	R–W	R–W	R–W	R–W	R	R–W
Drive-unit cost ($1982)	$500–2,500	$1,000–6,000	$50–80	$500–900	$1,000–1,500	$400–750	$700–1,200
Media cost ($1982)	$5–10	N/A	$0.5	$1	$15	$15–25	$10–15

*Frames/sec.

Audio cassettes offer a low-cost digital and audio storage medium in which digital information is stored using frequency-shift keying techniques.[1] While basically an audio recorder, more integrated versions offer an audio and digital track. These recorders generally do not have any search mechanism; rather, the tape counter is used as a manual indexing system. Because of its relatively low price ($50 to $80), only a modest decrease in cost can be expected. This storage medium has found most applications with personal computing systems.

Digital cassettes were introduced to provide higher density, greater reliability, higher speed, and longer cassette life. Mechanical improvements to the cassette shell and drives have increased information storage to 500 kbytes with transfer rates of up to 8 kbytes/sec. Current prices for digital cassette recorders are $600 to $900 and are expected to decrease with increased penetration. More importantly, improvements in recording tape technology and better drive designs are significantly increasing the quantity of digital information that can be stored on a (nonstandard, high-density) cassette. These drives now permit storage of up to 50 Mbytes of information.

Cartridge tape units offer considerably more storage than standard audio and digital cassette recorders. The tape width is 0.25 inch instead of 0.15 inch, and there are 4 recording tracks rather than 2. As a result a 300-foot cartridge tape may hold 2.5 Mbytes of data with a transfer rate of up to 70 kbytes/sec. As with digital cassettes, high-density recording techniques can increase storage by an order of magnitude to 20 Mbytes (and the price to $2,000). The outlook is for microprocessors to provide better drive reliability, replacing some existing mechanical functions (Allen, 1980). Thus prices are expected to decrease from the present range of $1,000 to $1,500 and, conversely, storage densitites are expected to increase at comparable cost.

Audio, digital, and cartridge tape systems all have miniversions that are geared toward lower-cost applications. Storage size and transfer rates are typically reduced by a factor of 5 compared to the standard-size models. Recorder prices are then comparable to the standard models of the same storage size (Manildi, 1980).

Videodiscs and videocassette tapes are primarily video storage media, although cassette units provide for audio recording as well. Digital information, when appropriately converted, can also be stored on such media. For example, a videodisc can store 2,500 Mbytes of data (one frame is roughly equivalent to 25 kbytes). Such units currently cost $300 to $750. The drawback of today's videodiscs is that they are read-only and thus are mainly suitable for stand-alone applications (Figure 8.9). They have, of course, a tremendous storage capacity, with a maximum single frame random-access time of 2.5 seconds. As a result, discs can combine full-motion video, slides, and voice (or music) without being played through from start to finish. The price of videodisc players is expected to continue decreasing as penetration increases. In addition, videodisc recorders will become available during the next 20 years, but are likely to remain too expensive for the home market.

Videocassette tapes also offer a viable storage medium. Cassette tapes come in 2-, 4-, and 6-hour recording densities, a 2-hour cassette tape being capable of storing 216,000 frames of video information. Cassette recorders have a parallel audio storage

---

[1] Frequency-shift keying: A frequency modulation method in which the frequency is made to vary.

FIGURE 8.9 RCA Videodisc Player. (*Source: RCA Corporation.*)

channel, and, in contrast to the videodisc, they have both read and write capability. Because the storage medium is tape, access time to individual frames is much slower. Prices of videocassette recorders range from $600 to $1,200, depending on features, and are also expected to decrease (Lachenbruch, 1980).

Twenty-Year Forecast

Many of the above storage devices are already well established in the home entertainment market, e.g., audio cassette recorders and, to a lesser degree, videocassette recorders and videodisc players. Other devices have more recently entered the home market, e.g., videodisc players, also for home entertainment, and floppy disc drives for home computer hobbyists. With the introduction of teletext/videotex systems and the development of new uses for these storage devices, mass production will permit lower prices in many cases (Table 8.9).

In addition to the above prices, there is also a substantial cost to the user to interface each device to his or her home terminal. For example, an interface unit to link a

TABLE 8.9 TWENTY-YEAR FORECAST: STORAGE DEVICE PRICES

	1980	1990	2000
Floppy disc drives	$500–$2,500	$500–$2,500	$500–$2,500
Hard disc drives	$1,000–$6,000	$500–$5,000	$500–$5,000
Audio cassette players	$50–$80	$50–$80	$50–$80
Digital cassette players	$500–$900	$100–$300	$50–$100
Cartridge tape players	$1,000–$1,500	$500–$1,000	$100–$500
Videodisc players (playback only)	$300–750	$250–500	$100–300
Videocassette recorders	$600–1,200	$350–800	$200–600

laser videodisc player to an Apple II computer is available from Discovision Associates of Costa Mesa, California, for $225 (Marshall, 1981). However, unlike the storage devices that rely on mechanical parts, interface facilities are typically provided by a microprocessor. As discussed earlier, microprocessor costs are declining rapidly, and thus it is likely that the basic teletext/videotex display unit will eventually incorporate all the necessary interface arrangements at a slight incremental cost. By 1990, therefore, the cost of additional storage is unlikely to be influenced to any significant degree by the cost of any interface equipment.

CONCLUSION

In this chapter we have attempted to convey the diversity of technological options that are available both now and in the future as building blocks for the videotex terminal. As we indicated at the beginning of the chapter, it is unlikely that there will be a single teletext or videotex terminal. We can already discern a number of overall distinguishing characteristics in the developing videotex terminals. These characteristics include home or business usage and integrated or modular design.

The requirements placed on the videotex terminal and the capital available to purchase it are very different in the home than in the business environment. The home user approaches videotex from the standpoint of familiarity with television, video games, and home computers. The business user, on the other hand, is more experienced with terminals, data processing, and timesharing services.

The various components of the videotex terminals that we have discussed can be packaged as an integrated terminal or provided in a modular arrangement. Integrated terminal units offer the prospect of lower costs and slightly simpler ease of use; modular arrangements offer greater flexibility in matching users' needs.

In examining current videotex terminals (Table 8.10), we note that the terminal may display text or both text *and* graphics. For example, Radio Shack's videotex terminal displays text only and thus is geared to existing database services such as Dow Jones, CompuServe, or The Source. Terminals that support text *and* graphics use one of the major videotex standards (Antiope, AT&T Presentation Level Protocol, Prestel, or Telidon). Since the standards include the ASCII coding of characters, these terminals can be used to obtain not only current text-based services but also new information services that use the coding standards.

While most current videotex terminals provide no locally available processing capability, future terminals are likely to provide local processing power and thus will resemble a personal computer. This processing power might arise from the introduction of a videotex decoder board into a personal computer (e.g., Norpak's videotex decoder board for the Apple II or III personal computer to be available in mid-1983) or from the introduction of a CPU board into a dumb videotex terminal.

In a similar way, the current videotex terminals can be enhanced in the future with the options discussed in this chapter to provide the technological support for the diverse videotex applications. The evolutionary paths followed by the various terminal arrangements will be determined by trade-offs between costs and the various attributes required by videotex services and by market demand.

TABLE 8.10 CURRENT VIDEOTEX TERMINALS BY TECHNOLOGY COMPONENTS

Videotex terminals	Integrated	Modular	Interface device	Display unit	Input device	Estimated single unit cost (Dec '81)
Radio Shack videotex terminal	X		Telephone modem	TV	Keyboard	$ 399
Norpak Mark III		X	Telidon decoder Telephone modem* Cable decoder*	Monitor	Keypad keyboard	$1,825
Electrohome EGT01	X		Telidon decoder Telephone modem	Monitor	Keyboard	$2,887
AEL Microtel VTX 202	X		Telidon decoder Telephone modem*	Monitor	Keyboard	$2,107
Matra	X		Antiope decoder Telephone modem	TV	Keyboard	$ 500
Plessey VUTEL	X		Prestel decoder Telephone modem	Monitor	Keypad	$ 800
INDAX home terminal unit		X	Cable decoder	TV	Keypad	$4–6 month**

* Not included in basic terminal cost.
** Includes interactive service charge.

CHAPTER 9

FUTURE ALTERNATIVES
FOR TELETEXT AND
VIDEOTEX

In the discussion so far we have examined teletext and videotex systems in terms of their technological components and the information services that might be provided over the systems. The aim now is to synthesize this information and describe alternative teletext and videotex configurations that are likely to emerge over the next 20 years.

Although we recognize that new technologies often shape consumer demand (the "science push" effect), we believe that in the longer term the configuration or configurations of teletext and videotex systems that emerge in the United States market will be influenced by one or more driving or predominant applications of the technology. Thus, in this synthesis we integrate the technology alternatives with the five generic classes of information services defined in Chapter 4 to provide "technology-application" scenarios. The pivotal criteria for matching technology with applications are the set of teletext/videotex system attributes described in Chapter 4 (Table 4.3) and expanded in Chapters 6, 7, and 8. The attributes define the level or quality of the information service on the one hand, and the minimum necessary technological design to provide the service on the other.

We recognize that within a 20-year-time horizon the technology is likely to evolve considerably from the present designs. To investigate the direction of this change we introduce the concepts of *basic* and *enhanced* teletext and videotex systems.

A basic system is the technological design obtained by postulating a minimal set of attributes for each information service and describing a least-cost (teletext or videotex) system using our 20-year cost forecasts. No distinction is made at this stage between home and business terminals, but an attempt is made to assess the major differences in costs among the alternative communication networks.

A series of likely technological enhancements, based on the attributes, is then postulated for each generic information service. While this distinction between basic and enhanced systems is convenient for discussion purposes, in reality the services will constitute a spectrum: at the lower end, there are simple text services such as captioning; at the upper end, videotex terminals tend to merge with personal computers and office work stations. In fact, turnkey videotex systems are competing in the marketplace with graphics terminal systems and minicomputers. It is worth noting that by focusing on attributes from a user's perspective, the differences between videotex, teletext, and other classes of services such as stand-alone devices, personal computers, and communicating terminals become more apparent.

BASIC TELETEXT AND VIDEOTEX SERVICES

The five classes of information services for teletext/videotex systems, namely, information retrieval, transactions, messaging, computing, and telemonitoring, are sufficiently broad to cover specific applications such as education, electronic games, computer-aided instruction, and electronic banking. We now define a basic system for each class of information services.

Information Retrieval

The most fundamental teletext/videotex service is information retrieval. There are three basic parameters that describe the type of information retrieval service:

- The specificity of the information being accessed. This can range from very general data on news, weather, and sports to information specific to a professional group or a small class of users.
- The volume of information stored.
- The currency of the information or the frequency with which it requires updating.

Thus, we have an array of information services defined in terms of eight permutations through the matrix

Specificity $\times$	Volume $\times$	Frequency of update
Specific	High	Infrequent
General	Low	Frequent

The upper and lower paths define two characteristically different basic information retrieval services. The upper path is for a specifically focused user group requiring large volumes of slowly changing information, e.g., the traditional encyclopedia or a weekly current affairs update. The lower path is for the general user, with limited detail but frequent updates; the print analogy being the daily newspaper and the broadcast analogy being the hourly news bulletin.

The basic system for *general information services* relies on a one-way transmission service. Pages or frames of information are displayed using color and low-resolution graphics to complement text formats. While black-and-white may be acceptable for many applications that produce solely textual page displays, consumers are psychologically conditioned to color and hence would require color not only in graphic/textual pages but also in textual display pages. The format that most closely resembles a sheet of paper is the 40 × 24 display. However, the resolution of existing TV sets may mean that initially a smaller quantity of information is displayed, e.g., 40 × 20 or even as low as 32 × 16.

There is widespread agreement among all industry groups that the basic display should be *at least* 40 × 20. The proposed European standard involves 24 rows but this is being disputed within the United States because the United States scan line raster television system has only 525 lines compared to the 625 in Europe. Recent evidence shows that 24 rows does not reduce quality on the U.S. system (Ciciora, 1981), and so we have suggested a 40 × 24 page format as a basic default option.

For page selection the user requires a numeric keypad to input the page number or the branch path on a hierarchical structured database. There is no requirement for local processing or storage. In its most basic form the service may be sent over the broadcast television system on the vertical blanking interval and thus can be viewed as an adjunct to television.

To display pages in 10 to 15 seconds, the effective transmission rate needs to be 300 bits/sec, either 100 pages cycling at 13,000 bits/sec in 2-VBI-line broadcast teletext systems or 300 bits/sec on telephone-based videotex systems. The reason for a waiting time in videotex systems is that the appearance of a text and graphic page as it is being received depends on the assembly technique. For text pages in an ASCII data protocol videotex system (e.g., CompuServe), the information can be read as it is received. For other protocols that include text *and* graphics, the page may need to be almost completely received before its information content is apparent.

This waiting time is, however, much longer than the 1 to 2 seconds deemed "maximum" in interactive computer systems (Miller, 1968). Thus users may well become impatient with the slow response. For specific information delivery, waiting times on the order of five seconds appear as a "maximum" limit in timeshared computer systems. Thus, while the waiting time problem is less severe than for general information services at a comparable transmission rate, users may also be less than tolerant of such a lengthy waiting time.

Specific information services require two-way capability. A transmission speed of 300 bits/sec from the user to the database is more than adequate. The user may require more sophisticated access mechanisms such as an alphanumeric keyboard instead of a numeric keypad. The information can still be displayed in a 40 × 24 or 40 × 20 text format without local processing or storage requirements. There may, however, be a need for a paper copy of the displayed information if the range of information is very large (classically a business application). From the user perspective it is immaterial whether services are provided by a single centralized database or a network of databases, provided access and search procedures are standardized.

Transactions

As transactions require the user to interact with at least one external computer, the basic service for transactions requires a two-way capability. A relatively slow transmission speed to and from the individual undertaking the transaction is adequate. Again a comparison with complex interactive computer requests suggests a 15-second maximum waiting time. At the basic level of service, the user needs only a numeric keypad to search for transactions and respond to the preformatted transaction pages. The exact size of the display unit is not important, and a text format of 32×16 is adequate. This application does not require color or high resolution. All computations, bill payments, transfers of funds, and statements of balance may be performed remotely, and no local storage or processing is necessary.

At this basic level, transactions might involve ordering and paying for products and services. Thus, the possibility of remote purchase of products and services is included within the term "transaction." In this case the purchaser requires good quality images in color of the product (the catalog shopping analogy). These attributes are included within specific information retrieval services.

Electronic Messaging

Messaging services have two distinct arrangements—one-to-one communication and one-to-many communication. In the former, which resembles telephone conversations or personal mail, there is generally a requirement for user identification, privacy, and store-and-forward communication. In the latter, there is the need for a common control system to provide services such as community bulletin boards and computer conferencing. The basic messaging service for the user reflects the minimal acceptable presentation level. It is not influenced by intent, i.e., to or from a person or to or from a bulletin board, as in both cases either free text or preformatted messages can be sent. The service must be two-way to allow both sending and receiving of messages. An average transmission speed of 300 bits/sec for communication with the remote computer is adequate, and a text format of 40×24 or 40×20 characters to a page closely matches the current density of personal mail. The basic service does not include color, although black, white, and shades of gray may be used.

The creation and editing of mail messages requires an alphanumeric keyboard and is supported by a remote computer that stores incoming messages until the addressee clears his or her "mailbox." The service resembles store-and-forward messaging currently available on computer timesharing and packet-switching networks (e.g., Telemail on Telenet). The basic service does not include provision for printing messages at the user's terminal.

Because users can begin reading as the message is received there is in reality no waiting time. In fact, a transmission speed of 300 bits/sec is equivalent to 360 words per minute and thus closely matches the average reading speed. In addition it is likely that messages will be numbered, and with the limited space assumed for message storage (2,000 characters), a menulike presentation similar to that in specific information services will suffice.

150

Computing

The computing service may be considered in its most basic form in terms of the physical location of processing, and possibly of storage, functions. The terminal may merely act as an online interface to a computing service, similar to existing computer timesharing services. Alternatively, down-loaded software supported by user-created programs would require an intelligent terminal with considerable memory to operate in an offline mode.

The basic computing service has two possibilities. The first requires two-way transmission at relatively low speeds, a visual display unit, and a numeric keypad. No processing or storage is necessary, as the computing is undertaken remotely with the user providing numerical input in response to formatted requests.

The second basic form of computing is a one-way service with high transmission speed, e.g., 2,400 bits/sec or higher, local CPU processing power, and a local storage device. Both black-and-white display and an alphanumeric input are adequate. The applications are limited only by the size of the CPU and the extent of local storage.

In remote computing, the user requires rapid response to inputs (e.g., two seconds). In the down-loaded situation, the acceptable waiting time to down-load a program into the user terminal appears to be as long as 15 to 60 seconds (Miller, 1968). In this latter case, of course, once the program is down-loaded, response times are virtually instantaneous. In the basic service, menu presentations of database materials are adequate.

Telemonitoring

The telemonitoring service provides a link between a host computer and a user sensor network and involves two distinct services. The first is the transmission of user sensory data to the host computer for detection of fire, security, and electrical appliances malfunctioning. The second is an active decision-oriented service that in addition to monitoring attempts to control various devices, e.g., an energy management system that aims at minimizing energy utilization by optimally setting thermostats.

The most basic form of the service is one that once in place does not require user intervention. It has a two-way capability with low transmission speed requirements (e.g., 300 bits/sec). The input device is a sensor, and the system requires a small amount of local CPU and memory to interface with the host computer automatically. If the service is to include any management functions, then there is the need for user intervention, i.e., a display and numeric keypad input.

Basic Systems Defined

It is clear that by viewing these services in terms of attributes it is possible to design special purpose units for each service. To some extent this represents the current state of teletext/videotex developments. The focus on attributes also provides a link to the set of alternative technological components identified in Chapter 4.

Each basic information service can now be summarized in terms of its attributes, as shown in Table 9.1. Then, drawing on the relation between the attribute and tech-

TABLE 9.1 ATTRIBUTES BY SYSTEM CHARACTERISTICS AND GENERAL APPLICATION; BASIC SERVICE

System characteristic	Information retrieval		Transactions	Messaging	Computing		Telemonitoring
	General	Specific			Remote	Down-loaded	
Network							
Communication interface	One-way 300 bps	Two-way 300 bps	Two-way 300 bps	Two-way 300 bps	Two-way 300 bps	One-way 2,400 bps	Two-way 300 bps
User terminal							
Display	40 × 24 Full color; low-resolution graphics	40 × 24 Full color; low-resolution graphics	32 × 16 b & w	40 × 24 b & w	40 × 24 b & w	40 × 24 b & w	Nil
Input	Numeric	Alphanumeric	Numeric	Alphanumeric	Numeric	Alphanumeric	Sensory
Processing	Nil	Nil	Nil	Messages created and edited remotely	Nil	Local CPU and memory	Local CPU and memory
Storage	Nil	Nil	Nil	Remote storage	Nil	Local storage	Nil
Database							
Search procedure	Menu/index	Keyword	Menu/index	Menu/index	Menu/index	Menu/index	—
Acceptable waiting time	2 sec	5 sec	15 sec	—	2 sec	15–60 sec	Immediate

nological options developed in Chapters 6, 7, and 8 and comparing these with Table 9.1, we can determine a set of technology components to provide the basic information services (Table 9.2).

While we have considered the basic service requirements for the five generic classes of teletext and videotex services separately, it is extremely likely that if these technologies become widespread, there will be one basic teletext system and one basic videotex system to support the generic application classes. It may be that the teletext system is contained within the videotex system, or there may be two separate compatible or even incompatible systems. From Tables 9.1 and 9.2, we see that the common attributes and corresponding technology components of such a basic system are as given in Table 9.3.

In describing a basic system, we must introduce a series of caveats. First, we have not included telemonitoring in the basic system. The reason is that it is fundamentally different from the other generic information classes. Telemonitoring is system initiated, while other information classes are user initiated. Furthermore, telemonitoring generally requires an uninterruptable link to the user's home or business. In the long term, however, electronic services to the home are likely to be multiplexed on a common communications link allowing concurrent use of each. But, at least initially, telemonitoring can be considered as a separate system.

In the basic system, we also have chosen to support only remote computing rather than include local computing as well. Because there is little difference to the user between local and remote computing, we opt for the technologically simpler configuration. A local computing capability is considered as an enhancement to the basic system. Many systems may offer local processing as a matter of course, since the decoder unit already has a CPU and memory to interpret and display the incoming signal.

As shown in Figures 9.1 and 9.2, the basic system can be provided using a number of alternative communication network configurations. These include both one-way communication technologies for teletext services and two-way technologies for videotex services. Using the previously developed data on technology costs, we can estimate expected user-borne technological costs of the various alternatives. These are shown in Tables 9.4 and 9.5.

What is evident from these tables is that by 1990 and beyond there will be very little cost differential between the various network configurations. In the short run, however, costs appear to favor the cable-based systems. The reason is that these networks are used to provide other services with the same equipment (e.g., entertainment) and thus the costs are shared with these more-developed services. Hence, in the cable environment videotex services are an add-on to existing services rather than sole services that must pay their own way.

In other environments, users might also never have to pay the full price for their own terminal equipment. It might be provided by information providers such as banks in return for a small monthly leasing charge and the use of their electronic services. In this way the information service provider can substitute relatively inexpensive electronic equipment for expensive labor and buildings.

TABLE 9.2 TECHNOLOGY OPTIONS TO PROVIDE BASIC INFORMATION SERVICES

Technology component	Information retrieval		Transactions	Messaging	Computing		Telemonitoring
	General	Specific			Remote	Down-loaded	
Communication network option	Broadcast TV, One-way cable	Telephone, Cable, Hybrid	Telephone, Cable, Hybrid	Telephone, Cable	Telephone, Cable, Hybrid	Broadcast TV, One-way cable	Telephone, Cable
User terminal option							
Display	TV	TV	TV (b & w), Small CRT terminal & keyboard	TV (b & w)	TV (b & w)	TV	Nil
Input	Keypad	Keyboard	Keypad Touch-tone telephone	Keyboard	Keypad, Touch-tone telephone	Keyboard	Sensor
Processing						8-bit CPU 4-kbyte ROM 64-kbyte RAM	4-bit CPU 4-kbit ROM
Storage						Floppy disc	

TABLE 9.3   ATTRIBUTES OF THE BASIC SYSTEM

	Teletext		Videotex	
Characteristics	**Attributes**	**Technology component**	**Attributes**	**Technology component**
Communication interface	One-way 13,000 bits/sec (2 VBI lines)	Decoder	Two-way 300 bits/sec	Decoder, possibly modem
Display	40 × 24 text format, full color, low-resolution graphics	TV	40 × 24 text format, full color,low-resolution graphics	TV
Input	Numeric	Keypad	Alphanumeric	Keyboard
Processing	Nil		Nil	
Storage	Nil		Nil	
Database	Menu index 15-sec display time		Menu index 15-sec display time	

Even though the cost of the basic system is derived from a linear summation of the component costs, we do not imply that the system is composed of separate entities. In deriving our cost forecasts for the technology components, we have assumed a modular design in the declining costs. For example, the future TV set will provide a series of built-in modular jacks into which various other components can be plugged. In addition, we have only indicated the interface cost of hybrid systems in which the user-to-database link is a touch-tone telephone. More elaborate hybrid systems might incorporate a modem-decoder unit in the telephone return link.

Another aspect of a comparison among the basic system alternatives is the sensitivity to telephone local area calling charges. Here, only the two-way cable-based sys-

FIGURE 9.1   Basic Teletext System.

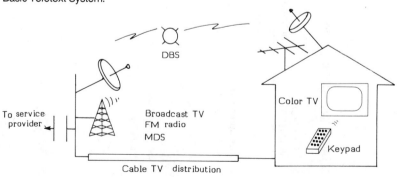

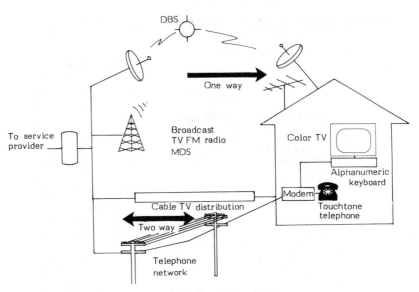

FIGURE 9.2   Basic Videotex System for Hybrid Configurations, Return Path via Telephone Network.

tem is unaffected by such charges, thus adding to its price attractiveness. However, this alternative is limited to the availability of two-way cable systems and the cable operator's willingness to implement the service. Even by 1990, cable penetration is likely to be only about 50 percent of households, of which less than 30 percent will have two-way cable capabilities. But where it exists, it is indeed an attractive alternative.

COMPARISON WITH EXISTING TELETEXT/VIDEOTEX SYSTEMS

In comparing our basic teletext systems with existing teletext systems (Table 9.6) we note there is a great deal of similarity in attributes. The graphic standards are quite different, but to the user the difference is perhaps not too noticeable. In fact, because of the limited number of pages, there is a pressure to keep pages as compact as possible. This constraint may often preclude taking advantage of some of the features of the alphageometric graphic standards.

The key differences among the systems—from the user perspective—relate to the quality of graphics and to the average waiting time per frame. The Telidon system and the enhanced British teletext or French Antiope with DRCS or alphaphotographic possibilities offer significantly better graphics than the basic alphamosaic system. In the limited-page teletext service, however, this qualitative difference may be hard to exploit and currently requires the user to pay a price in terms of average access time (see Table 9.6). In addition, the variable format systems, such as Antiope and Telidon, are more prone to transmission error than the fixed format, Ceefax-type system. This becomes an important consideration for the user who, because of interference in the variable format mode, might lose the total page.

TABLE 9.4   COST FORECASTS OF THE BASIC TELETEXT SYSTEM BY
TECHNOLOGICAL COMPONENTS

	Cost ($1982)		
	1982	1990	2000
Communication network decoder			
Broadcast TV (2 VBI lines)	200	20	10
One-way cable	300	100	30
FM radio	400	100	40
MDS	300	100	40
DBS	4,000	200	50
User terminal			
Color TV	400	400	400
Keypad	50	20	20

With existing videotex systems, however, there is much greater diversity (Table 9.7). A number of differences are readily apparent. In terms of display, the services that have evolved from computer timesharing (e.g., The Source and CompuServe) are solely text oriented, while other systems incorporate both text and graphics. Also, these solely text-based systems are black-and-white only.

In terms of the display transmission speed, the information telephone-based systems may run at either 300 or 1,200 bits/sec. As indicated, most users of The Source and CompuServe access the systems at 300 bits/sec. This is due to the relatively high cost of

TABLE 9.5   COST FORECASTS OF THE BASIC VIDEOTEX SYSTEM BY
TECHNOLOGICAL COMPONENTS

	Cost ($1982)		
	1982	1990	2000
Communication network decoder			
Telephone-based (includes modem)	850	150	30
Two-way cable	300	100	30
Hybrid			
Broadcast/touch-tone telephone	225	45	35
One-way cable/touch-tone telephone	325	125	55
FM radio/touch-tone telephone	425	125	65
MDS/touch-tone telephone	325	125	65
DBS/touch-tone telephone	4,025	225	75
User terminal			
Color TV	400	400	400
Keyboard	100	30	30

TABLE 9.6 RELATIONSHIP BETWEEN EXISTING TELETEXT SYSTEMS AND THE BASIC SYSTEM

Teletext system	Communication interface	Display	Input	Processing/ storage	Database index	Average waiting time (100 pages)
BASIC	2 VBI lines	40 × 24 full color, low-resolution graphics	Numeric	Nil	Menu	12.5
Ceefax/ Oracle	4 VBI lines	×	×	×	×	7.5
Antiope	×	40 × 20 full color, medium, resolution graphics	×	×	×	12.5
Telidon	×	40 × 20 full color, high-resolution graphics	×	×	×	20.0
Closed captioning	1 VBI line	Text only	Nil	×	—	—

Note: × = same as basic service.

1,200-bits/sec telephone modems. Other existing telephone-based services opt for a split transmission speed (1200 bits/sec for display and 75 or 150 bits/sec for input). The lower input transmission speed reduces the cost of the modem substantially.

If we examine the attributes of the basic system (Table 9.1) and consider a trade-off between transmission speed and waiting time (Table 6.4), we note that the waiting times are comparable for transactions and message applications for the basic system we have defined. For information retrieval, however, the difference is quite large. As a result, systems with transmission speeds of 300 bits/sec appear more suited to transactions and messaging than information retrieval. At 1,200-bits/sec, however, the average waiting time for information retrieval and that tolerated by users becomes comparable. Thus, at a 1,200-bits/sec display transmission speed, systems can offer an entire range of videotex services with some assurance of user satisfaction. This requirement is evidenced by the preponderance of systems with a 1,200-bits/sec display rate.

ENHANCED TELETEXT AND VIDEOTEX SERVICES

Modifications to the basic system characteristics discussed above are already taking place in all teletext and videotex systems, and by the year 2000 systems will be avail-

TABLE 9.7   RELATIONSHIP BETWEEN EXISTING VIDEOTEX SYSTEMS AND THE BASIC
SYSTEM

Teletext system	Communication interface	Display	Input	Processing	Storage
BASIC	300 bps	40 × 24 full color, low-resolution graphics	Alpha-numeric	Nil	Nil
Prestel	1,200/75 bps	×	×	×	×
Antiope	1,200/75 bps	×	×	×	×
Telidon	1,200/150 bps	40 × 20 high-resolution graphics	×	×	×
Viewtron	1,200/75 bps	40 × 20 full color, high-resolution graphics	×	×	×
The Source	300 (1,200) bps	80 × 16 characters per line b & w	×	Remote	Remote
CompuServe	300 (1,200) bps	32 × 16 b & w	×	Remote, home computer (optional)	Remote, home computer (optional)
INDAX	28 kbps	32 × 16 full color, low-resolution graphics	Numeric	×	×

Note: × = same as basic service.

able that include numerous additional facilities. Like stereo systems, microcomputers, and digital watches, the range of possibilities for teletext and videotex enhancements is limited only by the imagination of equipment manufacturers and the size of the consumer market. The key attributes described in Table 9.8 indicate the major dimensions of enhanced services for teletext and videotex.

For each of the five generic information services and the numerous permutations of derived services (such as information retrieval *and* messaging), there is a number of likely enhanced configurations depending on the particular application. For example, information retrieval may range from news or sports summaries for which basic teletext systems may be adequate to intricate diagrams and designs for an architect or an engineer. The analogy of the car market seems fitting; a wide variety of options exist, each class of which provides a marginally different service with different performance attributes.

The technological options for enhanced services may be explored in the context of

TABLE 9.8 USER ATTRIBUTES OF TELETEXT/VIDEOTEX SYSTEM

System characteristic	Attribute
Communication network interface	Transmission speed 1- or 2- way
User terminal display	Text format Graphics resolution Still frame video Hard copy Color Animation Speech Music Full-motion video
Input	Text Graphics Speech Video Sensory

Processing capability	Software: preprogrammed, user created
	Hardware: CPU power
	            memory size
	            volatility
	            read/write capability

Storage	Capacity: digital audio, video Replaceable or fixed storage medium Read or read/write capability
Database	Search procedure Waiting time

the five generic classes of information services along the lines adopted for describing basic teletext and videotex services. Based on the cost forecast data presented in Chapters 7 and 8 and drawing on potential teletext and videotex applications discussed in Chapter 5, it is possible to postulate alternative system configurations that may emerge during the next two decades.

Information Retrieval

The bulk of the potential future applications of teletext and videotex systems (Table 5.5) are concerned with information retrieval. What is evident from examining these applications is the broadening of the current text and graphics display to include pictures (e.g., multiple listing services), motion video (e.g., on-demand TV), music (e.g., electronic jukebox), and hard copy (e.g., specialized newsletters). In fact, the applications we have identified are probably but a few of the many enhanced services that will be offered.

Such applications require a much higher transmission rate to the home. In teletext and cable-based videotex, application services can easily use a full channel (8-Mbits/sec). In the telephone-based videotex environment, the transmission rate will have to be substantially higher than it is currently. We envision a transmission speed of at least 4,800 bits/sec from the system to the user but an upstream rate of only 300 bits/sec or less, e.g., 75 bits/sec. Enhanced videotex and teletext information retrieval services tend to be similar in all other respects. The attributes of an enhanced information retrieval service are listed in Table 9.9.

Even though teletext remains essentially a one-way service, an 8-Mbits/sec communications channel terminating on a high-capacity local storage device has the ability to provide interactive information services almost identical to those on a two-way videotex service. In this enhanced teletext service, large volumes of data can be broadcast (approximately 30 million pages per hour). Intelligent user terminals can be programmed to monitor these transmissions and select the particular pages of interest, storing them on a local floppy disc or videocassette recorder. A local printer could then automatically output the pages selected or, alternatively, the user could search these pages on an interactive basis, as in a videotex system, to locate the specific information required. The only difference in the videotex system is that the user is interacting with a *local* database rather than a remote one.

A further enhancement that will become increasingly common toward the end of the century will be speech recognition input and synthesized speech output. Users will be able to request information verbally and receive a spoken output if desired. The speech output will be indistinguishable from the human voice, and users will even be able to select from a range of different voice outputs, e.g., male or female. Another possible development is an interactive stand-alone system such as a videodisc with full-motion video as part of a videotex system. An obvious extension of this is to

TABLE 9.9   ENHANCED INFORMATION RETRIEVAL SERVICE ATTRIBUTES

	Teletext	Videotex
Communications interface	One-way full channel cable (8 Mbit/sec)	4,800/75 bits/sec (telephone) 8 Mbit/75 bits/sec
Display	Frame video Motion video Hard copy Voice	Frame video Motion video Hard copy Voice
Input	Alphanumeric speech	Alphanumeric speech
Processing	Fast local processor Large memory	Fast local processor Large memory
Storage	Video storage Digital storage	Video storage Digital storage

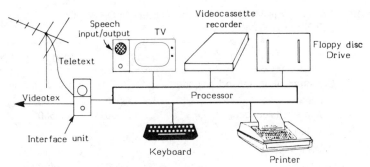

FIGURE 9.3   Enhanced Information Retrieval Terminal.

provide a two-way link, e.g., for information retrieval and for transactions. Figure 9.3 illustrates the terminal components of a typical enhanced information retrieval service.

The likely costs of the configuration illustrated in Figure 9.3 are given in Table 9.10.

Transactions

The basic videotex system described earlier in this chapter is able to deal with simple transactions such as the payment of accounts. However, the basic system relies on passwords for system security. As the use of videotex systems for financial transactions becomes increasingly widespread, e.g., electronic checkbooks, electronic funds transfer, and electronic credit cards, system security is likely to become a major concern to users. Such concern is likely to stimulate the use of more secure system enhancements, e.g., card readers and speech recognition. Smart cards, coded with a fixed amount of dollars, are likely to become especially important for activities that do not allow credit, such as gambling. In the cashless society, the home smart card reader not only serves as a means of identification but also becomes the means to obtain "ready cash"; cards are "recharged" by authorizing a transfer from the user's bank account to the special card account.

Enhanced transaction services are likely to be closely linked with information retrieval so as to include applications such as teleshopping that involve electronic catalogs. The attributes of a possible enhanced transaction service to support teleshopping are summarized in Table 9.11.

For the display of merchandise, a high-quality video display allowing the full range of still frame video, animation, full-motion video, speech, and music (i.e., a television receiver) will be required. A wideband communications channel (8 Mbits/sec) will be necessary to support this range of display attributes. With a narrowband communications channel (9,600 bits/sec or less), full-motion video is not possible, but merchandise could still be displayed using still frame video on a portion of the TV screen (as in the photographic enhancements to the Prestel and Telidon systems). The return communications channel, from the user to the videotex system

TABLE 9.10 TYPICAL TERMINAL COSTS: ENHANCED INFORMATION
RETRIEVAL SERVICE ($1982)

	1982	1990	2000
Communication network decoder			
Teletext	300	100	30
Videotex (telephone)	3,000	500	70
User terminal			
Display			
TV	400	400	400
Speech synthesis unit	200−400	100	20
Printer	500	200	200
Input			
Keyboard	50−300	30−250	30−250
Speech recognition	900−7,500	500−5,000	100−1,000
Computing			
32-bit CPU	150	10	1
1-Mbit RAM	2,000	200	40
Storage			
Floppy disc drive	500−2,500	500−2,500	500−2,500
Videocassette recorder	600−1,200	350−800	200−600

computer, need only be sufficient to support the input devices used in the transaction service, typically a keyboard or numeric keypad. In enhanced transaction services, these input devices are likely to remain the commonest means for initiating transactions, although graphic inputs such as a light pen may provide a more acceptable user interface. In either case, a slow-speed (less than 300 bits/sec) communications link from the user to the system computer will be adequate.

Because of the widely different requirements in data transmission rate in the two directions for some enhanced transaction services, hybrid communication network arrangements are likely to be the most suitable (Figure 9.4).

Enhanced transaction services are unlikely to require much local processing capability or large amounts of local storage. However, users may desire to keep a record of all their transactions (similar to a checking account record) for tax or other purposes. Inexpensive floppy disc equipment is likely to be the most suitable. The likely costs of the configuration illustrated in Figure 9.4 are given in Table 9.12.

Electronic Messaging

While the introduction of electronic messaging capability to the home is likely to result in a number of interesting applications such as referenda and consumer action, the requirements for electronic messaging are likely to remain relatively modest, at least for the next two decades (Table 9.13). The most significant development will be

TABLE 9.11 ENHANCED TRANSACTION SERVICE: ATTRIBUTES

Communication interface	2-way 4,800 bps-8 Mbps/75 bps
Display	Still frame video Animation Full-motion video Speech Music
Input	Alphanumeric graphics Speech authentication
Processing	Preprogrammed software Small CPU memory
Storage	Digital Replaceable Read-write capabilities

the introduction of digital speech and its corresponding need for a 4,800 bits/sec transmission rate channel to the home. It will then be possible for users to communicate verbally while at the same time taking advantage of store-and-forward messaging techniques. In addition, because of the personal nature of messaging (compared to the group nature of television, for example), it is likely that electronic messaging services will stimulate the growth of personal terminals (flat-screen, pocket-size, portable units), rather than remain focused on the home television receiver. The local processing capability required for digital speech is also likely to filter ''junk mail'' and enable users' terminals to communicate directly with one another, with messages being stored in a local mass storage device, such as a floppy disc, rather than in some central computer database.

FIGURE 9.4 Enhanced Transaction Service Terminal.

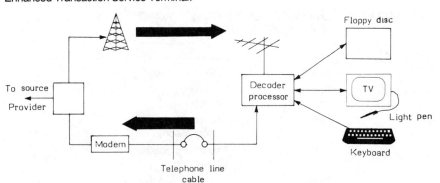

TABLE 9.12 TYPICAL TERMINAL COSTS: ENHANCED TRANSACTION SERVICE ($ 1982)

	1982	1990	2000
Communications network decoder			
Wideband receiver (e.g., low-power TV)	300	100	30
Telephone modem	150	50	10
User terminal			
Display			
TV	400	400	400
Input			
Keyboard	50−300	30−250	30−250
Light pen	50	40	25
Speech recognition	900−7,500	500−5,000	100−1,000
Storage			
Floppy disc drive	500−2,500	500−2,500	500−2,500

The ability to produce a printed copy of a message is likely to become one of the most demanded attributes of an enhanced videotex service, at least for the next two decades. This becomes especially important when greetings are sent by electronic means. The evolution of inexpensive plotters and color printers will allow users to customize their own greeting cards.

Message authentication will also be an important attribute of enhanced messaging services. Card identification (similar to today's credit cards) is likely to evolve initially, but ultimately authentication may be achieved by speech recognition techniques. These procedures will also be widely used for transactions. Figure 9.5 illustates the terminal components of a typical enhanced electronic messaging service.

Computing

The key requirement of an enhanced computing service on a videotex system is flexibility for the user. Some computing applications (e.g., inventory/stock control)

TABLE 9.13 ENHANCED ELECTRONIC MESSAGING SERVICE: ATTRIBUTES

Communication interface	2-way 4,800 bps/75 bps
Display	Hard copy Speech
Input	Speech authentication
Processing	Fast local processor Large memory
Storage	Digital (replaceable)

TABLE 9.14 TYPICAL TERMINAL COSTS: ENHANCED ELECTRONIC
MESSAGING SERVICE ($ 1982)

	1982	1990	2000
Communication network decoder			
Telephone modem	150	50	10
Radio receiver	400	100	40
User terminal			
Display			
Portable VDU/keyboard	2,000	500	200
Printer/plotter (color)	2,000	1,000	600
Input			
Card decoder	500	200	50
Computing			
32-bit CPU	150	10	1
1-Mbit RAM	2,000	200	40

require large amounts of storage while others (e.g., financial management) involve lengthy iterative calculations. An enhanced computing service would permit users to write their own software or down-load software from centralized library systems. Likewise, users would be able to input their own data or access remote data files. The user's terminal would be able to operate as a stand-alone personal computer or as a "dumb" remote timesharing terminal. The attributes of a possible enhanced computing service are listed in Table 9.15.

The keyboard is likely to remain the major input device of any computing service; however, graphic input devices such as light pens and graphic tablets will be essential enhancements for many specialized applications such as telework applications. These are also likely to require a higher quality of display than is possible with a standard television receiver. Enhanced computing services are therefore likely to emerge with dedicated video display units. There is also a very practical problem in using the home television set for computing applications. Television sets are typically placed in rooms used for entertainment; in multiperson households in particular, such rooms are not conducive to programming work.

FIGURE 9.5 Enhanced Electronic Messaging Terminal.

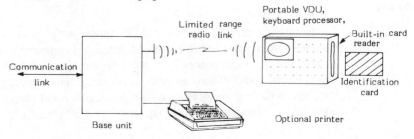

TABLE 9.15 ENHANCED COMPUTING SERVICE: ATTRIBUTES

Communication interface	2-way 4,800/75 bits/sec
Display	High resolution Hard copy
Input	Alphanumeric graphics
Processing	User-oriented software Large CPU Large memory
Storage	Digital storage Replaceable Read-write Fast access and transfer rate

Similar to today's hi-fi sound reproduction equipment, computing terminals are likely to emerge as a set of modules rather than an integrated whole. Users will then be able to balance the local processing and storage facilities required against those available in remote mainframe computers. However, it is likely that during the next two decades the processing power and storage of today's personal computers (8-bit microprocessors; 64-kbytes RAM's) will become the *minimum* available in enhanced videotex computing applications.

Local mass storage is likely to be initially provided by floppy discs and hard discs and eventually by videodiscs. Computing applications involving the bulk transfer of large volumes of data require high-speed communication links. Most other applications can readily be accommodated on links operating at speeds of 9,600 bits/sec or less. This means that enhanced computing services could be provided using either the telephone network or the cable television network. If a large volume of data in a centralized database is to be transferred to a user's terminal, it is likely that the user will be able to request temporary use of a timeshared cable channel. Figure 9.6 illustrates a typical configuration for an enhanced computing service.

The likely costs of the configuration illustrated in Figure 9.6 are given in Table 9.16.

Telemonitoring

Telemonitoring services, which include remote meter reading (user's premises to centralized computer) and telecontrol (centralized computer to user's premises), can be provided quite independently of any videotex system (which is true of other individual videotex applications as well). However, given the existence of a videotex system, telemonitoring services such as home security and energy management can be added at little incremental cost. The amount of information transferred between the

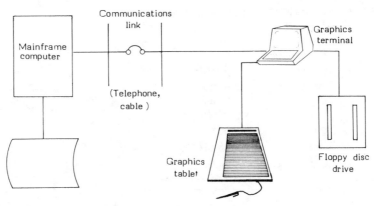

FIGURE 9.6   Enhanced Computing Service.

user and the centralized computer for any telemonitoring services is extremely small, and therefore the speed of transmission is not critical. In fact, systems working at the speed of a rotary telephone dial (10 bits/sec) would still operate satisfactorily. Thus, any type of two-way communication channel can be used. As discussed earlier in this chapter, 24-hour monitoring systems may require a dedicated communications link, in which case the existing switched telephone network could not be used (leased lines will generally be too expensive for widespread use). An alternative involves the use of automatic dialer units to call preselected numbers whenever a local alarm occurs. During the next two decades, voice plus data multiplexers are likely to become economically viable, thereby providing a dedicated link over the telephone network in

TABLE 9.16   TYPICAL TERMINAL COSTS: ENHANCED COMPUTING SERVICE ($ 1982)

	1982	1990	2000
Communication network decoder			
Telephone (4,800 bps)	3,000	500	70
Cable	300	100	30
User terminal			
Display			
Graphic terminal and keyboard	1,000–5,000	500–1,000	200–800
Input			
Graphics tablet	800	500	250
Computing			
32-bit CPU	150	10	1
1-Mbit	2,000	200	40
Storage			
Floppy disc drive	500–2,500	500–2,500	500–2,500

TABLE 9.17   ENHANCED TELEMONITORING SERVICE:
ATTRIBUTES

Communication interface	2-way 300 bits/sec (or less)
Display	Audio Visual Hard copy
Input	Sensory
Processing	Small CPU
Storage	Digital read-write capability

addition to a voice channel (dial-up links can be unreliable). The attributes of an enhanced telemonitoring service are listed in Table 9.17.

The main input device for a telemonitoring service is the solid-state sensor (or transducer); the output display is typically at some remote location (fire station, police station, utility company), although local visual and audible alarms could also be provided. Processing and storage are not essential for telemonitoring services but could potentially be used for record keeping and for controlling multisensor installations. A floppy disc is a possible add-on to be used to record continuous readings such as energy usage, which then could be read in batch mode. Figure 9.7 illustrates a typical configuration for an enhanced telemonitoring service.

The likely costs of the configuration illustrated in Figure 9.7 are given in Table 9.18.

CONCLUSION

In this chapter we have attempted to merge the technology forecasts of Chapters 6 through 8 with the present and future applications of Chapters 4 and 5. In this merging

FIGURE 9.7   Enhanced Telemonitoring Service.

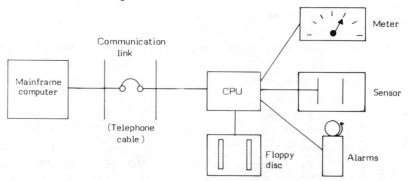

TABLE 9.18 TYPICAL TERMINAL COSTS: ENHANCED TELEMONITORING
SERVICE ($ 1982)

	1982	1990	2000
Communication network decoder			
Modem	150	50	10
User terminal			
Input			
100 sensors	100−200	1−20	0.1−10
Computing			
4-bit CPU	2.0	0.2	0.1
4-Kbits RAM	8.0	0.8	0.2
Storage			
Floppy disc drive	500	500	500

process we outlined the characteristics of a basic system and indicated some possible future enhancements. The specific details of any future enhancements will be dictated by the applications that are provided and that consumers find useful. While we have described five enhanced systems and their possible uses, there will no doubt be many other enhanced systems tailored to specific applications.

We also wish to reiterate that even though we examined the merging of the technology and applications in terms of the five generic applications (information retrieval, messaging, transactions, computing, and telemonitoring), we believe that successful videotex systems will have to support several or all of these generic classes of services. The reason for this assumption is that the U.S. society is so diversified in interest that few users will be identical in all respects. Further, the combination of many services within the same system increases its utility since the consumer will find it easier to learn one system than a number of disparate ones. Thus, the consumer is likely to make more use of the system, and, conversely, videotex systems are likely to achieve higher penetration of the consumer market.

The broadcast teletext systems, especially those that piggyback on an existing signal and thus utilize a small proportion of available bandwidth, are technologically different and offer a specialized class of information services. The key features of these systems will be their potential to be rapidly updated and to maintain a low average waiting time for successive frames.

In terms of the technology paths, our enhanced systems have used a telephone network, a cable network, or a hybrid network to connect to the home. We perceive that *all* of these technology paths will be used to provide videotex services. While the ultimate mix of these various technological paths will be determined by market forces, we can safely say that the relative costs of interfacing to each network will not be a factor by the 1990s. Rather, each network arrangement will support the applications for which its transmission characteristics are most suited and the transmission cost of the network.

CHAPTER 10

POLICY ISSUE
IDENTIFICATION

As is the case with most new technologies, there is no clear single path into the future for teletext and videotex. Instead, there are many alternative directions in which these systems can develop. The paths they take will result in part from public policy decisions. "Public policy" is taken in its broadest sense here as a conscious decision by a recognized authority to support a particular course of action in dealing with a matter of economic or social importance to the general public (as distinct from a special interest group). The policy decisions may, therefore, come from a public body, such as the FCC, or a private organization, such as a trade association agreeing on standards for a new technology. Within the public arena, policy decisions may emanate from regulatory, judicial, or legislative bodies at the federal, state, or local level. International organizations may also make policy that impacts teletext and videotex domestically. Finally, policy toward teletext and videotex may result from preexisting laws, regulations, and judicial decisions as well as from new laws, regulations, and judicial decisions. In short, the policymaking arena is filled with many potential actors and many possible ways of directing or diverting these technologies.

Given that these technologies by definition involve both an array of previously discrete technologies (e.g., television, telephone, and computers) as well as an array of application areas (e.g., news, mail, and banking), the potential policymaking spectrum becomes extremely broad. Not only are there separate public policy assumptions and precedents related to each of the technologies of teletext and videotex, there are also separate precedents in the applications areas.

This chapter reviews the existing policy environment in which teletext and videotex are emerging and then identifies a number of basic policy issues raised by teletext and videotex. Finally, a number of key policy issue theme areas are described. These form the nucleus of the subsequent policy analysis.

BACKGROUND

Teletext and videotex are emerging in the United States in a policy environment that has a tradition of separating various types of communication media (i.e., broadcasting, point-to-point communication, newspaper, and mail). Yet because teletext and videotex potentially represent the merger on a mass consumer basis of (regulated) communications and (nonregulated) computer technologies, they challenge traditional policy distinctions. The foundations of these traditional distinctions are worth reviewing. Let us briefly consider some of the component media technologies of teletext and videotex, their underlying assumptions for public policy purposes, and factors that challenge those assumptions.

Component Technology	Traditional Regulatory Assumptions	Challenging Factors
Newspapers	No government involvement; constitutionally guaranteed free press promotes multiplicity of voices	Electronic newspapers utilizing telephone lines and cable TV systems currently being tested
Broadcasting	Locally based radio-TV with licensee serving as public trustee; limited spectrum must be regulated to ensure public interest	Multiplicity of video sources including distant signals on cable, pay programming, videocassettes and videodiscs
Telephone	Telephone service provided on nondiscriminatory basis through monopoly common carrier; rate and rate-of-return regulation; no content regulation	Increasing competition for equipment and services
Computers	No direct government intervention; market-driven technology	Merging of computer technology and communications technology
Mail	A government monopoly for first-class mail; uniform rates; cannot be used for illegal purposes; limited subsidy for distribution of books, periodicals, and newspapers; universal service guaranteed	Electronic mail already commercially available

Changes in technology have already begun to blur not only the distinctions within telecommunications but also the distinction between nonelectronic media and electronic media. Thus the policy environment for teletext and videotex is one in which old rationales and old policies are under reconsideration. The fact that efforts have been under way since 1976 to rewrite the Communications Act of 1934 indicates the complexity of the issues and the strength of the stakeholders in the process.

Overall the regulatory environment in the United States has undergone major shifts in the last decade. Not only in telecommunications, but also in the airline, trucking, and securities industries, there has been a movement away from dominant federal government regulation. In telecommunications this change really began with the Carterphone decision in 1968 and has picked up increased momentum in recent years. For example, the 1981 Senate version of the Communications Act revision was known as the Telecommunications Deregulation Bill. For purposes of comparison, it is interesting to note the changes in the last decade and the implications they have on new technologies.

1960s through mid-1970s	Late 1970s through early 1980s
Monopoly common carrier	Emergence of competitive common carriers
Federal government actively involved in stimulating alternative services (e.g., public broadcasting)	Decreasing government role; let marketplace decide
Enforcement of existing regulation	Deregulation and "unregulation"
Constraints on new media (e.g., cable)	Minimize barriers to new entry

Of course, the above trends do not imply continuation into the 1990s and beyond. Just as a major shift occurred in the 1970s, so other shifts could occur in the 1980s and 1990s. Indeed, some shifts can be anticipated. For example, as the federal government unburdens itself of regulation, there may be an increased burden placed elsewhere, such as at the state and local level, through public utility commissions and local franchising authorities. Also, at the federal level, there may well be pressure for a kind of mid-course correction. That is, affected parties begin to feel that deregulation has gone too far, abuses are apparent, and a new middle ground is sought.

This policy environment is unique in the world. In other countries, governments not only are maintaining tight regulation over their existing media but have taken aggressive approaches towards supporting development and testing of teletext and videotex. The United States, on the other hand, has maintained a neutral policy stance vis-à-vis the introduction of these technologies. While they have not benefited from any government support, neither have they been impeded. This policy environment is considerably different from that which surrounded the growth of another new medium 10 years earlier. Cable television for many years was closely regulated and had artificial barriers placed on its development as a means of protecting broadcast television.

However, even though the government has not established a "videotex policy," development of the technology will be influenced both by existing laws and regulations and by policy decisions yet to be made.

THE CURRENT POLICY ENVIRONMENT FOR TELETEXT AND VIDEOTEX

In looking at any new technology and the policy issues it potentially raises, it is important to first understand the prevailing policy environment. There are at least 10 broad areas of government policymaking that impinge on the development of teletext and videotex. These areas range from the First Amendment to local franchising for cable TV. The following is a brief summary of each area, including an indication of the implications for teletext and videotex. They are arranged roughly in order from the broadest and most far-reaching to the most specific and local.

1 The First Amendment

Background   The First Amendment to the Constitution states that "Congress shall make no law . . . abridging the freedom of speech . . . ." By way of the Fourteenth Amendment's equal protection doctrine, the states cannot deny people freedom of speech either.

As interpreted by the Supreme Court, free speech protection does not apply to purely commercial messages (advertising), to false or fraudulent statements, or to libelous or slanderous speech. In addition, the court has singled out radio and television broadcasting as an area where limited frequency availability justifies government regulation. For reasons more historical (or political) than logical, such regulation has also been applied to newer media, such as CATV and satellites, although this is changing.

Implications   Teletext and videotex are hybrid technologies, combining aspects of print and electronic media. A fundamental policy decision will be to determine the model on which regulation of these new media is to be based. For example, it is generally conceded that newspapers enjoy greater First Amendment rights than broadcasters. Will electronic newspapers have the same First Amendment protections as print newspapers? Will the content of broadcast teletext be granted protection equal to that of the content of telephone-based videotex?

2 Federal Communications Act

Background   In 1934, Congress passed the Communications Act, which created the Federal Communications Commission and empowered it to regulate all "interstate and foreign commerce in communication by wire and radio." Title II of the act pertains to FCC regulation of common carriers. Title III of the act provides for FCC regulation of radio and television. The FCC enforces the act by making case-by-case decisions (e.g., on station licenses and common carrier tariffs) and by making rules (e.g., prime-time access rule and computer inquiries). FCC rules have the force of federal law.

Rewrite of the Communications Act New developments in communication technology have increased pressure for new legislation to update the 1934 act. For example, because cable television did not exist in 1934, the act contains no mention of it. (The FCC has justified its regulation of cable by treating it as a service "ancillary to broadcasting"—an increasingly inappropriate definition.)

Since 1976 Congress has been attempting to formulate either a new communications act or a revision of the existing law. Central to the debate over various versions of proposed changes has been the role of AT&T. Recent versions have focused particularly on a videotex application—electronic yellow pages—and what the dominant common carrier is allowed to do with it.

Implications There are several potential changes in the act that have bearing on the shape of teletext and videotex in the years ahead. How the common carrier sections of the act are amended and what new ground rules are spelled out for AT&T are key points of contention. AT&T wants to participate in the new competitive environment. Its opponents fear that AT&T could take over the market. A major debate in Congress in 1981 was on whether AT&T would be allowed to offer electronic yellow pages. Other areas of the legislation that may impact teletext and videotex include possible changes in broadcasting regulations. If, for example, Fairness Doctrine obligations were eliminated or decreased, such a policy would impact teletext use. In addition to specific aspects of the law, the general policy orientation will set the foundation for U.S. policy for years to come. Language of recent draft bills emphasizes the promotion of marketplace competition, deregulation, and reliance on the private sector to provide telecommunications services.

3 Antitrust Law

Background The U.S. antitrust law is essentially a creature of the federal government. Before World War I, Congress passed the Sherman Act and the Clayton Act. These statutes broadly proscribe "monopolies" and "acts in restraint of trade." Antitrust law has actually developed in the courts. Most (but not all) of the important antitrust cases have been brought by the Justice Department.

Antitrust law is concerned with three major kinds of violations that interfere with the free competitive market. The first is horizontal combinations or agreements, such as price fixing or restraint-of-trade agreements between companies. The second kind of violation is vertical arrangements, by which a single entity gains control of a commodity from production through distribution. This enables a vendor to control the terms of sale (e.g., tying agreements linking purchase of one item with another). The third kind, and perhaps the most relevant here, is structural violations involving harmful concentrations of economic power. An example would be the total domination of telecommunications by AT&T. It should be noted that regulated industries are generally exempted from antitrust law since they are presumably under regulatory control. Concern about AT&T arose because of believed misuse of the regulatory process.

The 1956 Consent Decree was a compromise settlement of an antitrust suit brought against the Bell family of companies by the government. Bell agreed to refrain from engaging in unregulated services, including the sale or lease of computer equipment or services, except as a direct part of its common carrier telephone service. Thus Bell could provide modems to connect computer terminals to the phone lines but could not provide the terminals themselves. Similarly, Bell could not sell timesharing services on its own computers or engage in providing cable TV service.

In 1974, the Justice Department filed another antitrust suit against AT&T, seeking to break the company into several smaller entities. In a settlement reached in January 1982, AT&T agreed to divest itself of its 22 local operating companies, which represent approximately two-thirds of the company's assets. In return, AT&T would be freed from the terms of the 1956 Consent Decree and would be able to enter areas of business previously off limits to it. Much remains to be seen about how this settlement is implemented and how Congress and the FCC respond to it.

Implications Antitrust problems in the videotex area have potentially greater sensitivity to First Amendment concerns than do other types of antitrust problems. Concentration of information control and distribution in the hands of a few large conglomerates could jeopardize the basic First Amendment value of a diversity of voices and free expression of ideas. The problem to be resolved is: What constitutes abusive concentration in the information and telecommunications areas? There are differing interpretations about how much and what type of concentration should be tolerated.

In the case of videotex, decisions on antitrust issues related to AT&T could have major implications for the future direction of this technology.

4 FCC: Common Carrier Regulation

The FCC, under Title III of the Communications Act, regulates common carriers. The FCC is empowered to review all "changes, practices, classifications, and regulations" made by common carriers to ensure that they are "just and reasonable." Because of its size and dominant position, AT&T has been the focus of much of the FCC's common carrier regulation.

One persistent problem in recent years has been the increasing difficulty in distinguishing between computing and communication. As the telephone system has been utilized to carry data communications and as computers have been used to support new forms of communication, the line between the two has blurred.

In 1971, the FCC attempted to settle this issue with a decision now referred to as Computer Inquiry I. CI1 stated that all forms of data processing (including so-called hybrid data processing) should remain outside of the FCC's jurisdiction. Because AT&T was banned by the 1956 consent decree from entering any unregulated business, CI1 effectively kept AT&T from offering any service involving data processing.

A few years later, the FCC reopened the subject by means of Computer Inquiry II. In its final decision, released in 1980, the FCC defined two classes of services:

"basic" and "enhanced." Basic service is the use of transmission capacity for information delivery. It is subject to rate and other common carrier regulation. "Enhanced service" is everything else besides transmission, such as data storage and retrieval. This area would not be regulated. Under CI2 AT&T would be allowed to enter certain parts of the data processing business through separate subsidiaries. Divestiture of AT&T's local operating companies may make CI2 irrelevant. However, the FCC will continue to have wide jurisdiction over AT&T and other common carriers.

Implications   The FCC's role in relation to common carrier regulation generally, and to the regulation of AT&T specifically, is in a state of flux. Its role will be influenced by court decisions and by congressional actions. To date, however, the FCC has played virtually no role in shaping the development of videotex.

5   FCC: Broadcasting Regulation

The mandate over broadcasting given to the FCC by Congress is extremely broad. Title III of the Communications Act gives the commission authority to see that all broadcasters operate in a manner consistent with "the public interest, convenience, or necessity." The FCC's principal power over broadcasters derives from its authority to grant and renew temporary licenses to operate radio and television stations.

Although the act forbids the FCC from making any rules governing the content of broadcasting, the commission has enacted several rules that influence content. Under FCC regulations, broadcasters are prohibited from broadcasting lotteries or obscene materials. Under the commission's Fairness Doctrine, television broadcasters are required to provide programming on controversial issues of public importance and to do it fairly. And by means of its relicensing, the FCC influences TV content toward public affairs, news, and community involvement and away from excessive commercialism.

Most of the controls on the content of television are based on industry self-regulation. Voluntary standards for TV and radio content are set by the National Association of Broadcasters (NAB) through its Television Code. The code includes provisions relating to kinds of products that may or may not be advertised, permissible advertising techniques, and the maximum amount of broadcast time to be devoted to commercials.

In addition to granting licenses to all broadcast stations, the FCC also sets technical standards for these transmissions. The FCC approves TV receiver standards, such as what type of terminal can be attached to a TV set. These decisions can influence the nature of teletext (and perhaps videotex) systems. The FCC is also responsible for authorizing teletext trials, since these involve use of the vertical blanking interval of the television channel.

In 1980, CBS petitioned the FCC to adopt a set of teletext standards. In October 1981 the commission decided against setting a formal technical standard, choosing instead to allow the marketplace to decide. (In the meantime, industry associations are actively involved in developing recommendations.)

Implications So far, the FCC has chosen to play a minimal, yet significant, role in the development of teletext. Its decision against setting a teletext standard was a clear indication of the FCC's inclination to favor the marketplace over detailed regulation. At some point, however, it will be necessary for the FCC to determine the extent to which regulations that apply to normal television broadcasting apply to teletext.

6 Federal and State Banking Laws

Background Banking is one of the most highly regulated industries in the United States. Its regulatory structure is also complex because multiple federal and state bodies are involved.

The following agencies regulate private sector banking:

• The Federal Reserve Bank supervises its member banks (typically all large banks).

• The Federal Deposit Insurance Corporation (and the Federal Savings and Loan Insurance Corporation) are independent agencies that protect depositors in virtually all banks and savings institutions.

• The Comptroller of the Currency regulates federally chartered financial institutions. This office is part of the Treasury Department.

• State banking agencies regulate state-chartered financial institutions. State banking laws cover federally chartered banks also. This coverage includes limits on branch banking, usury rates, and interstate banking.

• The Federal Trade Commission enforces several statutes relating to credit and lending.

Currently, a movement is under way to diminish the regulation of banking. Particular emphasis has been placed on eliminating the distinctions that have traditionally been maintained between different categories of financial institutions (e.g., different maximum interest rates on savings permitted to banks and to savings and loan associations).

Implications Videotex does not raise issues in bank regulation that have not come up before regarding electronic funds transfer, banking by telephone, and automatic teller machines. However, widespread use of videotex payment and funds transfer will make these issues even more important.

Videotex adds a further element of uncertainty to the established regulatory structure, because banking and fund transfers from videotex terminals blur or eliminate many of the traditional regulatory distinctions concerning bank and branch location, the distinction between intrastate and interstate commerce, and, in fact, the definition of what constitutes a bank.

7 Federal Trade Commission Regulation of Sales and Advertising

Background The Federal Trade Commission has the basic objective of preserving competitive enterprise and prohibiting unfair acts in commerce. It has antitrust responsibilities, overlapping those of the Justice Department, but with an additional emphasis on consumer protection.

The FTC has jurisdiction only over interstate commerce, however broadly interpreted. Each state, by common law and statute, also prohibits fraud or deception in commercial dealings. The FTC distinguishes itself from the standard state regulation of trade by enforcing a variety of federal statutes and its own related rules that bear on specific commercial abuses. Examples are the Fair Packaging and Labelling Act, the Fair Credit Reporting Act, and the Truth in Lending Act. The FTC puts a good deal of emphasis on eliminating false and misleading advertising and sales techniques. The FCC acts both on a case-by-case basis and by promulgating industrywide trade rules and regulations.

Implications Since videotex selling would presumably use an interstate means—the telephone network—the FTC would no doubt assert jurisdiction, much as it now monitors broadcast and print advertising and mail-order sales. The very concept of advertising may need restructuring in an on-demand user-controlled system.

8 Copyright Law

Background It is now well accepted that material held in nonprint media can be protected. Works protected under U.S. copyright law can also be protected abroad in countries having copyright relations with the United States.

The judicial doctrine of fair use was developed over the years as a defense to copyright infringement. It is now seen by user communities as the basis of their ''right'' to copy (but not to engage in resale). However, the ''Betamax'' case, which challenges the right of individuals to make videotape recordings of copyright broadcast material, represents the state of flux introduced by new technologies.

The Copyright Act of 1976 as amended in 1980 specifically protects computer software and information stored in an online database. The problem is the ease with which these copyrights can be infringed. From the publishers' point of view, direct user violations result in loss of revenue and, it could be argued, theft of material. The problem is two-edged as violations also enable widespread dissemination of what would otherwise be private information.

Implications The new technologies, including audio and videotape recorders, photocopying equipment, and computer terminals with microprocessor capabilities, make it increasingly easy to violate the law. A major issue will be whether this act carries over to videotex systems and how it, or some associated act, will be enforced, given a multitude of information dissemination networks and a proliferation of individual storage media.

9 State Utility Regulation

Background Under the U.S. Constitution, the federal government can regulate only interstate commerce. The courts have defined interstate commerce very broadly over the last 40 years, but the state legislatures and public utility commissions (PUCs) still have considerable authority over local communications common carriers—the operating units of the Bell system and other telephone companies. State regulation has emphasized tariffs, typically in an effort to keep down charges for local service to residential customers.

State regulation has historically ignored broader telecommunications issues. The PUCs have not concerned themselves with technical standards either, as long as traditional telephone service was adequate. Because of their more rapidly rising costs, electricity and natural gas, also responsibilities of the PUCs, have tended to be more controversial than telephone service. However, with AT&T's divestiture of its 22 local operating companies, state commissions are likely to take on a larger role in telephone regulation.

Implications While federal regulation of telecommunication services is decreasing, state regulation and oversight is on the increase. Evidence of this shift has already appeared: the 1980−81 case in which the Texas Public Utilities Commission had to rule on whether Southwestern Bell could implement a demonstration of electronic yellow pages in Austin, Texas. If state PUCs did become more actively involved, there could be problems of large discrepancies in services and rates from one state to another. As with cable, such a situation may precipitate lobbying for federal preemption of some of the state activities.

10 Cable Franchising and Regulation

Background Cable television is the only major communication medium that is regulated primarily on a local level. Franchises to build cable systems are granted by municipalities under the power of local governments to control the use of streets and grant rights-of-way. In return, cities usually receive a portion of cable revenues as a franchise fee.

Franchising has proved a difficult challenge for local governments. Apart from the abuses stemming from the financial stakes involved, many cities have been ill-prepared to deal with the complex technical issues of cable service. On the other hand, service demands made by cities have escalated rapidly. Franchise bids in major cities today routinely promise sophisticated interactive services including videotex (*Electronic Publisher*, 1982). Whether cable operators will be able to make good on these promises remains to be seen.

State-level regulation of cable varies widely across the country. Eleven states regulate cable on a comprehensive basis through a state agency. Three states—Massachusetts, Minnesota, and New York—have established separate commissions concerned solely with cable; other states regulate through state utility commissions. Five states—Alaska, Connecticut, Hawaii, Rhode Island, and Vermont—have preempted local control in favor of state franchising and regulation.

At present there is little FCC regulation of cable. Federal regulation reached its high point in 1972 when the FCC created an elaborate set of rules requiring a minimum number of channels, mandating access channels and two-way capacity for systems in major markets, and placing restrictions on the signals these stations could carry. Subsequently these rules have largely been eliminated by court decisions or by the commission itself. The FCC has also reduced its regulation of state and local franchising. As part of its emphasis on deregulation, the FCC has moved toward encouraging development of other competing media such as low-power TV stations, multipoint distribution systems (MDS), subscription television (STV), and direct broadcast satellites.

Implications   Under existing rules a cable company could provide an information service and control access to that service. Similarly, the cable company could offer a home security service and deny the use of its cable to other local security and burglar alarm companies. (The telephone company, on the other hand, could not discriminate against any alarm company.) As cable penetration increases, there may be pressure for some access policy concerning cable systems. Information and service providers who are excluded from major markets may ask policymakers for relief.

TELETEXT AND VIDEOTEX POLICY ISSUE IDENTIFICATION

Where do teletext and videotex fit in the existing policy structure just described, given that they involve existing media that are regulated under different policy assumptions? This fundamental question is perhaps best illustrated from the home perspective. Figure 10.1 shows current communication links to the home and indicates whether they are regulated monopolies, semiregulated monopolies, or not regulated at all. By *regulated monopolies*, we mean organizations that are given exclusive franchise to deliver a service—e.g., the U.S. Postal Service and the Bell system. The rates they charge customers and their profits are regulated. By *semiregulated monopolies* we mean aspects of the particular service are under the aegis of a particular government agency, e.g., broadcasting and cable television. However, unlike regulated monopolies, there is competition, and there is no rate-of-return regulation. In addition, Figure 10.1 identifies "gray" areas where policy has yet to be determined, e.g., satellites, and unregulated areas, e.g., print, videocassettes, and video discs. The fact that teletext and videotex are carried by several of these media and potentially can incorporate the functions of several others is why its regulatory status is so uncertain.

Policy issue identification, especially that of future policy issues, represents the middle stage in this technology assessment. To this point, we have been answering the questions: What is the technology? What might it be used for? Where is it heading? What is its current policy environment? We are now at the point of knowing the technology and its current and potential applications. Having this foundation, we are able to isolate specific issues arising now and likely to arise in the future both from within the context of the existing policy environment and from outside this context.

A policy issue arises when disagreement occurs. If there is no disagreement or controversy, no policy issue exists. An initial set of policy issues was derived from a

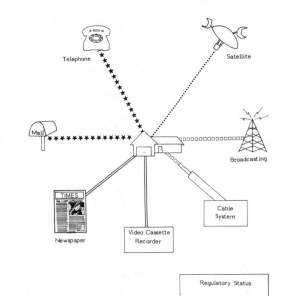

FIGURE 10.1 Regulatory Status of Communication Channels to the Home.

series of three policy-oriented futures workshops. One aim of these sessions was to identify likely policy issues arising on the path toward and as a result of widespread adoption of a number of teletext and videotex services such as electronic mail, electronic transactions and teleshopping, and electronic yellow pages.

The workshops began with participants generating an extensive list of potential applications. These are described in Table 5.6. Eleven of these applications were selected as being particularly important, and a series of policy questions pertaining to each was raised in the workshop discussion sessions.

Our aim at this stage was merely to identify potential issues—not to suggest what should (or could) be done about them. The applications areas were chosen to represent the five generic classes of teletext and videotex applications. The emphasis on information retrieval was predictable as almost every videotex service has an important information retrieval component. The workshops generated more than 100 specific policy questions related to these applications. These are summarized in Appendix 2. This number could be increased manyfold by considering the policy implications of other limited applications. However, our aim in this study was to identify the *major* recurring policy themes raised by the emergence of videotex and teletext. These policy themes are defined in the next section.

The 11 applications and the policy issue themes associated with them were as follows:

APPLICATIONS AND ISSUE THEMES

Applications	Other generic classes involved	Policy Issue Themes
I Information retrieval		
1 Electronic yellow pages	Transactions Messaging	Guarantee of access Competition Content regulation Consumer protection Industry structure
2 Education/ retraining	Computing	Content regulation Copyright Privacy and security Equity of access Industry structure Employment
3 Storefront medicine (medical information accessed by professionals or consumers)	Messaging Transactions Computing	Content control Privacy and security Consumer protection Industry structure Employment
4 Multiple listing service (e.g., real estate, travel packages)	Transactions	Guarantee of access Privacy and security Consumer protection Industry structure Employment
5 Real-time job market	Messaging	Copyright Privacy and security Equity of access Consumer protection Industry structure Employment
6 Consumer information (comparative pricing and performance ratings)	Computing	Consumer protection Industry structure
7 Electronic library	Messaging	Copyright Equity of access Industry structure

APPLICATIONS AND ISSUE THEMES (cont.)

Applications	Other generic classes involved	Policy issue themes
II Messaging		
8 Electronic mail		Standards Guarantee of access Competition Privacy and security Equity of access Consumer protection Industry structure Employment International trade and communications
III Transactions		
9 Home banking and shopping		Standards Content regulation Privacy and security Consumer protection Industry structure Employment
IV Computing		
10 Video games and interactive gambling	Transactions	Standards Content regulation Privacy and security Consumer protection Industry structure
V Telemonitoring		
11 "Electronic mother" (electronic information, support, and monitoring from cradle to grave)	Information retrieval	Privacy and security Equity of access

KEY TELETEXT AND VIDEOTEX POLICY THEMES

In looking at present and future instances of issues emerging from videotex and teletext, it is clear that not all policy issues are on the same level. Some issues are of immediate concern and the battle lines are already being drawn; some are to be faced further down the road, given a scenario of widespread penetration. These two kinds of issues have been called *developmental* and *consequential*.

Developmental issues are the set of issues that arise *as the technology is emerging*. Examples are the questions of what standards (if any) should be set, control over videotex content, and full and fair competition. Although these issues arise in the near term (some, such as the role of AT&T, have already come up), their resolution will not necessarily be simple or quick. How they are decided will have a significant impact on how the technology develops, what industry configuration emerges, and what applications are included.

Consequential issues are the broader issues likely to arise *as a result of widespread penetration* of videotex. Many of these issues are familiar—e.g., potential threats to privacy, copyright problems, and impact on employment. At this level, videotex is less likely to raise new issues than to accelerate or accentuate trends already in evidence from other technologies and from the developmental issues.

Based on our analysis of applications and interest groups, we have identified 11 major policy issues. Five of these fall under the category of developmental issues and six under the category of consequential issues:

Developmental issues	Consequential issues
1 Standards	1 Privacy and security
2 Guarantee of access	2 Equity of access
3 Competition	3 Consumer protection
4 Content regulation	4 Industry structure
5 Copyright	5 Employment
	6 International trade and communication

The five developmental issues are discussed in detail in Chapter 11; the consequential issues are discussed in Chapter 12. In addition, Chapter 14 looks at a series of possible broad societal impacts of the technology that, though potentially significant, are less easily analyzed in terms of shorter-term policy implications.

In brief, the five developmental issue themes we have identified are as follows:

1 *Standards.* Mass-market products achieve great economies of scale through standardization. In teletext and videotex, standards can influence how the future technology takes shape. For example, the standards debate at present has focused on the degree of sophistication necessary for graphic displays. The trade-off is, the more sophistication built into the system at the outset, the higher the initial cost versus adopting an outmoded (or soon to be dated) technology.

2 *Guarantee of access.* Whether information providers could be excluded from a teletext and videotex service is a potential policy problem. Under existing policies, the same information provider might be guaranteed access in one medium and excluded in another. When information providers are arbitrarily excluded from a service because of a competitive threat, then it is likely that they will seek relief through legislation or regulation.

3 *Competition.* Even before teletext and videotex emerge as commercial services, there is already much discussion on how this marketplace will be structured to ensure full competition. Because it involves regulated media, decisions are looming as to who will be allowed to enter this new marketplace and under what circumstances. Of particular concern is the role to be played by the dominant common carrier, AT&T.

4 *Content regulation.* A number of FCC regulations affect broadcast content, such as the equal time rule and the Fairness Doctrine. There are no controls, however, on the content of newspapers or of nonbroadcast video programming (e.g., cable-only networks, prerecorded videocassettes and videodiscs). What model teletext and videotex will follow may influence whether there will be any content regulation.

5 *Copyright.* Teletext and videotex, together with other new computer/communications technologies, make the possibility of down-loading, manipulating, and repackaging software very real. Creators of software may want added protection for their products—both from copyright law and from contracts with user organizations. Some argue, however, that by the nature of the technology, it is impossible to monitor and to guarantee integrity to the original work.

The six consequential issue themes are the following:

1 *Privacy and security.* Under the different generic application areas, potential privacy issues are raised. What restrictions are placed on the vast amounts of personal information collected from videotex systems is a major policy issue. Defining an individual's right to privacy in this electronic environment is a complicated task, but one that has considerable public interest. Closely associated with privacy is the issue of security. Electronic transfer of personal and financial data creates the potential for illegal access and electronic fraud and theft. The policy question centers on responsibility for the integrity of the data or the information service.

2 *Equity of access.* This becomes an issue area in two ways. First, with widespread penetration and the comprehensiveness of information services carried electronically, it may be deemed that teletext and/or videotex, like the U.S. Postal Service, should be a service available without discrimination to the general public. The second aspect is a social convenience question: Does the new technology offer unfair advantages to the "haves" at the expense of the "have nots"?

3 *Consumer protection.* Interactive videotex allows people to do impulse shopping electronically. The policy problem here is to determine the implications of protecting the consumer against reckless impulse spending, questionable sales tactics, and misrepresented products or services, as well as to determine how to deal with disputes and guarantee quality of product. In particular, do these new media create special problems for consumer protection or will extensions of existing consumer policies be sufficient?

4 *Industry structure.* Teletext and videotex bring new combinations of industries together to offer household and business services. Existing industry structures, for example, traditional newspaper publishing, are challenged. The new vertical and horizontal alignments affect the antitrust legislative fabric. In this broad category we have also included associated economic impacts on industry.

5 *Employment.* Widespread implementation of teletext and videotex could impact the work force and work styles in a number of different ways. By definition, it will create a new class of information providers. It is not clear, though, whether those information providers will be commercial extensions of large existing organizations or whether there will be a significant role for new enterpreneurs as information providers. New job categories will be (and already have been) created, i.e., videotex artist and videotex editor, and existing categories may be replaced. At the same time other jobs may be eliminated. If videotex in its future evolution becomes a significant extension of the office (e.g., through closed-user groups), these work styles will be heavily impacted.

6 *International trade and communication.* Prestel International, the early name given to an international test of Prestel, linked multinational corporations in six different countries. The technical connection of the geographically dispersed offices of a particular company to a common database is not unusual. What this trial demonstrates is the possibility of videotex users routinely accessing and interacting with foreign databases. Transborder data flow issues arise here as well as issues of national sovereignty and international trade. The United States has traditionally argued for the free flow of information across national boundaries; other countries have become concerned about privacy vis-à-vis foreign databases and about the danger of becoming overly "information dependent" on other countries such as the United States.

There are numerous other individual and social concerns of new technologies. They include rights of minors, impacts on family structures and on political participation, and challenges to societal values. These issues are examined in Chapter 14.

POLICY SYNTHESIS

We have argued in the technology section that there are three alternative videotex configurations (telephone, cable, and broadcast hybrid) and two general teletext configurations (full channel—including software down-loading—and limited VBI service), based on the medium used to carry the service. We have also noted that these different media have traditionally fallen under different regulatory schemes. Table 10.1 indicates which policy issues are likely to arise under each of the five configurations. An examination of the table shows that:

1 The policy issues raised under all three videotex configurations and under full-channel teletext are much the same, although the stakeholders, the implications of the issues, and how they may be resolved may differ for each configuration; and

2 VBI teletext is significantly different from a policy standpoint: it raises fewer issues, but those issues may arise first because of the more rapid penetration of the service. How those issues are resolved may, in turn, affect how some videotex-related issues are settled.

TABLE 10.1 RELATION OF POLICY ISSUES TO TRANSMISSION MEDIUM

General policy issue	Videotex			Teletext	
	Telephone	Cable	Hybrid	Full channel	VBI
Developmental					
Standards	×		×	×	×
Guarantee of access		×	×	×	
Competition	×	×	×	×	×
Content regulation	×	×	×	×	
Copyright	×	×	×	×	×
Consequential					
Privacy and security	×	×	×		
Equity of access	×	×	×	×	
Consumer protection	×	×	×	×	×
Industry structure	×	×	×		
Employment	×	×	×		
International trade and communication	×	×	×	×	×

Finally, we can look at the policy issues in relation to the five generic application areas (Table 10.2). By examining the policy issues in this way we clearly can see the policy similarities and differences of incorporating a range of services in a teletext or videotex system. Among the points to be made about this table are:

1 Some issues (e.g., guarantee of access, standards privacy) are raised under all the applications. The generality of these issues gives them special significance.

TABLE 10.2 RELATION OF POLICY ISSUES TO APPLICATION AREAS

General policy issue	Information retrieval	Messaging	Transactions	Computing	Tele-monitoring
Developmental					
Standards	×	×	×	×	×
Guarantee of access	×	×	×	×	×
Competition	×	×	×		
Content regulation	×		×		
Copyright	×			×	
Consequential					
Privacy and security	×	×	×	×	×
Equity of access	×	×	×	×	
Consumer protection	×		×		
Industry structure	×	×	×		
Employment	×	×	×		
International trade and communication	×	×	×		

2 Some application areas raise more issues than others. Virtually every issue comes up in one way or another in terms of information retrieval, while only a relatively few are raised by telemonitoring.

3 There are no issues raised by only one application. Where these issues are, in fact, approached broadly, the effect may be to bring about a new policy that cuts across previously separate domains.

CHAPTER 11

DEVELOPMENTAL POLICY ISSUES

In Chapter 10, we classified policy issues according to the stage at which they emerge during the implementation of a new technology. Those that emerge primarily during the technological development and early stages of testing are referred to as *developmental issues*. Although market forces will ultimately determine whether or not a new technology succeeds or fails, the public policy environment also influences the direction that the technology takes. The major concern of our public policy analysis is not the normative question of whether a particular policy decision should be made, but more pragmatic questions: What are the major policy choices to be made, and What are the implications of choosing a particular policy option or set of policy options?

The five developmental issues we have identified may be considered under three broad impact areas. First, there are issues that influence the specific technological path. The most significant issue affecting the technological paths of teletext and videotex in the short term is *standards*. The adoption of a particular videotex standard (or standards) will determine the quality—and cost—of the service available in the United States. A decision on standards is also important for resolving the "chicken-egg" problem for manufacturers of hardware; i.e., when costs decline, a mass market is likely to emerge, but until a market has been clearly defined, costs will not decline.

The second key impact area concerns the structure of the teletext/videotex marketplace and hence the competitiveness of the industry. Even in an environment of minimal regulation, the question will arise as to whether any government intervention is needed to prevent undue concentrations of power and to ensure diversity and competition. Two broad issues are discussed. The first involves the possibility of providing some *guarantee of access* for information providers. The second issue, which we have called *competition*, involves who will be allowed to enter this marketplace and under what circumstances. The specific concerns include the possibility of separating carriage and content and the role of the dominant common carrier.

Two additional developmental issues have been identified that will influence the content of the services or the range of applications provided by teletext and videotex systems. First, there is the issue of *content regulation*. Currently, content regulation varies among different transmission media—e.g., broadcasting, cable, and telephone. Whether these distinctions are maintained for teletext and videotex may influence which medium predominates, quite apart from that medium's technical characteristics. On the other hand, the fact that the same service can be carried by several hitherto distinct media may force a rethinking and reformulation of current regulations to bring about greater consistency. Second, there is the question of *copyright* and the protection of intellectual property. Here the issue is primarily one of enforcement, as this technology greatly increases the possibilities for down-loading, copying, manipulating, and reselling information. How this issue is resolved may influence pricing policies and the conditions under which information services are offered.

Each developmental issue is discussed in terms of its key impact areas, and a range of policy options for dealing with the issue is outlined. Two broad policy profiles are described—a deregulation profile and a proregulation profile—in terms of likely policy options to be selected for the various issues. These profiles are compared for each of the three general areas: technology path, teletext/videotex marketplace, and the range of applications. The two profiles are then carried over into Chapter 12 in our analysis of "consequential issues" arising from widespread penetration of teletext and videotex systems in the United States.

AREA 1:   THE TECHNOLOGICAL PATH

Standards

Standards setting is a powerful tool for shaping the development of a new technology. Standards enable manufacturers to produce with confidence; they allow users to purchase with confidence; they allow service providers to market new services with confidence. But the adoption of any standards also implies a choice among available alternatives. In choosing between competing interests, some are likely to benefit while others may be disadvantaged. In addition, standards setting for new technologies inevitably faces an environment of uncertainty—the effects of standardization on international trade, product innovation, system penetration, etc., are largely unknown at the time standards are adopted. In view of this uncertainty, it is common for different groups to adopt divergent positions in the standardization debate, reflecting their particular interests. A key public policy question, then, is: What are the implications of setting or not setting standards for teletext and/or videotex?

Teletext and videotex face the classic problem in achieving a critical mass for the service, i.e., a large enough customer base to support the initial capital investment of the system operator. Users are reluctant to subscribe to the service if they fear that the equipment they purchase may be incompatible with future services. Standards can help resolve this dilemma by instilling confidence in manufacturers, system operators, and users alike, thus enabling a steady and reasonably predictable growth rate.

A counterargument suggests that formally specified standards for teletext/videotex systems are not necessary. Historically, the marketplace has played an important part in standards development. Standards-setting processes, especially at the international level, have typically taken many years—seven to eight years is not uncommon in both the International Telegraph and Telephone Consultative Committee (CCITT) and the International Organization for Standards (ISO). During that time, systems have been established and de facto standards have emerged. By the time a formal standards agreement is possible, participating interest groups have often invested so heavily in their own systems that they are unwilling to compromise, and the standard then becomes a formal ratification of a number of alternatives. The two videotex recommendations approved by the November 1980 Plenary Assembly of the CCITT (S.100—"International Information Exchange for International Videotex" and F.300—"Videotex Service") were of this type. Proponents of the "marketplace standards" approach cite examples such as these to support their view that standardization in formal institutions is essentially a waste of time and effort.

We recognize, therefore, that there are sound general arguments both for and against standardization—hence, its selection as a key developmental policy issue.

No single organization at either the national or the international level has a clear mandate to set new standards for the emerging teletext and videotex technologies. All the existing standards-setting organizations in the telephone, television, and computer industries have roles to play in setting teletext and videotex standards. The following organizations are the major potential actors in setting standards at the four broad levels of activity: international, national, industry, and organization.

International
International Organization for Standardization (ISO)
International Electrotechnical Commission (IEC)
International Telegraph and Telephone Consultative Committee (CCITT)
International Consultative Committee for Radio (CCIR)

National
Federal Communications Commission (FCC)
American National Standards Institute (ANSI)
State Department through:
   1 U.S. Organization for the CCITT (U.S. Org. for CCITT)
   2 U.S. Organization for the CCIR (U.S. Org. for CCIR)

Industry
Electronic Industries Association (EIA)
Institute of Electrical and Electronic Engineers (IEEE)
Society of Motion Picture and Television Engineers (SMPTE)

Organization
American Telephone and Telegraph Company (AT&T)
International Business Machines (IBM)
National Bureau of Standards/Institute for Computer Sciences
and Technology (NBS/ICST)

TABLE 11.1 POTENTIAL DECISION MAKERS IN TELETEXT AND VIDEOTEX STANDARDS

Organization	Teletext/ videotex	Authority	Interest	Current activities
ISO	Both	Intl. agreement	Intl. trade	Yes
IEC	Both	Intl. agreement	Intl. trade	Yes
CCITT	Videotex	Intl. agreement	Intl. interconnection	Yes
CCIR	Teletext	Intl. agreement	Intl. interconnection	Yes
FCC	Both	Statutory	Public interest	Yes
ANSI	Both	Voluntary	Business, manufacturers	Yes
U.S. Org. for CCITT	Videotex	State Dept.	Carriers, manufacturers	Minor
U.S. Org. for CCIR	Teletext	State Dept.	Carriers, broadcasters, manufacturers	Minor
EIA	Both	Voluntary	Manufacturers	Yes
IEEE	Both	Voluntary	Manufacturers	No
SMPTE	Teletext	Voluntary	Manufacturers, broadcasters	No
AT&T	Videotex	Market size	Economic	Yes
IBM	Both	Market size	Economic	Minor
NBS/ICST	Both	Statutory	Federal	Minor

The standard-setting authority of these organizations and their activities in teletext/videotex standards setting are summarized in Table 11.1.

Standards may influence the nature of services available, the quality of these services, the interchangeability of user equipment, the cost of user equipment, the interconnectability of different systems, the ease of system use, and so on. Therefore, many other organizations, groups, and individuals with a stake in teletext and videotex have an interest in the standardization issues. These groups, which are also involved in most of the other developmental issues, include:

- users (home, business, special interest, and government)
- communication network providers (local distribution and long haul)
- system operators (e.g., Dow Jones, HomeServ, CompuServ, Viewtron, Comp-U-Card)
- information service providers (e.g., advertisers, newspapers, banks, retailers, travel)
- software and hardware vendors (e.g., Prestel, Antiope, Telidon, Sony, Zenith)

At the formal policymaking level, groups with a potential interest in teletext and videotex standards include those responsible for formulating national information policies as well as those responsible for their implementation:

- Congress
  House Subcommittee on Telecommunications, Consumer Protection, and
    Finance
  Senate Communications Subcommittee
- Administration
  White House Office of Science and Technology Policy
  National Telecommunications and Information Agency (NTIA)
- State regulatory commissions
- National Association of Regulatory Utility Commissioners (NARUC)

Where Do Standards Issues Arise?   There are four main components in a teletext/videotex system: (1) users' terminal equipment, (2) communication links, (3) computer database(s), and (4) information provider's terminal equipment (Figure 11.1). These are not components of some newly invented technology but rather a new application of existing technologies. Basic elements of users' terminal equipment,

FIGURE 11.1   General Teletext/Videotex System.

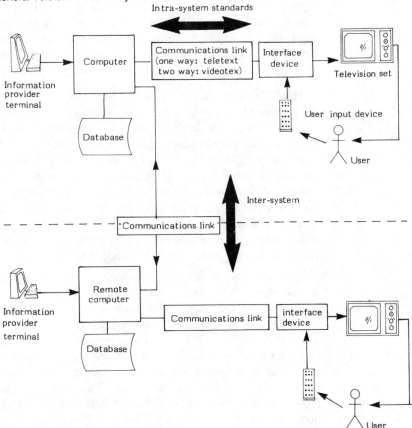

such as the telephone and television set, have been in nearly all American homes for three decades. The communication links providing telephone services (local loop) and television service (primarily over-the-air broadcast but increasingly coaxial cable) are likewise well established. Nor are computer databases new to the American public, especially the business sector. The information provider's terminal is also nothing new; remote job entry terminals are now commonplace in many business operations. However, videotex systems link all these components together in a new way that requires standards to ensure all the components work together properly.

Several different types of standards may be required. Standards could potentially be developed for each component of a teletext/videotex system on its own as well as for the interfaces between components. All such standards, however, relate to a particular system and therefore can be considered as *intrasystem* standards. On the other hand, to connect a videotex system computer to a remote database or to another videotex system requires *intersystem* standards (Figure 11.1).

Neither type of standard necessarily raises public policy issues. But, since broadcast teletext systems involve the use of a scarce resource, the frequency spectrum, the allocation of this resource is clearly a public policy issue. In addition, whether or not a particular teletext and videotex standards question becomes a public policy issue depends on who is involved in providing the service. Private videotex systems or systems in which all components are provided or controlled by one single interest group are unlikely to generate standards issues. On the other hand, systems in which the various components are provided by different interest groups or systems that are to be interconnected with other systems are more likely to generate standards issues.

How May Teletext And Videotex Standards Issues Be Classified?   The essential purpose of teletext and videotex systems is to transfer information between centralized or distributed databases and remote user terminals. In teletext systems, information transfer is unidirectional; in videotex systems, it is bidirectional.

Recent work in the American National Standards Institute (ANSI) and the International Organization for Standardization (ISO) has resulted in a framework for thinking about and developing standards for complex computer communication systems (known as the Reference Model for Open Systems Interconnection—OSI). This framework contains seven independent functional layers of standards. Each layer, which comprises an agreed upon set of protocols, performs one well-defined set of functions. Briefly, each layer is described in terms of teletext and videotex services.

Application layer (layer 7). Protocols of this layer directly serve the end user by providing the distributed information service appropriate to each particular videotex application, i.e., the application layer defines the information service itself, e.g., Prestel, Viewtron, The Source, captioning.

Presentation layer (layer 6). The purpose of this layer is to provide the set of services that may be selected by the application layer to enable it to interpret the meaning of the data exchanged. These services are for the management of the entry, exchange, display, and control of structured data. For videotex systems, the presentation layer protocols define the character repertoire (sets of usable characters and their interpretation), text and graphics coding schemes (alphamosaic, alphageometric, alphaphotographic, and dynamically redefinable character sets), and attribute coding (serial/parallel attribute). Standard protocols at this layer allow applications in an Open Systems Interconnection environment to communicate without unacceptable costs in interface variability, transformation, or application modification.

Session layer (layer 5). The purpose of this layer is to assist in the support of the interactions between cooperating presentation entities, i.e., procedures for log-on, user identification, billing, and statistics gathering.

Transport layer (layer 4). This layer provides transfer of data between two videotex systems or between a user's terminal and the videotex computer. Its task is to optimize the use of available communications services to provide the performance required for each intersystem connection at a minimum cost; e.g., if the communications network is a packet-switched network, the transport protocol defines the routing algorithm and the flow control procedures.

Network layer (layer 3). This layer provides functional and procedural means to exchange data between two videotex systems (or a terminal and the videotex computer) over a communication network connection; e.g., for packet-switched networks this protocol defines how data is assembled into packets (CCITT Recommendation X.25, level 3).

Data link layer (layer 2). The purpose of this layer is to provide the functional and procedural means to establish, maintain, and release data links in a communication network, e.g., SDLC (synchronous data link control), HDLC (high-level data link control).

Physical layer (layer 1). This layer provides mechanical, electrical, functional, and procedural characteristics to establish, maintain, and release physical communications between two videotex systems (or a terminal and the videotex computer); e.g., EIA RS-232-C, lines 15 through 18 of the vertical blanking interval.

The above seven layers must be defined before standardized equipment can be manufactured and before a completely standardized service can be offered. However, because each layer of the model has a clearly defined function, the development of standards can proceed relatively independently at each of the architectural layers. This may result in a single standard for one layer and multiple market-determined standards at another.

The seven layers can be divided into three classes: *application* standards (layer 7) concern the functions of the services provided, *dialogue standards* (layers 5 and 6) concern the protocols for the encoding and display of information, and *transmission standards* (layers 1−4) concern the transport of information between computer databases and user terminals.

Application standards define the way in which a user will be served by the teletext or videotex system. They concern the functions or services available to the user. They include the control functions (to clear an unwanted entry), the service functions (to select an application provided by a videotex service), the retrieval functions (e.g., to obtain direct access to a page, to progress from frame to frame, to return to a previous frame, to repeat the same frame, to retrace the progression of the user's action), and the display options (24 rows or 20 rows, 40 or fewer columns).

The way in which application standards are implemented can have a major effect on the possible range of services. A standardized terminal would ensure functional compatibility, but this would be unacceptable from a marketing point of view; i.e., not all users will want the same range of functions. Some users will be content with a text-only service, while others will require sophisticated graphics. Likewise, fully standardized databases are unlikely to meet the diversity of existing or future information needs; e.g., existing keyword-structured databases could not be accessed using a simple numeric keypad.

One possible solution to this dilemma is to define a set of standard functions and then precede all information interchange with a terminal recognition request. That is, the terminal would inform the database of its capabilities before any information is sent; if the terminal has limited functions, the database would only send information that could be understood (e.g., graphics may be deleted). In the worst case, the database would simply inform the terminal that the information cannot be transmitted.

Dialogue standards include session and presentation layer considerations. These layers or levels, in particular the presentation level, are the most critical in the videotex debate as they relate to the protocols or procedures used to provide the service. The protocols define data formats and data transfer procedures and are generally but not always invisible to the immediate user. They include graphics sophistication, serial/parallel attribute codes for graphics, bit-error rates, data transmission rates, and fixed or variable format transmissions (discussed more fully in Chapter 3). Those that are visible to the user are session layer concerns and include log-on, user identification, and billing protocols.

Presentation standards pose a dilemma. Teletext and videotex systems have emerged as low-cost information retrieval systems based on the home television set as the display device. If both services are to remain centered on the television set, there are economic advantages to the user in adopting a common presentation standard, e.g., character generator and frame memory functions can be shared by the two services, thus reducing the cost of the decoder.

The quality of the teletext/videotex service is determined to a large degree by the way in which information is accessed and the way in which it is displayed. A presentation standard therefore becomes important to information providers and information consumers (users). Multiple presentation standards would lead to a duplication of not only user terminal equipment but also source databases.

On the other hand, teletext and videotex could develop as two distinct classes of service. For example, teletext could remain as a mass medium, general purpose information service while videotex could evolve into a special purpose information/transaction service. The display standards for the two types of service could therefore be quite different. A sophisticated display standard for one service may impose undue costs on users of the other service; on the other hand, a limited display standard for one may inhibit the range of services available on the other.

Transmission standards are primarily the concern of the communications network provider, although for the over-the-air broadcast services, the FCC has a statutory responsibility to authorize these services since the frequency spectrum is considered a scarce public resource. The transmission requirements for digital data over the three principal teletext/videotex communication media—telephone, broadcast television, and cable television—differ, and therefore different standards are required for each. These standards would be transparent to the actual service provided.

Transmission standards do not constitute a major public policy issue. Although standards are likely to be necessary within a particular communication network, they do not need to be common to all networks. The forums for setting these standards reside within the respective industries, i.e., AT&T and CCITT in the telephone industry, NCTA in the cable television industry, and the FCC in the broadcasting industry.

Current Status   FCC inquiries and rule-making procedures represent a well-established mechanism for setting a national standard for *broadcast teletext*. Acceptance of these procedures was demonstrated by the many parties who submitted comments to the FCC in response to the initial petition for rule making filed by CBS in July 1980. In October 1981, however, the FCC declined to set a standard for teletext, choosing instead to let the marketplace and industry deal with the standards problem. Industry groups, such as the Electronic Industries Associations (EIA), continue to address the relevant technical issues (Interim report, Broadcast Television Systems Committee, 1981).

For *videotex*, standards-setting procedures are less clearly defined, especially at the national level. The number of interested parties is larger and more diverse than for teletext standards, but most will have little if any influence on the standards selected. The dominant actor, AT&T, has already made its position clear by publishing presentation level and session level videotex standards (Videotex Standard, 1981). No formal mechanisms exist for most other actors in the United States to contribute to or to debate this standard (except perhaps to lobby Congress or litigate through the courts to prevent AT&T from entering the videotex business at all). The American National Standards Institute supports the AT&T standard in principle and is working jointly with Canada toward a recommendation on a North American standard. At the international level, however, there is an opportunity for other interests to be considered.

Two recommendations on videotex standards—S.100 ("International Information Exchange for International Videotex," which deals with the characteristics of coded information and display formats) and F.300 ("Videotex Service," which describes the standard parameters for a public videotex service)—were ratified by the CCITT in October and November 1980. In May 1981 the European Conference of Post and Telecommunications announced the definition and adoption of a verified European videotex standard (CEPT, 1981). The standard, which is a presentation level protocol defining a common alphamosaic system, includes Prestel and Antiope options. It has been ratified by 26 European countries, but this sytem and the AT&T presentation level system are not entirely compatible in three areas, namely, the handling of mosaic attributes, the default level display area (40×24 for CEPT and 40×20 for AT&T), and the AT&T−Telidon-derived alphageometric capability.

In the CCITT, AT&T's "standard" has the status of a proposal until accepted by all member countries. Given that the competing European standard will also be seeking the support of the CCITT, the final outcome may well be some compromise between the two. But this debate will be restricted to only *one* interest group—the communication common carriers—since they alone have standing (and voting power) in the CCITT.

Policy Options Standards policy decisions are likely to be most effective at the three highest layers of the ISO reference model; i.e., application, presentation, and session levels. The protocols adopted at these layers have a direct impact on the user, whereas lower-level protocols tend to be totally transparent to the user.

For each of these three high-level protocols, a set of five policy options for teletext and videotex standards in the United States was identified. It should be noted, however, that each layer must be examined independently, as the consequences of each option are likely to be different for each layer. The options are as follows:

1 A single national standard, common to all broadcasting, telephone, and cable-based teletext *and* videotex services.

    Key bodies responsible for development of option: FCC, influenced by CCITT, CCIR, ISO, NBS, ANSI.

2 A single national standard is set for teletext services only. Videotex services emerge for specialized users only; the standard used for each service depends on the particular application, for example, AT&T uses its presentation level standards (plus relevant session standards), Cox Cable markets INDAX, CompuServ goes cable.

    Key bodies responsible for development of option: FCC, ANSI.

3 A single national standard is set for videotex services. Teletext services remain unregulated—the standards are set by the marketplace.

    Key body responsible for development of option: AT&T.

4 Many different teletext and videotex services emerge using numerous market-determined standards; the marketplace is large enough to allow these different standards to coexist, although, in general, systems remain incompatible. A de facto standard may ultimately evolve for each technology.

    Key bodies responsible for development of option: Market-determined input by NBS, AT&T, ANSI.

5 The cable industry, outside the realm of the FCC, develops standards for teletext and ultimately videotex, both of which eventually become de facto national standards.

    Key body responsible for development of option: NCTA.

These options relate to national (or, more probably, North American) standards. They may or may not be compatible with European standards either for videotex, which is moving toward a common European standard, CEPT, or for teletext, which is not.

A second level consideration is whether or not any standards are compatible with those set by international bodies (CCITT, CCIR, and ISO). Thus, the options exist to have a single international standard, nationally determined standards that may or may not be compatible, or marketplace-determined standards.

AREA 2: STRUCTURE OF THE TELETEXT AND VIDEOTEX MARKETPLACE

The regulation of mass communications in this country has developed on the basis of two fundamental, and often conflicting, principles—freedom from government interference, and the need to ensure diversity of expression in society. The balance between these two has been struck differently at different times and for different media, but the tension between the two principles has defined much of the ongoing debate over communications policy in the United States.

Freedom of the press from government interference is, of course, guaranteed by the First Amendment, whose language is about as absolute as any legal statement can be. However, few would argue that freedom of expression ought to be unlimited by any bounds or restrictions. One such modification is the concept of diversity. The Supreme Court has held that the First Amendment "rests on the assumption that the widest possible dissemination of information from diverse and antagonistic sources is essential to the welfare of the public" (Gillmor and Barron, 1974). This has been interpreted to mean that a certain amount of competition among different sources must be ensured if freedom of speech is to be maintained.

The policy issue area is that of the emergent market structure or role model for teletext and videotex. Since current regulations on competition and market structure in the communications industry generally are medium-specific, there are numerous possible precedents for teletext and videotex—broadcast, telephone, cable, and print (Table 11.2). In addition, videotex involves a number of generic applications, each of which raises a number of different issues and carries with it different traditions of

TABLE 11.2 COMPARISON OF REGULATORY FRAMEWORKS FOR POTENTIAL TELETEXT/VIDEOTEX DISTRIBUTION NETWORKS

	Rate regulation	Content regulation	Structural regulation
Broadcasting	None	FCC: Fairness Doctrine Equal time rules Ban on cigarette advertising, lotteries, obscenity Requirement for local programming	FCC: Limit on number of stations per owner Limits on TV-newspaper cross-ownership
Telephone	Tariffs set federally (FCC) and by state PUCs	No "mass media" services by AT&T (S.898)	Common carrier rules on access FCC: Computer inquiries Courts: Antitrust suit Congress: Communications Act
Cable	Local franchises or state regulation	FCC: Broadcasting-type rules applied to "origination cablecasting" by cable operator; "must carry" rule on local signals Local franchises may require specific program services (e.g., local access)	FCC: Ban on cross-ownership of cable and TV station in same market Ban on network ownership of cable Ban on telephone company ownership of cable in urban markets
Print	None	Case law on obscenity, libel, national security	None

structural regulation. Thus, it should be borne in mind that videotex is not merely a communications medium for information but also can be a vehicle for carrying banking transactions or providing a security service. As with other policy areas it is the inclusion of this range of services in a single videotex system that poses a particular challenge to policymaking.

There are two major developmental policy issues whose resolution will determine the structure of the market environment for videotex and teletext. First, there is the issue of access for information providers. Second, there is the issue of competition in this marketplace. It is worth noting that here we are concentrating specifically on the actors in the teletext and videotex service. In Chapter 12, we explore the broader impacts of teletext and videotex on the structure of U.S. industry.

Guarantee of Access

The major policy question is well summed up by Madden (1979):

> Is the videotex highway, whether cable, telephone, or hybrid (broadcast-telephone, cable-telephone) accessible to all information providers who choose to use it? If not, who should make the choice? Who should decide on barriers to entry? What criteria should be used?

Currently "guarantee" of access is determined by the type of delivery system or communication network used for a teletext or videotex service. Common carriers must provide interconnection to all who apply. Thus, in principle one or more information providers may organize a videotex service that in turn is offered to the end users over the telephone system. The videotex system operator in this case takes a critical pivotal role in the debate. If the videotex system operator is not the communication provider, then there would seem to be no legal need for guaranteed access since it is already ensured.

Broadcast system operators, on the other hand, offer no guarantee of access other than the requirements for equal access for political candidates. A similar situation holds for cable. Unlike the telephone system, cable has not been defined as a common carrier and so is under no obligations to provide access to its network to outside suppliers. As the federal government has reduced its regulation of cable, the authority for cable policy has shifted to the state and local levels. At present, cable operators are able to choose the services they wish to offer their subscribers. The development of cable-based videotex-type services (e.g., Warner Amex's QUBE, Cox Cable's INDAX) has been conducted by the cable operators themselves, who have retained the right to select the outside information providers who participate in these services.

The "access" discussed so far has been for videotex system operators to the communication networks. An equally important concern is access for information providers to the videotex system. It is worth bearing in mind that the information services could include transactions, messaging, computing, and telemonitoring. The possibilities exist to have multiple competing videotex systems and/or a single umbrella service that becomes a gateway for information providers. It is in this latter case where guarantee of access for information providers is not ensured.

The guarantee of access, and conversely the barriers to entry, for information providers have been discussed in terms of the major communication networks. There are a number of general economic considerations that may act as barriers to entry for information providers to the videotex industry. These include:

Government regulatory restrictions that exclude firms from certain markets. (FCC regulations, for example, have at various times restricted entry into the telephone interconnect market, cable television market, and specialized telecommunications common carrier market.)

"Absolute" cost advantages and unavailability of major resources. (This occurs if there are exclusive contracts for existing information providers, if patent rights make a particular resource inaccessible, or if a resource is totally controlled by one organization, for example, a cable operator.)

Noncompetitive pricing. (The classic cases are those of cross-subsidization, where transfer prices or prices of one of the products are below the "market price," and predatory prices, where prices are held below costs to capture a market share or force out a competitor.)

Advantages of scale economies. (Certain forms of business behavior are prohibited. Monopolizing trade is illegal under the Sherman Act and the Clayton Act makes it illegal to acquire a competing corporation if the effect may be substantially to lessen competition or tend to create a monopoly.) (Braunstein, 1981).

Policy Options   Congress and the FCC are the major policymakers. The options considered, which are not mutually exclusive, are:

1 To rely on the existing common carrier model to provide for access to the telephone network. There would be no guarantee of access for information providers to any particular videotex service, but nothing would preclude them from developing and offering a separate service.

2 Continue present cable-franchising arrangements or even move further to a "let-the-market-decide" attitude toward allocation of cable channels. In this case there would be no guarantee of access for information providers or system operators.

3 To reinstate the policy of requiring cable systems to make leased channels available on a nondiscriminatory basis. Access would then be guaranteed for anyone wishing to provide a videotex service over cable.

4 Apply the common carrier model to the videotex service itself to guarantee access for information providers to that service. For example, in the case of a cable-based videotex service that includes home banking, all banks that wish to participate would be allowed access to the videotex system. In this case cable operators offering videotex services would be treated as common carriers.

Competition

The key question here is: Does the fact that a single entity plays more than a single role in a videotex system give that entity an unfair advantage vis-à-vis its competitors?

Typically, this concern arises in terms of combining the provision of information content with the provision of the communications network to disseminate that information. The structure of a videotex system introduces a third key factor—the teletext or videotex system operator. Thus, the policy debate includes consideration of issues involved in separating content, service, and carriage (Table 11.3). The major reasons for carriage, service, and content separation are to stimulate the variety and scope of services, to encourage fuller competition among information service providers, and to minimize the possibility of abuse by the carriers.

Once again, the debate is affected by the different transmission media involved. Separations policies have not been applied to newspapers, cable, or broadcast industries. A newspaper is totally vertically integrated, controlling the editorial content, advertising, rate setting, production, and distribution. With cable there are no federal regulations with regard to limiting the roles of the cable company. However, local franchising bodies may, and in some cases do, act to prohibit operators from providing programming on their own systems.

Broadcasters have control over both the transmission and the content of their programming, but they are subject to content regulations such as the Fairness Doctrine (balanced coverage of controversial subjects), the personal attack rule (permitting on-air response to certain types of commentary), and equal time rule (applying to appearances by candidates for public office). Although the FCC exercises no direct authority over national television networks, it has developed a set of rules to limit their dominance. One of these is the Prime Time Access Rule (PTAR), which requires local stations to be responsible for programming at least one half hour per night of the prime time period, when viewing levels are highest.

The telephone companies, with the exception of AT&T, have been allowed to offer unregulated information services over their own networks. For example, GTE owns and operates the Telenet timesharing service over the telephone lines. Even

TABLE 11.3 MARKET STRUCTURES

Information provider	System operator	Communication network provider	Example
Same*	Same	Same	INDAX/Cox
Same	Same	Different	The Source
Same	Different	Same	Viewtron
Different	Same	Same	Prestel (No U.S. example)
Different	Different	Different	No videotex service but computer time-sharing provides examples

*"Same" implies that an information provider is also the system operator and/or network provider for a given system.

AT&T has been allowed to offer a number of ''Dial-It'' prerecorded message services (e.g., Dial-a-Joke).

The possibility of cross-subsidization is one major objection to the communication provider also acting as service provider and/or information provider. For example, it has been argued that an arms-length subsidiary to provide unregulated services over the telephone system can still give AT&T an anticompetitive economic advantage over its competitors by selective pricing policies (*Videoprint,* 1980).

On the other hand, there are distinct advantages in terms of uniformity of service, indexing, access, and billing if the communication provider is also the service provider. As Table 11.3 indicates, this is the model adopted by British Telecom, which plays these two roles for Prestel but does not act as an information provider on the system. To date, however, no United States service has adopted this model.

In the cases of both broadcast teletext and cable-based videotex, the network provider is likely also to play the roles of the sole system operator and a major information provider. This has been the structure utilized in current teletext trials and in cable services such as QUBE and INDAX. On the telephone system, there is more likely to be multiple system operators who, in turn, may be major information providers (e.g., The Source, Viewtron). Significantly, in the one case in which AT&T planned to fulfill all three roles—in its proposed 1981 EYP trial in Austin, Texas—so much opposition was generated that the company decided to abandon the project in favor of more limited participation.

The role of the dominant common carrier, AT&T, is sufficiently important to warrant special attention. Videotex on the telephone system is but one example of the much larger issue of ensuring full and fair competition for the entire spectrum of telecommunications services provided (or potentially provided) by the nation's telephone industry and, in particular, by AT&T. Concern arises, of course, because of Bell's position as the nation's dominant common carrier. In the past, detailed regulation of tariffs and services, at both state and federal levels, has been used as a mechanism to guard against possible abuses of AT&T's monopoly power. As deregulation proceeds, increasing attention is being focused on developing structural safeguards to ensure full and fair competition. The Justice Department's antitrust suit, the FCC's two computer inquiries, and congressional attempts to define Bell's structure through legislation are all attempts to deal with this issue.

The 1982 settlement by the Justice Department gives AT&T greater freedom to offer videotex services without as much concern about its monopoly over the communications network. In this environment, AT&T will be just one of many service and information providers, though it is certain to be a formidable competitor. The local operating companies will maintain their monopoly position and it remains to be decided how much freedom they will have to act as system operators and information providers.

Policy Options The broad outlines of communication policy for cable, broadcast, and common carrier are set by Congress, which delegates the interpretation and enforcement of those policies to the FCC. The structure of videotex trials conducted by telephone companies in specific areas requires the approval of local PUCs. In the case of cable, the franchise granting body can specify the services to be made available and the degree of competition.

The basic policy options include:

1 Have no carriage-content separations policy for teletext and videotex services but let the industry structure and groupings of conglomerates evolve "naturally." In this case, there would be the existing framework to deal with unfair practices, antitrust, and discriminatory pricing.

2 Allow existing carriage-content separations policies to apply to common carriers providing videotex services, but give all other teletext and videotex communication providers complete jurisdiction over their role in information provision. There may be some provision for public access and government information services on the VBI or the other networks.

3 Determine carriage-content separations policy for each videotex system, for each class of services offered. In a remote area, for example, a telephone company may be permitted to offer telemonitoring services if the service is not offered by any other vendor.

4 Prevent all videotex communications network providers and system operators from controlling content. A modification would be to exempt some services, e.g., messaging and white pages, from the regulation.

AREA 3: THE RANGE OF VIDEOTEX APPLICATIONS

There are two final developmental policy issues that influence the range and nature of teletext applications and, especially, videotex applications. These issues, content regulation and copyright, are not unique to videotex but have been discussed widely in the proliferation of home-based and business electronic technologies.

Content Regulation

The major policy question is: Given the multiple roles involved in providing a videotex service, who is responsible for the content and quality of the service?

Content rules and regulations are primarily directed at information retrieval and electronic publishing services on teletext and videotex. At present, content rules apply primarily to broadcasting and to a lesser extent cable services but not to newspapers, telephone services, or cable channels that are not controlled by the cable company (see Table 11.2). As teletext and videotex may be distributed in a number of ways via a number of communication networks, content rules, unless universally applied to all networks, could be both unfair and anticompetitive.

The extent to which government becomes involved in content regulation will depend on (1) how the content is classified and (2) how much weight is given to the medium over which videotex is transmitted. The classification of content is important because the courts have held that advertising does not enjoy the same far-reaching First Amendment protection as editorial content. Since 1938, the Federal Trade Commission has been empowered to prohibit "false advertising," and this authority could certainly extend to videotex-based advertising. One potential problem here is the tendency for videotex to blur the line between advertising and editorial content, raising the danger that excess regulation may be imposed on editorial content, or that advertising content may escape proper regulation. A more difficult problem may arise in distinguishing between content control issues for information retrieval services and the relevance of these controls for other services such as messaging and transactions when all such services are combined into a single system.

The medium by which videotex is transmitted and the role model that finally emerges for videotex are potentially significant as we have demonstrated. Broadcasters have historically enjoyed less First Amendment protection than print publishers primarily because of the scarcity of broadcast channels. Courts have upheld the FCC's authority to impose various public interest requirements on broadcasters in return for their license to use a portion of the broadcast spectrum. Whether these obligations will apply to teletext has yet to be decided. If the VBI is judged to be part of a broadcaster's basic service, it is possible that obligations such as the Fairness Doctrine might be extended to the content of VBI. Alternatively, the FCC may deem teletext to be an ancillary service. It could then be offered free of broadcast content regulations such as the Fairness Doctrine and the Equal Time Provision (Brandon, 1981).

Without some major shift in the FCC's regulatory stance, the basic framework of broadcast regulation could carry over to broadcast teletext. Thus, the content of a station's teletext "magazine" would be the prerogative of the station, which would be responsible for the content. The station would also have the right to sell advertising on the service and receive revenues from it. While teletext is currently being developed (in the United States) primarily as a local service, it is possible that a network might provide its affiliates or other networks with a national teletext service occupying some or all of the available VBI lines. If this service proved overly dominant, the FCC might invoke some remedy along the lines of the PTAR.

There is an alternative precedent for regulating teletext, however. The FCC's Subsidiary Carrier Authorization (SCA) for FM radio broadcasting allows an FM licensee to lease a portion of its signal to another entity for a private service such as Muzak. A similar arrangement could be envisioned for broadcast teletext, whereby a station would lease its vertical blanking interval to a separate information provider. Who would ultimately be responsible for its content is an open question. Another area of uncertainty surrounds what can be done with teletext material in the VBI when a broadcast signal is retransmitted by a satellite carrier or a cable system. Do these entities have the right to strip the stations' teletext service and insert their own signal? Such a case occurred in 1981, and a federal judge ruled that the carrier, United Video, did have the right to replace the station's teletext service with its own (*International Videotex and Teletext News*, 1981*b*). The case, which is being appealed, was decided on copyright grounds, however, which illustrates how various policy issues and areas of regulation interact.

It is also possible for a full broadcast channel to be utilized to carry a one-way teletext service. Such a service would not be integrated with conventional programming and is likely to be quite different from the VBI-based service, which is limited to a few hundred pages of text. In the case of over-the-air subscription television (STV), the FCC has shown its willingness to allow broadcast stations to be used for purposes quite different than conventional television. The same might be true of full-channel teletext as well. Many of the rules that would be appropriate to a limited VBI service would not be relevant.

Although the FCC and ultimately Congress can intervene directly, there is a tendency to favor industry self-regulation rather than direct intervention. The NAB's Television Code, which includes detailed content guidelines, is an example of this. So is the recent FCC ruling on low-power television. These stations, which have a limited range of 15 to 20 miles, will not have to comply with standard broadcasting rules governing commercials, news and public affairs, and pay TV programming. They offer yet another regulatory model for either teletext or videotex, or both.

The question that remains, however, is how responsibility will be divided among information providers, system operators, and communication network providers. In the Prestel model, British Telecom, which is acting as both system operator and network provider, determined it would exercise the least possible control over the content supplied by its information providers. In turn, information providers agreed to accept legal liability for their content. Under British law, British Telecom cannot escape all legal responsibility. It can be held "partially liable for any legally offensive information, both in civil and criminal proceedings, but can greatly reduce the risks in law if it can be seen to take reasonable steps to withold information until it has been tested. [Therefore,] British Telecom reserves the right to exclude (temporarily) any material suspected of being a breach of law" (A. Smith, 1980).

There is some question about how relevant this model is for the United States. There is no equivalent of British Telecom. By contrast, videotex services developed to date have not drawn a clear demarcation between system operator and information provider. In many cases, the system operator *is* the information provider (Dow Jones, QUBE); in other cases, the system operator is actively involved in selecting the information providers and developing their content (OCLC, Viewtron).

If videotex and even teletext system operators are seen as publishers, then they are likely to share the liability for information providers' content and will be allowed—or required—to exercise some degree of editorial control. If system operators evolve into a role more like that of a common carrier along the lines of Prestel, then they may well be freed of legal responsibility for content and precluded from exercising content control. A parallel here might be with the role of cable operators toward public access channels, which some offer on a first-come-first-served basis and over which they exercise no censorship.

One industry-led initiative that has already been taken in Europe is the establishment of national associations of viewdata information providers (AVIP). An international body has also been formed. In the case of the United Kingdom, for example, the association has outlined a set of principles governing the quality, pricing, and responsibility for information on Prestel and has defined a code of practice for advertising. This code builds on current rules and regulations and attempts to preempt some of the possible "videotex specific" issues (Viewdata Code of Practice, 1980).

The general issue of ethics and responsibilities has been addressed by numerous professional groups such as engineers, physicians, and pharmacists. These societies offer models in which there are barriers to entry to the profession, such as licenses, and rules of acceptable conduct. Such models may supplement any industry self-regulation.

In summary, although the federal government is unlikely to exercise substantial direct control over videotex content generally at this stage, it is likely to become involved through extension of existing regulation over such things as advertising. In addition, it is likely to act at some point to establish structural relations among the various providers, either through court decisions assigning legal liability, through legislation spelling out industry structure, or through FCC action similar to the Computer Inquiries. As teletext emerges as a full-fledged commercial service, the FCC will have to decide how much of its broadcasting-related rules should be applied to the new service.

Policy Options   The decision-making bodies include the courts, which decide the First Amendment questions, determining legal liability on a case-by-case basis; the FCC, which determines broadcaster responsibility for VBI and establishes rules for use; the FTC, which enforces rules against false advertising; and Congress, which can decide whether videotex should be structured as a common carrier service.

The general policy options include:

1 No government regulation of content for electronic information retrieval on teletext or videotex systems

2 Selective government regulation by application of existing laws to specific kinds of content (obscenity, advertising)

3 Government application of different degrees of content regulation depending on the medium of transmission

4 Self-regulation by information providers

Copyright

The basic question posed here is: How can rights to intellectual property be protected in the teletext/videotex environment?

There are no copyright issues that are unique to teletext or videotex. The Copyright Act of 1976 was a general omnibus act pertaining to all works of authorship including individual database entries. The problems, as with copyright of print material, are related primarily to enforcement. Just as photocopying machines have created problems for protecting books, and home video recorders for broadcast programs, so too the widespread use of home information systems poses a potential threat to electronically stored material.

There have been widespread copyright problems of both input and output from modern communications and information systems. The input problems have been those of the definition of property rights under traditional categories of legal protection; the output problems with the passage of H.R. 6933 in November 1980 (an amendment to the Copyright Act), which made it explicit that all works of authorship, including software and data bases, are protected by the Copyright Act in both computer-readable and human-readable form. On the output side, since service providers can control screen displays, information providers can be remunerated for them. But offline duplication or printing, which users may well want, would be very difficult, if not impossible, to control. Videotex service providers could make some allowance for such copying when setting monthly connect charges; permission to make limited copies could even be explicit (Peyton, 1981).

Section 107 of the Copyright Act does allow recipients to make "fair use" of copyrighted information. This poses interpretation problems when applied to the recording of information from databases on home computers (and of course, off-air video taping).

Renumeration itself is a difficult issue. A Copyright Royalty Tribunal was established to determine appropriate disbursements, but its rulings have been appealed. Further, royalty structures are often geared to connect time, a criterion that makes little sense in an environment of high-speed dedicated lines or broadband signals for down-loading information and intelligent terminals for storing and processing the information locally and then, perhaps, illegal use. This has led the information industry to consider pricing on the basis of intended use of down-loaded information. The whole area of economics of product pricing may need to be revised, as the first copy cost now is in the structure of the database and the marginal cost to an end user is virtually zero (Hamilton, 1981).

In spite of the 1976 act and the 1980 amendment, publishers argue that further steps are needed to protect intellectual property rights with these new media. There does not seem to be a clear consensus as to what steps are feasible and effective. In fact, the very concept of copyright may be obsolete and the development of new marketing and pricing strategies may be more effective than additional legal protection.

Although copyright and patents are the principal tools for encouraging the creation of information and for protecting intellectual property rights, new electronic technologies for disseminating information goods and services, such as videotex, pose considerable problems. Information goods, for example, computer software, do not readily fit within the traditional categories of legal protection. Further, online access to databases allows original works to be copied or altered, recorded, and disseminated without the knowledge of the creator. An alternative policy approach to copyright protection is for the government to *subsidize* the creation of information. This overcomes the problems of protection of intellectual property rights but increases the opportunity for governments to exercise censorship and control over information content (Yurow, 1981).

Policy Options The network organization of teletext and videotex makes first-access billing for information services relatively easy. The difficulty arises in illegal use and access to duplication of electronic material. Four broad options for Congress, the major decision-making body, include:

1 Do not protect copyright; allow the marketplace to resolve the problem.
2 Provide better application and enforcement of existing copyright law.
3 Create a new law specifically tailored to deal with electronic information services. The major focus would be on subsidies for information creation rather than copyright protection.
4 Create a new enforcement agency or mechanism to pursue copyright claims.

Developmental Policy Option Profiles

We have now examined five major developmental policy issues for teletext and videotex. For each issue, we have identified the key decision-making bodies and a series of policy options for dealing with the issue. These issues and options are summarized in Table 11.4.

The options that are selected will help shape the evolution of the technology and the services offered (this is, in fact, our definiton of a "developmental issue").

One strategy for a decision maker at this stage would be to define a desirable "end state" for the technology, then ask the question: What policy options should be selected to bring this end state about? A second strategy would be to attempt to anticipate the likely regulatory environment over the next decade and to choose from the array of policy options in Table 11.4 those that would be most appropriate to that environment.

It is not the purpose of a technology assessment to determine which policy options are "best" or "most desirable," nor is it our objective to predict the most likely policies. However, it is our intention to explore the implications of alternative developmental policy options for widespread penetration of the technology. In order to assess these implications, we have chosen two broad alternatives to information policy, based on two different political and regulatory philosophies. One approach, called *proregulation*, involves active participation by government in directing the development of the technology. In the other approach, called *deregulation*, a laissez faire government policy permits the technology to be shaped by market forces. The specific options chosen under each of these approaches represent feasible but contrasting policy profiles. In a general way, the two profiles that we have selected represent boundaries for videotex and teletext policy.

Consider first the *deregulation* profile. The technology path will be predominantly influenced by market factors and the positions taken by major actors. This means multiple videotex systems, for example, cable-based systems, telephone-based systems, and even broadcast-hybrid systems that may or may not be compatible. For teletext it implies that vendors will attempt to sell particular systems—Ceefax, Antiope, or Telidon—to broadcast and cable outlets. If standards do emerge, they are likely to be voluntary industry-determined.

The regulatory structure that emerges is likely to be one that includes little or no policies regarding access or separations. Communication providers, information providers, and system operators will adopt positions dictated by economics and comparative advantage rather than regulations. In the case of limited-bandwidth teletext this will mean broadcasters controlling content and carriage. For videotex systems and full-channel teletext, there will be a number of structural arrangements among carriers and providers, none of which guarantee access for any organization, such as a bank or a fire and security monitoring company. Corporations will compete on non-common-carrier videotex services for right of access.

The services will be market determined and market monitored. Service quality will be enforced by industry self-regulation. In addition, property rights will be enforceable primarily by using pricing mechanisms. The videotex services that are offered will be those for which it is possible to determine relatively clear demarcation lines

TABLE 11.4 DEVELOPMENTAL POLICY OPTIONS: SUMMARY

Policy issue area	Range of options				
	1	2	3	4	5
Technology path					
Standards	Single national standard for teletext and videotex.	Single standard for teletext. Multiple standards for videotex.	Single standard for videotex. Multiple standards for teletext.	Market-determined standards for both teletext and videotex.	De facto standards for teletext and videotex set by cable industry.
Market structure					
Guarantee of access	No action: existing common carrier model for telephone networks.	Continue present cable franchising arrangements. No guarantee of access for information provider on cable.	Require cable systems to have leased channels available.	Make videotex system operators common carriers.	
Competition	No carriage content separations policy. Let the market decide.	Existing carriage-content separations for common carriers. No separations policies for other system operators.	Separate carriage-content policy for each class of videotex systems.	Network providers and system operators prevented from controlling content.	
Range of applications					
Content regulation	No government regulation of content for teletext or videotex.	Selective application to teletext and videotex of content laws.	Content regulation linked to medium of transmission.	Self-regulation by information industry groups.	
Copyright	Let the marketplace solve the problem—no protection of copyright.	Strong enforcement of existing laws.	Create a new law specifically tailored to electronic information services—focus on subsidies.	New agency to enforce copyright claims.	

among various responsibilities. Most disputes that arise will be settled on a case-by-case basis, with a set of legal precedents building up gradually over time.

The *proregulation* profile we have selected contrasts with the previous profile, yet it is also feasible. In such an environment it is likely that single (although not necessarily identical) standards will emerge for both teletext and videotex. The market structure that emerges will include regulations to protect competition and to provide for diversity of information services. There will be rules providing for competitive videotex systems on cable systems and hybrid alternatives (e.g., one-way cable with a telephone for voice transmission on the upstream end) as well as on the telephone network. The existing antitrust laws will be vigorously applied to restrictive trade and anticompetitive practices, and separations policies will be applied to limit the degree of vertical integration. The rules for limited-bandwidth teletext will allow the broadcaster to be the teletext system operator and the sole information provider, although some public interest requirements will be imposed. Restrictions will be placed on the number of teletext franchises that a single broadcaster may own. The major developments toward guaranteeing access to those offering services on non-common-carrier systems will be to require cable systems to make channels available for independent videotex services. However, it will still be up to the individual videotex system operator to determine whether or not an information provider has access to his or her videotex system—in the same way newspapers may refuse advertising or news items.

Finally, in this policy climate the government will be involved in the establishment of content control regulation and copyright protection. Existing regulations will be strengthened and content regulations developed for each of the transmission media (telephone, cable, and hybrid). Amendments to the copyright law will be tailored specifically to electronic information retrieval and transaction services.

These two developmental policy profiles are summarized in Table 11.5. It is outside the scope of this project to comment on the desirability of either profile. Our intention is merely to highlight the characteristics of the various developments. It is realized that as the videotex policy debate unfolds this "clear distinction" between alternative profiles will be replaced by a combination of market and government regulatory actions.

Impact of Policy Option Profiles

The policy option profiles that emerge will influence the technological path for teletext/videotex, the structure of the teletext/videotex marketplace, and the range of services that are offered on the various systems.

Technology Path   A number of impact measures can be used to assess the relative impacts of the developmental policy profiles. The evolution of the technological path, for example, may be described in terms of

- the likely time horizon before a commercial service emerges
- the extent of incompatibility among the various teletext and/or videotex systems
- the mix of communication network technologies providing the service; and the technical sophistication of the systems (such as graphics and text capabilities)

TABLE 11.5 DEVELOPMENTAL POLICY OPTION PROFILES

Policy issue area	Deregulation	Proregulation
Technology path		
Standards	Market determines standards for teletext and videotex—multiple, possibly incompatible coexisting standards.	Emergence of a single national standard for teletext and for videotex.
Market structure		
Guarantee of access	Content and carriage policies for common carriers. All other teletext and videotex system operators and communication network providers allowed to control both content and carriage.	Access to videotex system mandated. Separations policy or its equivalent, for each class of videotex applications. Limits on on the number of teletext systems that can be owned by a single broadcast station.
Competition		
Applications		
Content regulation	Self-regulation by information industry groups. Market-determined copyright protection policies. Pricing structures reflect likelihood that material will be copied.	Content regulation linked to the medium of transmission. Strengthening of existing regulation. Copyright law designed specifically for electronic information services.
Copyright		

The technology paths likely to emerge under deregulation and proregulation are quite distinct. Under proregulation, a widespread commercial service emerges earlier, as there are no regionalized competing standards for providing teletext or videotex. With a single national system, all teletext and videotex services are compatible. The dominant communication network mix for providing services will depend on the particular details of the standard or standards and the costs to the user of the services. Under the deregulatory profile, it is likely that the various networks (broadcast, cable, and telephone) would ultimately implement standards primarily designed to offer the services for which they had a comparative advantage—for example, messaging on the telephone network, rapidly changing and general interest information on broadcast teletext, telemonitoring on cable systems, and transactions, on both telephone and cable systems. Finally, the market-determined environment may result in the most sophisticated technical developments being adopted in some or all of the systems, whereas future technological breakthroughs are more difficult to incorporate within a single standard system. On the other hand, a dominant firm's inferior technology may be adopted as an industry standard or industries in the market may all make a series of suboptimal standards decisions (Braunstein and White, 1981).

Marketplace   The teletext/videotex marketplace may be characterized in terms of:

- the coalitions of information providers, system operators, and communications network providers that emerge (horizontal integration)
- the degree of vertical integration within the marketplace
- the extent of economic and pricing barriers to entry

The structure of the teletext/videotex marketplace is also likely to differ under the regulatory profiles. There are already a number of major coalitions of print, retail, communication, broadcast, and financial organizations in current trials. In a deregulatory environment with limited separations policies, there would be both a high degree of vertical integration within systems and a high degree of horizontal integration across systems. The absence of any guarantees of access, with the exception of common carriers, would mean that competitive cable and telephone videotex services could have vastly different information service providers. The complete separations policy under proregulation, on the other hand, eliminates obvious scale economies for existing vertically integrated organizations. In both profiles, the current antitrust laws could be applied to minimize barriers to entry. Under deregulation, however, the new alignments of organizations offering transactions, information retrieval, and messaging services may not correspond with the current industry structure (see Chapter 12).

Range of Applications   The range of applications may be described in terms of:

- the diversity of information services that are offered
- the market focus of the services (business, consumer) and,
- the general cost of the services.

In general, the deregulatory environment is more likely to provide a market focus (consumer, business) that best matches demand, thus providing a greater range of teletext and videotex services. The cost of the services to the consumer will depend on economy of scale of production and on extent of competition. A single standard enhances the likelihood of large-scale markets, but separations policies reduce the efficiency for vertically integrated corporations in the marketplace. It is not clear under which model the cost to the user could be less.

One further impact measure, a function of the technology path, the structure of the teletext/videotex marketplace and the range of applications, is the international market potential for U.S. business interests. This question of international trade poses an interesting dilemma. While the adoption of an internationally compatible standard may open the world market to the American information service providers, system operators, and hardware and software manufacturers, it also makes the local U.S. market that much more vulnerable to external competition. This issue is discussed in more detail in the following chapter.

CHAPTER 12

CONSEQUENTIAL
ISSUES ANALYSIS

Societal adoption of a new technology frequently has consequences beyond those identified during its development and initial implementation. Such issues, referred to as consequential issues, arise because of the type of applications supported by the technology and the nature of the underlying technology itself. In the case of teletext and videotex there is a wide range of applications—information retrieval, transactions, messaging, computing, and telemonitoring. The way the issues arise and are resolved will depend, to a large degree, on the regulatory climate at the time.

In this chapter, we examine six consequential issues—privacy and security, equity of access, consumer protection, industry structure, employment, and international trade and communication. We describe each issue in broad terms and discuss the issue in relation to the five generic classes of videotex services. Then we use the developmental policy profiles to compare alternative approaches to resolving each issue.

PRIVACY AND SECURITY

The public policy issue related to privacy and security is simply—given that personal and financial data will be transmitted between terminals and databases and in many cases sold on a subscription or per-page basis—how will personal privacy be maintained and security against unauthorized interception guaranteed? This issue is not unique to videotex; it has been debated since the introduction of computer timesharing over a decade ago. However, because of the potential pervasiveness of the technology, and its focus on the home, privacy and security issues have a much wider impact.

Legislation governing privacy of consumer information has been implemented on a piecemeal basis, e.g., separate acts for medical information, financial information, and so on. Currently, the issues are partially covered by Section 605 of the Communications Act, which prohibits unauthorized interception of some broadcast signals, and by Title III of the Omnibus Crime Control and Safe Streets Act of 1968, which provides criminal sanctions against interception of wire communications and regulates wiretapping by law enforcement authorities (Neustadt et al., 1981).

These laws, however, were not written with electronic data transmission in mind. In the case of the Omnibus Crime Control Act, ''intercept'' is defined strictly in terms of aural ''acquisition of the contents . . .'' and thus probably excludes textual messages. Further, as the law defines wire transmission as transmission by a common carrier, cable television may not be covered (these issues are discussed in detail by Neustadt et al., 1981). Section 605 of the Communications Act is primarily limited to over-the-air services, and, while it should bar the role or use of decoders for unauthorized interception of subscriber-supported over-the-air teletext (or videotex), it does not apply to broadcasting intended ''for the use of the general public.'' If encoded signals sold to subscribers are not deemed to be broadcasting, then this section of the act will offer legal protection to subscriber-supported teletext. Recent decisions regarding unauthorized reception of over-the-air pay television signals would seem to support this interpretation.

There are a number of other laws addressing the issue of privacy. The Privacy Act of 1974 addresses the rights of the individual with regard to personal data contained within federal data banks. The Fourth Amendment's protection against unreasonable searches and seizures and the Fifth Amendment's guarantees of due process and freedom from compelled self-incrimination give individuals protection against government and other collectors of information that is in their (the individual's) possession (National Telecommunications and Information Administration, 1981). In general, however, the federal courts have ruled that individuals have no inherent legal interest in records concerning themselves that are owned by others (Neustadt et al., 1981), the implication for videotex operators being that they can do whatever they want with customers' records.

The implications for privacy and security of widespread implementation of teletext and videotex are somewhat different for each of the five generic applications:

Information Retrieval   Under present laws, the communication providers and/or videotex system operators can monitor all information requests by users. This information may be stored at the household level, if household access numbers are used, or at the individual level. Preference patterns can be created and used, for example, to focus personalized advertising. These patterns, or subsets of these patterns, may be furnished to the government or sold to credit bureaus, investigation agencies, banks, retailers, political parties, product manufacturers, and so on.

There is not the same problem with one-way information services, although it is worth noting that in the teletext trials in Los Angeles and Washington, D.C., all users are being carefully monitored by remote attachments to the TV. Such monitoring could in theory be part of any commercial teletext service, especially if information services are sent scrambled down a one-way channel to addressable converters.

A related issue for the user is the distinction between frames that are advertising and those that are information. The privacy issue in this case is that of undesired reception of objectionable or unwanted information.

Transactions   A profile of consumer spending habits, savings, investments, and credit arrangements can easily be created for all videotex users. Credit ratings and credit card information can be updated daily and available instantaneously to anyone, as the Right to Financial Privacy Act has not been extended explicitly to include electronic funds transfer (EFT). Information service providers and system operators can maintain a complete record of each frame addressed and each transaction performed, thus allowing "customized" direct advertising and solicitation. The consumer is also threatened with the possibility of unintended access to personal records resulting from inadequate system checks. Unauthorized transactions and transfers are a real possibility. The Electronic Fund Transfers Act (1978) limited consumer liability for such infringements and will apply to videotex-based transactions.

The National Commission on EFT (1977) recommended that the "depository institution should be liable for erroneous, unauthorized, or fraudulent use of the EFT account . . . ." The provision of security for the electronic transfer of funds between consumers and financial institutions is seen as being the responsibility of the institution, which must either assume the risks or develop scrambling devices to protect transfer of the information.

The problem, however, is not simply one of legal rights but of early warning. The teletext operator may not be able to detect unauthorized interception. Videotex operators may not be able to determine who is illegally accessing databases. This leads to a key concern for all applications—namely, what information should be denied to whom? Computer audits may identify potential problem areas but not prevent interception.

Messaging   As with transactions, the consumer is faced with an uncertainty as to who has access to messages that are sent or received. Legal messages, formal requisitions, electronic votes, health reports, and personal correspondence could all be scrutinized by present and future employers, insurance companies, banks, lending societies, political parties, and separated spouses. Key concerns are intrusion, interception, and ultimate misuse of the information. The potential exists to construct social and political attitudinal profiles of consumers and households.

Computing   The concern here is similar to the concerns arising from information retrieval and transactions. It is possible to record all programs accessed from databases or transmitted to user facilities unless the information is freely available on a one-way system, and to monitor any personal or commercial information that is stored in any system database.

Telemonitoring   This is a similar situation to transactions except that it is the energy company or the insurance company that, potentially, has access to the household behavior patterns, e.g., consumption of energy or security. A major economic consideration of monitoring household expenditure, energy use, and transactions on any wide-scale basis is that it becomes possible to redirect tax policies from an *income* basis to *consumption*, i.e., to subsidize particular user groups or to ration the consumption of particular goods. It needs to be balanced against legitimate business and government needs to collect, use, and disclose information.

In summary, privacy is neither easily definable nor easily protectable in an electronic age. The above issues are reflected in five fundamental privacy and security concerns derived from Plowright (1980), Nash and Smith (1981), and the National Telecommunications and Information Administration (1981):

- The right to know about or have access to information stored about an individual.
- The use of information about an individual's purchasing habits or viewing preferences. Redress in the event of misuse of personal data and steps to prevent such misuses.
- The extent of access to private textual communications between individuals using a service provider's computer. Prevention of unauthorized access (especially of financial accounts).
- Implications of aggregated "household profiles" constructed from electronic transactions.
- The time period after which information stored on an individual (e.g., a traffic misdemeanor, bad debts) should be erased. Guarantees of such erasure.

Impact of Policy Profiles on Privacy and Security

The issues of privacy and security affect every individual within society. The key public policy decision-making agency is the federal government, although as with most other policy issues, an alternative to formal regulation is the establishment of self-regulating bodies (e.g., various national and international videotex information providers associations) or codes of ethics for handling records and access to private data.

Associated with the self-regulatory mechanism is the provision of consumer education about these new media. This represents an important forerunner to any national debate on the *appropriate* level of privacy, a level that may differ for various applications. For example, individuals may demand strict privacy about financial status or voting habits but tolerate much greater flexibility with respect to solicited or unsolicited requests to purchase products or services. Privacy protection may evolve as an economic service—the greater the privacy required, the greater the cost to the user.

The possible policy responses are influenced by the developmental policy profile. In a *deregulatory* context, the key elements are the multiplicity of systems and the fact that issues are primarily left to be resolved in the marketplace. A range of system-dependent protocols for security will emerge. Some general responses within this broad context include:

- Apply existing protocols such as the Privacy Act, for privacy and disclosure. Allow participants the choice of being included on mailing lists.
- Revise or strengthen Section 605 of the Communications Act on security so that it is "transmission medium free." From a practical standpoint, any legislation has to

be technically feasible for conducting business. As such, let those providing the services develop cost-effective protocols such as audits and scrambled signals. Any formal government involvement should only be to ensure that the penalties are appropriate for the crimes committed.

- Teletext/videotex industry associations establish a code of ethical behavior to guarantee privacy and security.
- Assume that computer security, like burglary, is a risk. Users may take out insurance against that risk.

In a *proregulatory* environment there is now a greater possibility of integrating information on users from a variety of systems, but there are also likely to be greater penalties for such activities. Common carrier status for cable systems offering videotex, for example, offers participants the same protection as that offered now by the telephone company and the U.S. Postal Service.

The general policy responses within this environment include:

- Make explicit to the user what information will be collected, how it could be used, and who could use it, and allow the user to determine, possibly at some cost, the degree of privacy required. Mechanisms for enforcing such a policy would be required. This follows from the recommendation of the 1977 Privacy Protection Study Commission—created by the Privacy Act of 1974—which provide the basis for a set of policy option guidelines for protection of privacy.
- Allow individuals to see records concerning themselves, to copy their files, and to correct errors, amend, or rebut information. Inform individuals when an adverse decision, such as refusal of credit, is based on information contained in the videotex system (e.g., a credit reporting service).
- Restrict government access to personal records to a formal legal process; whenever practical the individual is to be notified and given an opportunity to contest the access.
- Prevent third-party access to information unless explicitly authorized by the consumer.
- Assign responsibility for security of transmission on any teletext/videotex system. This is in line with the National Commission on EFT recommendations for electronic funds transfer but is more complicated. As a videotex system involves primary information providers, packagers and distributors of information services, system operators, and communication network providers, the demarcation of responsibility will be a substantial task.

EQUITY OF ACCESS

The general area of equity of access for individual users centers around universality and availability of information. The issue is well summarized in a recent National Telecommunications and Information Administration publication (NTIA, 1981):

> Access policies provide an important means of enhancing the availability of information . . . . Laws or regulation that authorize persons to obtain information from public or private institutions generally serve the same societal goals as laws requiring information

dissemination. Public access to such information permits popular oversight of decision-making in government and private sector organizations. Access to information is essential if the public is to participate effectively in political, economic, and other societal decisions. Access policies also permit the public to obtain information, which, when put into circulation, contributes to the creation of an information rich society . . . . Equality of access is important to an information oriented society in which superior ability to gain access to information can become the ultimate advantage.

Widespread implementation of teletext/videotex systems raises one fundamental access issue: Are there to be any guarantees of access for individuals to some or all of the services provided by the technology? A corollary of this question becomes: If so, who is to provide this access and how is it to be provided?

Consider each of the generic services against a backdrop of widespread penetration of teletext and videotex.

Information Retrieval   The numerous dimensions of the access debate, when narrowed to information retrieval, may be considered in terms of government information and the problem of disadvantaged groups.

In the case of government information, the Freedom of Information Act requires government agencies to disseminate and publish information relating to office organization, procedures for interaction with the public, explanations of all procedures, and statements of general policy (National Telecommunications and Information Administration, 1981). The distribution of government information through the videotex marketplace allows intermediate vendors to repackage and sell the information even though it is obtainable at little or no cost. This results in a "double payment" (taxes plus add-on), and, if it is the only source, it guarantees that a limited subset of the population will have access to the information. On the other hand, widespread public access allows teletext and videotex systems to be used for direct government dissemination of public services for both special user groups such as farmers and for general user groups.

A second area concerns disadvantaged groups including the physically handicapped, minority groups, and the poor. If all information retrieval services are privately owned "for-profit" databases, those least able to pay will be excluded. The government is a prime advocate for these minority groups; it can either subsidize services, guarantee access, or provide its own information service.

Transactions   A general issue relates to product pricing for those consumers with access to electronic ordering. The concern is not that the product may be cheaper because electronic purchase may be more efficient, but that price may be cross-subsidized. This becomes a policy concern if those outside the videotex system tend to be the disadvantaged minorities.

Messaging   The basic issue is that of universal access. It arises if electronic messaging becomes a service provided by private organizations that do not have any common carrier obligations to provide nondiscriminatory access.

Computing and Telemonitoring The access issues for these videotex services are related to the broader information-inequalities argument. In the case of computing services, for example, equality of access becomes an issue if electronic transmission becomes the major way of providing educational material. With respect to telemonitoring, an issue arises if those who don't have access to such services are discriminated against in prices for nonelectronic services.

Impact of Policy Profiles on Equity of Access

In this case the contrast between the policy responses under the two broad profiles is very clear.

Under *deregulation* it is reasonable to hypothesize that:

• Videotex systems are regarded as but one of many sources of information and, as such, will be available at costs to be determined in the marketplace. There are no guarantees of equity of access, although government may use teletext and videotex to disseminate public information.

Under *proregulation* a number of possibilities exist. The government provides public terminals in "public locations" such as schools, malls, and bus stops so that all citizens can access government or public information services for a nominal fee or free of charge. This may be extended to include distribution of terminals, either free, leased, or sold at subsidized prices, to certain groups or the provision of other forms of information subsidy. Specific policy responses include:

• Subsidization of low-income households.
• Rate regulation to ensure that rates are low enough to be affordable for certain income classes.
• Requiring electronic mail services, if non-common carrier, to offer delivery services to all residents in a given area irrespective of ownership of a videotex system. Thus, even if the user does not have a videotex system, mail may be obtained or sent from a public terminal.

CONSUMER PROTECTION

The Federal Trade Commission is the primary agency responsible for enforcing regulations prohibiting false or misleading advertising and sales practices. There are currently Truth in Advertising regulations and rights of return for products ordered by catalog, as well as acts giving consumers limited rights of notice regarding adverse credit or insurance decisions taken against them (the Fair Credit Reporting Act and the Equal Credit Opportunity Act).

The specific issue of consumer protection in the electronic marketplace has been considered primarily in terms of electronic funds transfer. For example, the rights of the consumer were considered by the Board of Governors of the Federal Reserve Board in Regulation E (1980). Under this regulation, which has proved difficult to manage, financial institutions must:

- Provide receipts for each electronic banking transaction.
- Provide a periodic statement for any month in which an electronic transaction took place.
- Provide a positive notification system for acceptance of preauthorized credits.
- Implement a resolution process for alleged electronic funds transfer (EFT) errors within 10 days of oral or written notice.
- Credit a preauthorized transfer into a savings or checking account on the day of the transfer (Gussman, 1980).

Consumer protection issues may be considered in each of the five generic application areas for teletext/videotex services.

Information Retrieval   Information arriving on a user's screen may be a value-added service offered by one information provider through an umbrella service and transmitted on the network of a third party. There are four specific areas of consumer protection that arise. The first concerns the legal liability for the reliability of the information, e.g., how is "defective" or unsatisfactory electronic information to be "returned" and appropriate restitution made to the consumer? Are standards developed for physical products appropriate for intangible information? As there may not be a paper record, and the database is continually changing, it may prove difficult to ascertain what precisely was in the database at any point in time.

Second is the issue of censorship and good taste. Censorship or enforcement of community standards to prohibit minors from access may be very difficult via a gateway-designed videotex network. The aspect of good taste is less clear-cut and relates to information in the database that may contravene public decency standards.

A third issue is that of electronic disclosure of warnings, disclaimers, and warranties—for example, for pharmaceutical and food products. A practical problem here may be the restriction on the amount of information conveyed because the videotex "frame" is so limited in size.

The fourth issue relates to the loss of clarity in defining what constitutes an advertisement. Teletext and videotex frames of information blur the traditional lines among advertisements, information, and editorial content. By definition, an advertisement is a biased message intended to persuade rather than merely inform. However, these messages may be interspersed with news and other reports and presented in such a way that the difference is not easily discernible.

Transactions   The electronic funds transfer aspect has been discussed earlier. Specific consumer protection concerns (Gussman, 1980; Colton and Kraemer, 1978) include:

- protection of personal privacy
- protection against unauthorized or fraudulent use of accounts or devices and liability for such occurrences
- stop-payment rights in the event of defective, wrong, or unwanted merchandise
- machine-printed or verifiable receipts for transfer of funds or purchase of products

- mechanisms for resolution of errors or disputes over the quality or prices of products and services
  - adequate disclosure statements
  - legal responsibility for delivery of an electronically ordered product and return of an unwanted product
  - guarantees against tampering with electronic services
  - protection for impulse buyers
  - price and service discrimination based on previous buying or utilization habits (it is possible to record all "returns" or "no shows" by a consumer)
  - length of time before an electronic "bad debt" record is erased and the extent to which this information and other data on purchasing or transaction habits may be disseminated

Messaging The key issues here are protection against unwanted electronic mail and the default-payment option. Both of these issues have their counterparts in the print system. It is not legal to send an unsolicited letter that requires payment for a product or service. In the case of unwanted, unsolicited messages, the receiver would need to have a detailed index page or synopsis of each piece of electronic mail. It may be possible to program an electronic word scan that automatically removes certain unwanted material—for example, advertisements for obscene material.

Computing and Telemonitoring These applications are essentially covered under transactions. Computing and telemonitoring are electronic services purchased by the consumer. From a practical perspective, the issue relates to the reliability of the service and the legal responsibility in the event of breakdown, failure, or improper performance of the monitoring device.

Impact of Policy Profiles on Consumer Protection

There is no single body of consumer protection legislation. Numerous laws and regulations have been proposed and passed at both state and federal levels, but their focus has usually been issue specific. The Federal Reserve Board, for example, has been responsible for regulation on electronic funds transfers.

The key attribute of the developmental profile influencing the shape of consumer protection legislation is the extent to which teletext and videotex emerge in a market-determined world. Policy responses to the *deregulatory profile* are characterized by:

- The transfer of most existing print, broadcast, and financial consumer protection regulations to teletext and videotex, i.e., the focus being the function performed (which is *not* new) rather than the way of performing the function (which *is* new). This transfer may be accompanied by a gradual reduction of paper record support for all transactional services (e.g., it becomes an optional extra).
- Trade associations defining responsibilities and liabilities for themselves and their customers.
- Sporadic "case-by-case" precedents in areas of transactions and telemerchandising, e.g., delivery of faulty products and returns of unwanted or unrequested merchandise.

The *proregulatory profile* is characterized by a more institutionalized approach to developing consumer protection policy, including:

- Regulation that requires advertisers to clearly differentiate between information and advertisements. (In Germany, each frame that contains an advertisement is identified by the word *Verbung*).
- Requirements for information providers (or system operators) to produce details of authenticity on demand.
- Default options with mailing lists so that consumers need not receive electronic solicitations from unsavory organizations or for unwanted products.
- Financial transaction legislation derived from the Federal Reserve Board's Regulation E, for dealing with fraudulent use, stop payments, legal responsibility for delivery and return of products, protections against impulse buying, and erasure clauses (in the case of bad debts).

INDUSTRY STRUCTURE

In a recent report, the National Telecommunications and Information Administration indicated that this would be a key concern of federal policy:

> While the principal subject or object of policy in this case will be suppliers, the larger implication is that industrial organizations will largely determine what services are available to whom and at what price. Under this rubric, the principal concerns are economic concentration along both horizontal and vertical lines, restrictive practices by suppliers, and the effects of government participation in the market as a seller or buyer of services (NTIA, 1981).

In Chapter 11, we examined the issue of competition and teletext and videotex market structure. Here we consider the broader impact of videotex on the industries that are or will be involved. This issue is important because when videotex (and to a lesser extent teletext) achieves widespread penetration, it will serve as the vehicle for a range of services currently provided either by other media (e.g., newspapers and telephone) or through face-to-face contact (e.g., banking and retail sales). The question is, as more of these services are conducted electronically from the home, what will be the impact on the structure of the key industries? Will the availability of videotex lessen or increase productivity, costs of doing business, and the quality of products and services provided by these industries? From a policy standpoint, the most significant question is whether videotex is likely to increase industry concentration and decrease the extent of competition within industry sectors.

The breadth of potential involvement in videotex is suggested by Table 12.1, which lists some of the major organizations that are currently involved in or have announced plans to be involved in videotex trials or services. Among the points to be made about the listing are:

- The early involvement of the largest organizations in the telephone, television, cable, banking, publishing, and retailing industries.
- There is no clear demarcation of roles along industry lines. Information providers may become involved as service providers (e.g., Dow Jones, Knight-Ridder, and

TABLE 12.1   SELECTED ORGANIZATIONS INVOLVED IN TELETEXT/VIDEOTEX SERVICES

Information service providers

Financial institutions	Publishers	Retailers	Others
American Express	Dun & Bradstreet	B. Dalton Booksellers	American Airlines
Banc One	Dow Jones	Comp-U-Card	AT&T
Bank of America	Harte-Hanks	Federated Department	Associated Press
Chemical Bank	Knight-Ridder	Stores	Hallmark Cards
Citibank	New York Times	J.C. Penney	Internal Revenue Service
First Interstate Bank	Readers Digest	Sears Roebuck	New York Stock
Merrill Lynch	Time Inc.		Exchange
United American	World Book		Grand Union
Bank			

Videotex system operators (include packaging and networking)

AT&T
CBS
CompuServe
Continental
 Telephone
Cox Cable
 Communications
Dow Jones
First Bank System

Keycom Electronic
 Publishing
Online Computer Library
 Center
Source Telecomputing
Time Inc.
Times Mirror
Viewtron (AT&T and Knight-Ridder)
Warner Amex Cable Communications

Communication network providers

Broadcasters	Common carriers	Cable
CBS	AT&T	Cox Cable Communications
NBC	Continental Telephone	Sammons Cable Communications
Westinghouse	GTE	American Television & Communications
Broadcasting	GTE Telenet	Times Mirror Cable
WFLD	Tymnet	Warner Amex Cable Communications

Terminals/Hardware/Turnkey systems

Apple Computer
Atari
IBM
GEC
Honeywell

Matra
Norpak
RCA
Sony
Tandy
Texas Instruments

Tocom
Videodial
Western Electric
Zenith Radio

*Source: *Business Week*, 198lb, Institute for the Future.

Citibank), and communication companies may take on the role of information provider as well as service provider (e.g., Cox Cable, CBS, and AT&T).

• Numerous organizations are competing to provide all classes of service. Companies may become involved in providing new information services, thereby competing in new markets. At the same time, other companies may enter their markets and provide new competition for them. In particular, the electronic delivery of information services to the home is likely to erode the advantage of geographical location for such things as financial transactions and purchase of goods.

• The extent to which availability of videotex fosters or inhibits competition will depend on the structure of the videotex market and, in particular, on the barriers to entry into that market. These barriers may be economic or technological. Finally, whether videotex increases or decreases competition may vary from industry to industry.

In this context, we will briefly consider the likely impact of videotex on the major industry groups involved in the five generic classes of applications, then consider how the alternative developmental policy paths might interact with this issue.

Information Retrieval   The initial model of Prestel, as developed by the British Post Office, envisioned a system that would lower barriers to entry for publishing and would greatly increase the diversity of available information sources:

> By enabling (in theory if not yet in practice) any information supplier, big or small, new or established, to communicate information of whatever kind he chooses to any audience, big or small, that he chooses, it offers a range of choice, a freedom of expression, a flexibility, and a responsiveness between suppliers and users, that is unique in electronic media, and conceptually at least is a rival or complement to the massively variegated output of traditional print publications (Winsbury, 1981).

The theory Winsbury states is based on the premise that the system operator and network provider, the British Post Office, makes the service available to all at a low cost and exercises minimum control over content. Ironically, this common carrier model did not exist previously in Britain; it was imported from the United States and incorporated into the structure of Prestel. However, there has been little indication that videotex in this country will follow the model. In no case has a U.S. videotex system operator decided to abstain from the role of information provider. To date, it is primarily the largest publishing companies that have participated in providing videotex information retrieval services. This suggests at least the possibility of vertical integration, which might inhibit competition by smaller entities in the electronic information marketplace.

A second issue related to information retrieval and industry structure is the potential competition for advertising revenue. It is likely that advertising will play a role in subsidizing the cost of consumer electronic information services, just as advertising currently supports newspapers and magazines. As the cost of printing and distributing paper publications continues to rise, newspapers and magazines may have increasing difficulty in competing with videotex-based advertising. Moreover, the potential of videotex to "customize" delivery of advertising based on profiles of individual preferences may make it an even more attractive vehicle. If videotex becomes a major medium for local, classified-type advertising, it could threaten the revenue base of the newspaper industry. In fact, it was the perception of this possibility that motivated the American Newspaper Publishers Association to lobby for a ban on AT&T's entry into electronic yellow pages.

Transactions   The widespread use of videotex for home banking and home shopping could have major implications for both the banking and the retailing industries. In terms of banking, the availability of videotex may make the actual physical location of one's bank irrelevant. Traditionally, the convenience offered by having many branches located throughout a bank's service area has been a major competitive factor in attracting customers. While the total number of banks in the United States has grown very slowly over the past 20 years, increasing just 8.6 percent since 1960, the number of branches has grown over 250 percent since 1960. The widespread availability of home banking on videotex could dramatically reduce the signficance of branch locations.

Current banking regulations make a distinction between interstate and intrastate transactions. This distinction becomes increasingly difficult to sustain in a videotex environment. Economies of scale would suggest the emergence of national "super banks" that are able to offer more services and lower charges to their customers. In fact, these institutions may not be banks at all: such companies as American Express and Sears have recently acquired large brokerage firms as part of an apparent strategy to become "financial supermarkets" offering an entire gamut of banking, investment, insurance, and tax services in a single package. As our case study on transactions demonstrates (see next chapter), bankers are already protesting that existing rules unfairly handicap them from competing with institutions that, because they are not classified as banks, can offer similar services with fewer limitations.

Since the retailing industry is generally unregulated and highly competitive, fewer public policy issues are likely to arise in this area. Although the nation's leading retailers (e.g., Sears, Penneys, K Mart, and Montgomery Ward) are very large, they have struggled in recent years to sustain their growth. While they are likely to have a presence in videotex-based home shopping, so are other organizations—as long as access to videotex services is not restricted. However, a certain amount of technical sophistication is likely to be necessary for survival in the electronic marketplace. A new specialized industry may well emerge to provide this expertise to smaller merchants and other organizations.

Messaging   The organization most threatened in this country is the U.S. Postal Service. It is likely that widespread electronic messaging will eventually diminish the volume of mail handled by the USPS. The result may be a decline in the quality of mail service and/or accelerated increases in postal rates.

Other battles for a share of the electronic messaging business are likely to occur. AT&T will undoubtedly offer such a service through an unregulated subsidiary. At the same time, there are likely to emerge a variety of specialized electronic messaging services competing for a share of the business and home communication markets.

These services may not be universally compatible and are likely to be concentrated along high-volume corridors. There may be a continuing need for government to monitor this market to see that competition is maintained.

Computing/Telemonitoring Home-based computing and telemonitoring are growing rapidly and are likely to emerge as significant industries. As of now, the development and sale of software designed for personal computers is a highly diversified business. Assuming again that barriers to entry are not erected, videotex could represent an ideal medium for distributing this software (including games) to end users.

If telemonitoring for remote energy management becomes widespread, it could potentially result in an overall reduction in peak-load demand for energy, thereby diminishing the need to build new power generating plants. In the case of a renewed energy crisis, government might develop incentives or requirements for such systems in homes and other buildings.

The purpose of home security monitoring systems is, of course, to reduce the likelihood of burglary. While these systems presumably will provide a certain measure of protection to homes equipped with them, it seems unlikely that their widespread use would have an effect on overall crime rates. More likely, they would shift the impact of burglary to homes not as well protected. In addition, a reduction in burglary of videotex-equipped homes might well be offset by an increase in the burden imposed directly or indirectly by various forms of electronic crimes.

Impact of Policy Profiles on Industry Structure

Because the existing industry structure differs for each of the five application areas, we have summarized the broad implications of the two developmental policy profiles in terms of each application area (Table 12.2).

Although we have argued that the development of a single videotex standard encourages the development of the technology, Table 12.2 indicates that if this standard is associated with a proregulatory environment, the net regulatory effect will be to retard growth.

EMPLOYMENT

There are two dimensions to any employment or job substitution effect as a consequence of widespread penetration of teletext and videotex. The first is the creation of new jobs in the teletext/videotex field—teletext and videotex, as information assembly industries, require videotex editors, graphic artists, database managers, and system engineers. The second dimension is the influence of the technology on the structure of the existing work force. From a public policy perspective, it is primarily the second dimension that needs to be considered.

As background, the steady expansion of the labor force over the last two decades is expected to continue through 1990, although the rate of growth will decrease markedly during the 1980s. The annual growth rate in the 1980s will be about half the

TABLE 12.2   IMPACT OF DEVELOPMENTAL POLICY PROFILES ON INDUSTRY STRUCTURES

	Deregulatory profile	Proregulatory profile
Information retrieval	Accelerated growth of multi-media companies—integration of publishing with other businesses.	Restrictions on growth of media conglomerates. Controls on content. Slow entry of print publishers into electronic information services.
Transactions	Distinctions disappear among types of financial institutions (e.g., banks, brokers; interstate and intrastate entities). Merging of banking and other transactions services. Market provides levels of security.	Detailed rules to protect existing banks from competition—slower growth of electronic funds transfer.
Messaging	Private companies allowed to compete with USPS and with each other.	USPS protected; controls on videotex messaging systems restrain growth.
Computing	Distinctions between computing and communications disappear—videotex becomes medium for "publishing" software.	Communications remains more restricted than computing—faster growth for stand-alone computer.
Telemonitoring	Market drives security and energy monitoring applications. Emergence of a new industry integrated with videotex system operators.	Government provides incentives for monitoring in case of energy crisis.

level of previous years (Table 12.3). This declining growth rate is due to a reduction in the number of young people entering the labor force (the number of people aged 16 to 24 will drop from 37.5 million in 1970 to 31.6 million in 1990) and the stabilizing of labor force participation by women.

The composition of the labor force has also been changing, a change that could be further exaggerated by widespread penetration of teletext/videotex (Figure 12.1). Until the early 1900s the largest work force component was in agriculture; until the 1960s it was the industrial sector; and it is currently the information sector, a sector that includes those working in industries that produce information machines or market information services as well as those who are engaged in information work such as planning, scheduling, and marketing activities (Porat, 1976).

Consider the general employment patterns and some impacts for selected industries within the five generic teletext and videotex application areas.

Information Retrieval   This is an area in which teletext and videotex have been predicted to make their greatest initial impacts. The growth rate in employment in the

TABLE 12.3 CIVILIAN LABOR FORCE (millions)

	Total labor force	Average annual growth rate
1960	69.6	
1970	82.7	1.7
1980	106.6	2.5
1990	121.8	1.3

Source: Institute for the Future, 1981.

printing and publishing industry over the last decade has been small—about 50 percent of the overall annual employment growth rate (Table 12.4).

To calculate a broad estimate of job displacement effects by teletext and videotex within 15 years, say, it is necessary to make some rather general assumptions. First, consider the information retrieval services presently coming into the home—newspapers, periodicals, and books. Assume that the value of output for each service is approximated by the expenditure on these services in 1980 dollars. Assume further that employment effects in a world without videotex are nil; that is, gains in employment due to output growth in a richer, better-educated, and more populous United States are balanced by employment losses due to productivity improvements. This ignores the employment impact due to changes in the scale of production, but it offers an explicit baseline from which to estimate the employment impacts of new information retrieval technologies. We can then estimate how much of the expenditure (and hence employment) on these products (newspapers, periodicals, and books) will be replaced by teletext/videotex.

FIGURE 12.1 Four-Sector Aggregation of the U.S. Labor Force. (*Source: Bell, 1980.*)

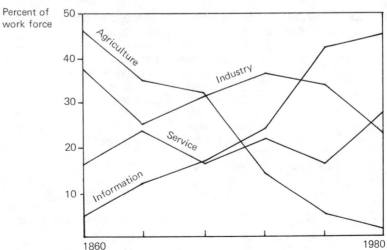

TABLE 12.4 EMPLOYMENT IN INFORMATION RETRIEVAL RELATED
INDUSTRIES (total employed [000])

Selected industries	1970	1979	Average annual growth rate in %
Printing and publishing	1,104	1,243	1.3
Newspapers	373	421	1.4
Periodicals	75	82	1.0
Books	101	103	0.2
Commercial printing	354	412	1.7

Source: U.S. Bureau of the Census, 1980.

Drawing on the videotex penetration forecasts in Chapter 5, we have household penetration in 1995 ranging from 12 million (low penetration) to 20 million households (high penetration). Suppose further that teletext and videotex information retrieval services replace 30 percent, 20 percent, and 10 percent of household expenditures on the three print-based services—newspapers, periodicals, and books, respectively—in those households equipped with teletext or videotex. In Table 12.5 we estimate a broad 1995 employment impact (displacement and need for retraining) scenario for these three components of the printing industry using the above penetration and replacement figures to estimate a videotex substitution effect. The values in Table 12.5 represent the number of people who would otherwise be employed in that year without a teletext/videotex industry.

Teletext/videotex will of course create some employment, but it is likely to be small relative to the number of positions replaced. For example, the national teletext services in the United Kingdom, BBC and ITV, each employ a total of about 30

TABLE 12.5 FORECAST EMPLOYMENT IMPACT OF VIDEOTEX ON SELECTED
INFORMATION RETRIEVAL INDUSTRIES, 1995

	Hypothesized videotex substitution in 1995		Estimated reduction in the number of positions		Job composition affected
	Low penetration	High penetration	Low penetration	High penetration	
Newspapers	4%	6%	14,000	21,000	50% production 20% clerical
Periodicals	2%	4%	2,000	3,000	10% professional 10% administrative 10% other
Books	1%	2%	500	1,000	

managers, editors, subeditors, and computer engineers. It is estimated that a full-channel teletext service could be provided by adding another 40 or 50 people. Transferring this to the U.S. context and adding three people to the staff at each existing broadcast television station to provide local information, the total number of positions created is less than 2,500. This number increases if the same rationale is applied to each of the 4,000 cable systems and to each of the 4,000 to 5,000 potential low-power TV stations that could be on stream by the end of the century.

Transactions   Industries that involve transactions between the consumer and the provider of a product or service employ a large share of the work force. Generally they are also the industries in which electronic transaction systems will make the most impact. In addition, employment in many of these industries has been exceeding the national average annual growth rate of 2.5 percent (Table 12.6).

Drawn from the videotex employment impact framework described for information retrieval, the general implications for selected transaction-related industries are shown in Table 12.7. In this case it was assumed that for households with videotex, 60 percent of banking transactions and 10 percent of retail transactions were undertaken using videotex .

Messaging   During the decade 1970 to 1980, employment in the U.S. Postal Service has been falling at a rate of about 1 percent per annum to its current level of around 650,000. During this time the volume of mail handled increased from 84.9 billion to 99.8 billion articles, an average annual increase of 1.6 percent.

The introduction of electronic messaging services is expected to increase this rate of decline of employment in the USPS as hard copy is replaced by electronic

TABLE 12.6   EMPLOYMENT IN INDUSTRIES UNDERTAKING TRANSACTIONS

Selected industries	Total employed (000)		Average annual growth rate in %
	1970	1979	
Wholesale trade	3,993	5,170	2.9
Retail trade	11,047	14,966	3.4
General merchandise stores	1,672	2,265	3.4
Food stores	1,731	2,281	3.1
Apparel and accessory stores	761	936	2.3
Furniture and home furnishings	472	614	3.0
Banking	1,044	1,489	4.0
Credit agencies	360	549	4.8
Security and broker services	205	205	—
Insurance carriers	1,033	1,213	1.8
Insurance agents	288	431	4.6
Real estate	661	940	4.0

Source: U.S. Bureau of the Census, 1980.

TABLE 12.7   FORECAST EMPLOYMENT IMPACT OF VIDEOTEX ON SELECTED TRANSACTION INDUSTRIES, 1995

	Hypothesized videotex substitution in 1995		Estimated reduction in the number of positions		Job composition affected
	Low penetration	High	Low penetration	High	
Banking	7%	12%	63,000	108,000	70% clerical 30% administrative
Retail (excluding restaurants)	1%	2%	98,500	197,000	80% clerical 20% supervisory, administrative

transmissions. By 1995, with a household videotex market of less than 30 percent, electronic messaging will not be the major carrier of home-to-home messages. However, this constitutes a very small volume of the total mail delivered (Table 12.8). A considerable volume of business-to-business mail, however, could be distributed electronically, and the business-to-and-from-household mail supporting electronic transactions could also be transmitted electronically.

Assume the same videotex household penetration and substitution factors for financial transactions, i.e., a low of 12 percent and a high of 20 percent of households doing 60 percent of their messaging electronically, and an estimated business market for electronic transactions of 30 percent of current business-to-business transactions. Further, assume that productivity gains in the USPS continue and that, without videotex electronic messaging, the volume of mail handled continues to grow steadily. We estimate that the work force needed to support the hard-copy U.S. mail service will decline at an even faster rate than at present, causing a further reduction from 75,000 to 50,000 positions.

TABLE 12.8   PERCENTAGE DISTRIBUTION OF MAIL FLOW, USPS MAIL

	Receiver		
Sender	Business	Household	Other
Business	21.0	48.7	1.9
Household	9.6	6.5	1.4
Other (government, nonprofit)	2.7	6.8	1.4

Source: University of Michigan, 1980.

Computing Teletext and videotex are add-on industries to the existing main-frame, personal computer, terminal, software, and data processing industries. As such they will create jobs in these areas. If we take the same 1995 range of household videotex penetration assumed earlier (12 million to 20 million households) and a $50 average cost for adding the necessary components to convert a television into a videotex terminal (1982 dollars), the size of the videotex hardware market ranges from $600 million to $1 billion. These values, however, represent a lower bound for the market, as many videotex users will use an enhanced rather than a basic terminal configuration. At an average value added per worker of $40,000, even these base figures represent an addition of 15,000 to 25,000 jobs in the semiconductor and hardware manufacturing industries, assuming 100 percent domestic production.

At least initially, teletext is likely to provide a larger hardware market and hence greater employment prospects. Assuming a 1995 household penetration of 40 to 80 percent and an average $25 cost for add-ons to the television set, the market size for teletext ranges from $1 billion to $2.5 billion. This translates to between 25,000 and 63,000 new jobs in the electronic industries.

In addition to employment increases in the semiconductor and computer hardware industries, teletext and videotex will create jobs in software areas such as programming and database management. These industries are already predicted to experience significant employment growth without widespread teletext/videotex. The Bureau of Labor Statistics predicts that while the white-collar work force in general will be increasing at about 3 percent per annum over the period 1978 to 1990, employment in computer specialist occupations (computer operators, computer programmers, and computer systems analysts) is projected to increase at 5 percent per annum during that time (Carey, 1981).

Telemonitoring It is extremely difficult to estimate employment impacts in this area. If telemonitoring includes fire and burglary protection and energy management and meter reading, then it could have significant direct and indirect employment effects. Tasks such as meter reading are beginning to be undertaken electronically; the indirect effects being that record management and billing can also be electronic, thus possibly reducing the need for clerical support.

Electronic fire and burglary protection systems may change the distribution and type of protective service workers required—if fires are detected earlier and thefts prevented. However, unlike information retrieval, transactions, or messaging, it is not obvious that there will be any net work force reductions or additions of protective service workers due to videotex.

Impact of Policy Profiles on Employment

Under *deregulation*, as we have argued in the previous section on industry structure, a number of large, vertically integrated organizations providing videotex services are likely to emerge. In this situation small, local, or regional organizations unable to obtain access to videotex systems may eventually be driven out of business. Within organizations, the interface positions—clerks, cashiers, and tellers—are the most vulnerable as electronic services such as transactions substitute for face-to-face interactions.

There will be a need for skilled personnel to design networks and videotex databases and to create frames, and a need for electronic marketing, information brokering, information analysis, and reformatting software. Positions will also be created in adjacent fields; for example, there will be an increased need for the physical distribution of electronically ordered products at international, national, regional, and community levels. Job creation will be encouraged by market incentives, and eventually formal academic training programs will emerge. Any worker relocation or retraining, possibly using videotex programs, is the responsibility of the individual, although corporations may wish to contribute to the retraining process. Powerful professions may be able to exert "occupational protectionism" on management in spite of the potential that electronic technology has for their professions.

Under *proregulation*, the regulated growth of videotex systems implies an employment impact more evenly spread across the board as the videotex medium becomes a major equalizer for large and small, national and local industries. Governments may acknowledge the need at least to stimulate retraining programs, which may utilize videotex technology. By influencing appropriations for education, federal and state governments can unburden themselves of some of the retraining responsibility and encourage institutions of learning to undertake training programs in those new skill areas where experienced personnel are scarce.

INTERNATIONAL TRADE AND COMMUNICATION

Teletext and videotex are emerging in the United States partly through the aggressive development and marketing efforts of foreign-owned companies. Most early U.S. trials have been based on technology developed in foreign countries. The French, Canadians, and British clearly intend to become major telecommunication suppliers to future world markets. These countries have government aid to stimulate their electronics and telecommunication industries in the videotex and teletext area. While Japan has not taken as strong a lead as these other countries, it is nonetheless waiting on the sidelines to provide the hardware needed for large-scale development of videotex and teletext.

Unlike in Canada and Europe, there have been only a few isolated examples of government-sponsored trials in the U.S.—the Green Thumb trial and the WETA teletext trial. The U.S. position has been to allow competition to develop with minimal government intervention. Foreign investors have as much opportunity to enter the U.S. market as domestic entrepreneurs have.

The issue of international trade and comparative advantage for the United States needs careful exploration for videotex. Although the data is not strictly comparable, it is useful to consider the trends in exports and imports of computer and telecommunication equipment (Table 12.9). The significant point here is that the crude trade balance in these areas has moved markedly toward imports. Much of this can be

TABLE 12.9   EXPORTS AND IMPORTS OF SELECTED COMMODITIES (in millions of current dollars)

	1973	1974	1975	1976	1977	1978	1979
Exports							
Electronic computers, parts and accessories	$1,717	$2,198	$2,228	$2,588	$3,264	$4,359	$5,671
Other office machines	368	501	411	349	621	647	804
Telecommunication apparatus (including TV, radios)	1,040	1,361	1,574	1,997	2,360	2,689	2,957
Total	3,125	4,060	4,213	4,934	6,145	7,695	9,432
Imports							
Office machines	915	1,007	1,052	1,341	1,568	2,254	2,500
Telecommunication apparatus (including TV, radios)	2,070	2,281	2,077	3,655	4,954	6,136	6,175
Electron tubes, transistors, semiconductors, and other parts	704	1,033	899	1,223	1,490	1,942	2,719
Total	3,689	4,321	3,925	6,219	8,012	10,332	11,394
Net export over imports	−564	−261	+288	−1,285	−1,867	−2,637	−1,962

Source: U.S. Bureau of the Census, 1980.

explained by focusing on telecommunications apparatus, the dominant component of which is TV sets. In fact, back in 1965 the trade balance for telecommunications apparatus was slightly positive—a clear indication of how the international market has changed as TV set manufacturers rely wholly or in part on foreign-produced components.

The issue of videotex and international trade could become serious for the United States if certain conditions were ripe: teletext and videotex achieve wide penetration; foreign manufacturers produce adapted TV sets at a lower cost than domestic manufacturers; foreign turnkey systems are implemented in major U.S. markets with residual payments to the foreign developer; and foreign software houses provide the teletext and videotex systems for microcomputers, minicomputers, and mainframes. In short, the above scenario portrays the United States as lagging increasingly behind both foreign hardware manufacturers and foreign teletext/videotex system developers. While the provision of information would still remain a predominantly U.S.-based industry (the U.S. online computer database industry accounts for 80 percent of total world revenue for online databases), the combined effects on balance of payments may still be negative.

In a technology assessment of personal computers, Nilles (1981) discusses the present and likely world market shares of the semiconductor market, the computer market, and TV set manufacturing by U.S. industry. Although the United States dominates the computer market and has stabilized its long-term share of semiconductors at about 60 percent of the market (Nilles estimate), the U.S. share of production of color television receivers for U.S. consumption has declined from 80 percent to less than 30 percent in a decade.

The teletext/videotex industry includes databases, software, terminals, transmission and switching equipment, computers, and semiconductors, all of which are currently being assembled in Asia, Canada, and Europe in addition to the United States. The possible implications have already been noted: "Telecommunications must continue to be at the cutting edge of American world trade. Sound fiscal and tax policy, and a powerful encouragement of investments in research and development must be implemented if we are to maintain this edge. The Japanese and others have clearly targeted the American electronics markets" (Wirth, 1980).

A basic question for the United States vis-à-vis the potential world market for videotex and teletext is: How will this new technology contribute to the U.S. balance of payments? Forecasts vary widely. Some skeptics compare the massive British investment in Prestel to that government's investment in the supersonic transport. Others argue that these countries have been more foresighted than the United States and are committed to retooling for a new information-based era. The U.S. position as a major provider of information databases has traditionally been to promote the free flow of information across borders. This position, however, has been rebuffed by other countries that have attempted to control the type and amount of information that crosses borders. In a scenario of widespread penetration of videotex and teletext, there are numerous ways in which international communication would be affected. For example, a teletext service from a direct broadcast satellite might allow a foreign information provider direct access to a domestic market—the emergence of the international information provider with national and even regional markets. This raises questions of personal privacy and national sovereignty—especially where industrial and social development is governed (or at least influenced) by the decisions of interest groups residing outside the United States. The long-term effect could be the unintended transfer of technology in this high growth area. International electronic messaging could be done through gateway services that bypass PTT regulations and other domestic restrictions. Similarly, international transactions could take place electronically, though in the case of teleshopping the actual delivery of goods could still be controlled by customs officials. How the ground rules are established for foreign organizations wanting to promote their products and services in the United States will be largely decided in Congress.

Impact of Policy Profile on International Trade

The options for U.S. policymakers vis-à-vis international trade range from minimal U.S. government intervention to direct legislation controlling and protecting the U.S. market, including fair-trade laws, import quotas, tariffs, and trade promotion.

The *deregulation* development profile, with a variety of standards, results in an uncoordinated U.S. market structure for international trade of products and information services. It places American industry at a disadvantage relative to the nationally coordinated efforts of the Japanese, Canadians, and Europeans for the international market. With easily accessible satellite communications and cheap local storage, the U.S. dominance of the database industry may ultimately be threatened. If the United States espouses total freedom of communication, as it has in the past, then it must be prepared for intense foreign competition (much of it government subsidized). In this environment there are few, if any, restrictions on transborder data flows into the United States.

The *proregulation* profile is characterized by policy responses that could include:

- The creation of a federal Department of Information and Telecommunications to develop policy toward these technologies at the domestic and international levels.
- Centralized and coordinated international trade planning at the federal level.
- National policy on inflow of data. May include tariffs or subsidiaries for local industries.
- Formal and informal measures to assist U.S. companies' market in countries that have gained significant footholds in the United States.

How mass information systems relate to one another in the world communications environment will be determined in various international bodies. The adoption of a single national standard within the U.S. may make the hardware market more accessible to foreign competition. There may well need to be special assistance provided to U.S. companies to give them a stronger footing in the emerging international information and telecommunications arena.

CONCLUSION

Consequential issues are shaped by the evolution of the technology, the applications or markets in videotex services, and the developmental policy paths. The two developmental policy paths that were examined reveal the likely policy dilemma with this new technology. At the societal level the unregulated market is usually a more economically efficient way to provide goods and services—with some notable exceptions. At the individual level, however, formal government regulation appears to be the most effective way to ensure privacy, equitable access for diverse groups within society, and general consumer protection.

The consequential issues seem to be primarily directed at videotex and possibly at one-way full channel broadcast services. Limited VBI teletext does not pose serious consequential policy issues. The key application area from a policy impact perspective is that of transactions. The potentials for loss of privacy and lack of consumer protection are coupled with major industrial structural impacts and potentially significant impacts on employment. This application is now investigated as a policy case study.

CHAPTER 13

TELEBANKING AND TELESHOPPING—A VIDEOTEX POLICY CASE STUDY

We now approach videotex policy questions from the perspective of a major *application* area for the technology—telebanking and teleshopping. This case study is intended to demonstrate some of the ways in which videotex issues are likely to actually arise and to provide a more focused analysis of some of the issues raised in the preceding chapters. Telebanking and teleshopping are analyzed because of their perceived importance in videotex systems and the potentially far-reaching consequences of widespread home-based transactions for individuals, as well as for the nation's retailing and financial industries.

CONTEXT

Teleshopping Home-based telephone and mail-order purchases of products and services (direct marketing) represent a large and rapidly growing fraction of all sales (Sawyer, 1982):

- More than $110 billion worth of direct sales took place in 1980.
- Direct mail advertisement volume increased from $2.87 billion in 1970 to $10.0 billion in 1980.
- Direct response advertising expenditures increased by 14 percent from 1979 to 1980, to nearly $22.5 billion.
- Over 45 percent of the total population (or 85 percent of the adult population) ordered at least one item through the mail in the last 12 months.
- Telephone expenditures for "telemarketing" rose by 15 percent during 1980 to $9.8 billion.
- Mail order is growing at a rate of 15 percent per year, or nearly twice the rate of retail sales.

Several videotex trials have included shopping services: Warner Amex's interactive QUBE permits selected home-based purchases. Comp-U-Card and Times Mirror in Los Angeles, California, and Adams-Russell and J. Walter Thompson in Peabody, Massachusetts, are offering shopping channels on cable systems. Cox Cable Communications' INDAX system and Knight-Ridder's Viewtron are offering home-based shopping services that permit consumers to select, order, and arrange for payment of merchandise in a single transaction.

Telebanking   By the late 1970s, Americans were estimated to be undertaking in excess of $250 billion cash, check, and credit card transactions at a cost totaling more than $14 billion annually (computed from Arthur D. Little, 1975). The environment in which these transactions are taking place is changing in several major ways. First, banks and financial institutions are moving into the electronic age. At the beginning of 1982, there were approximately 25,000 automatic teller machines in place around the country. Banks are now actively organizing national electronic networks for delivering a wide range of banking services.

Second, in March 1980, the U.S. Congress passed the Depository Institution Deregulation and Monetary Control Act—an act that guarantees financial deregulation. The key elements are the gradual elimination of interest ceilings on deposits by 1986 and the bringing together of all financial institutions under the control of the Federal Reserve. In addition to the likely reduction in the number of financial institutions, which currently number about 40,000, comprising 14,700 banks with 39,700 branches, 4,700 savings and loan associations, and 22,000 credit unions (U.S. Bureau of the Census, 1980), some further implications of this deregulation include (*Business Week*, 1980*b*):

• "True cost" banking requiring consumers and corporations to pay higher borrowing costs (currently, most banks do not charge customers the real cost of processing a check, which averages about 40 cents and is estimated to be as high as $1.60 for some interstate transactions.)

• Higher rates paid on checking and savings deposits of consumers, with all institutions free to pay what they wish by 1986 at latest, when all interest rate ceilings are abolished.

• Less differentiation between commercial banks and thrift institutions.

• "One-stop" banking for consumers and, ultimately, for corporations.

• Less market segmentation in types of services offered, although some institutions may be forced to "specialize" in order to survive.

• Less geographic segmentation as electronic funds transfer makes money move across today's arbitrary barriers. Eventually, the reality of the marketplace and the tendency toward "federalization" of regulatory oversight will hasten the adoption of some form of interstate banking in the United States.

• More conflict among depository institutions and such nonbanking interlopers as Merrill Lynch, Sears Roebuck, and American Express, with depository institutions struggling to retain market share. This poses serious questions as to the viability of separating commercial banking (deposits and loans) from investment banking (underwriting and selling securities), which is currently required under the Glass-

Steagall Act (1933). In the interests of competition, the potential conflict may require a revision of the 1956 Bank Holding Act, which limits bank ownership only to companies involved in banking-type activities.

Recently, the householder has been offered a number of limited at-home banking and telephone bill paying (TBP) facilities that are operated by either voice communication or by entering a set of digits on a touch-tone telephone keypad. However, videotex systems offer a more substantial outlet for home-based transactions, including both banking services and teleshopping. Trials of home-banking services are widespread. The Source, CompuServe, and Comp-U-Star, for example, provide personal computer owners a range of at-home banking and shopping services using national and regional suppliers of products. A summary of these trials is contained in Table 13.1.

STRUCTURE OF POLICY WORKSHOP

The findings of the case study are based on a policy workshop on transactions and videotex. Organized as part of the technology assessment project, the workshop was attended by a variety of financial, corporate, government, and academic participants selected because of their involvement in this subject. The structure of the workshop followed the broad structure of the policy analysis in this technology assessment. First, a set of policy issue questions associated with widespread use of videotex systems for transactions was identified. These issues were then ranked in terms of their importance for the development of videotex systems for transactions. Second, for the five issues identified as most important, a range of policy options was defined. Third, a deregulatory profile and a proregulatory policy profile were constructed. Finally, broad business and societal impacts of widespread penetration of videotex systems were outlined.

To set the scene for the policy analysis, a view of the possible evolution of videotex for transactions over the next 20 years was offered:

Videotex reaches widespread household penetration (i.e., in excess of 35 percent of households) in large part due to its transaction capabilities. At first, this activity is predominantly retail/discount shopping/ticket purchasing, but eventually it moves heavily into banking and financial service activities. Purchasing from the home is stimulated by rising fuel costs, inflation, increasing female participation in the full-time work force, greater emphasis on personal "leisure time," and wider electronic "product availability" with lower costs and increased product description/information.

Just as VISA and Master Charge were first the exclusive domain of separate banks, so too the initial services provided by banks were the domain of separate videotex services. That is, one had to use CompuServe to have access to a certain bank's services. Eventually the exclusivity agreements broke down. All banking services became accessible either through a terminal directly or through a videotex service and its gateway links.

Whether they use the service or not, all banking customers have electronic passbooks, checkbooks, and authentication cards using personal identification numbers (PINs). Branches are still available for personalized services, but they are much less pervasive. Customers tend to do their business electronically—through home and public terminals and through other telecommunication links.

TABLE 13.1   HOME BANKING TRIALS IN THE U.S., 1981

Primary sponsor	Financial institution	Location	Project status	Services offered	Participants	Others involved
Citibank	Citibank	New York City	Home Base, Summer 1981	*Banking* Bill payment Balance inquiry Funds transfer Order travelers checks	100	Unannounced
Chase Manhattan	Chase Manhattan	New York City, New Jersey, Connecticut	Chase Home Banking, June 1981	*Banking* Balance inquiry Bill payment Funds transfer Record keeping Electronic mail	300	Data Measurement Corporation (DMC) Telecredit
		National	Chase bill payment services, November 1981	Transaction processing	300 (banks)	
Chemical Bank	Chemical Bank	New York City	Pronto, November 1981	*Banking* Balance inquiry Checkbook reconciliation Funds transfer Bill payment Statement Validation Home budgeting *Informational* Theater tickets Sports News, weather Educational	200	Atari Inc.
Cox Cable	California First Bank, First Interstate Bank, Security Pacific, Home Federal Savings, San Diego Federal Savings,	San Diego, California	INDAX, New Orleans, Vancouver, Canada; scheduled for 1982–83	*Banking* Bill payment Funds transfer Balance inquiry *Informational* News Business news	300	Homeserve Viewmart The Source KPBS-TV San Diego State University Norpak

TABLE 13.1   HOME BANKING TRIALS IN THE U.S., 1981 (continued)

Primary sponsor	Financial institution	Location	Project status	Services offered	Participants	Others involved
	Manufacturers Hanover, Term Plan			Sports information Energy tips Shop-at-home Personal finance		
First Interstate Bank of CA	First Interstate Bank of California (formerly United California Bank)	Los Angeles San Fernando	September 1981	*Banking* Balance inquiry Funds transfer Bill paying Certificate rates *Informational* News Weather Sports information	250	The Source
Automated Data	25–50 banks to be selected nationally	Seattle, Washington	Home banking interchange, April 1982 start, to run 2 years Bill paying services for owners of personal computers; November 1981 start	*Banking* Bill paying Funds transfer Balance inquiry Check reconciliation Bank card cash advances Interest rate information New accounts applications Order travelers checks Order C.O.D.s Home budgeting *Informational* News Weather Sports Financial Travel services Catalog shopping	5,000	AT&T

				Services		
Banc One	Banc One	Columbus, Ohio	As a result of October 1980 90-day test, will begin franchise in 1983	*Banking* Balance inquiry Interest rate information Bill payment *Informational* News, weather Encyclopedia Computer games	200 (90-day test)	Micro-Computer Ventures, Inc. Decision Research Corp. Ohio State University O.C.L.C., Inc.
United American Services Corp.	United American Bank	Knoxville, Tennessee	1980 start	*Banking* Express statement Bill payment Bank-at-home	400	CompuServe, Inc. Radio Shack, Inc.
Microtel, Inc.	Erie Savings Buffalo	Buffalo, New York	2d quarter 1982	*Banking* Cash management for small business Bill payment Funds transfer		Infomart (Canada) Genesys Group
Viewdata Corp.	Southeast Bancorp	Coral Gables, Florida	Viewtron phase II, February 1981; Expanded services in mid-1983	*Banking* TBP Personal financial planning *Informational* News, weather, sports Message transmission Entertainment Advertising White and yellow pages Stock market Travel Consumer advice	200	AT&T Knight-Ridder

TABLE 13.1 HOME BANKING TRIALS IN THE U.S. 1981 (continued)

Primary sponsor	Financial institution	Location	Project status	Services offered	Participants	Others involved
Communications Technology Management, Inc.	5 to 10 banks to be selected in Baltimore-Washington area	Vienna, Virginia	May 1982 interconnect	*Transactional* Home banking Home shopping *Informational* National news Video games Directory Messages	1,000	NABU
First Bank System	First Bank System	Minneapolis, Minnesota; and Fargo, Wahpeton, and Valley City, North Dakota	May 1982	*Transactional* Home banking Clearing house for retail services *Informational* News, weather, sports Commodity reports Agribusiness information	250–300	Tymshare, Inc. Minneapolis Star and Tribune Intelmatique

Sources: TransData Corporation, reprinted from *National Study of Home Banking Services in the USA*, 1981; Telidon report no. 8, January 1982; Intelmatique press release, October 1981.

POLICY ISSUE IDENTIFICATION

The policy issues that were raised in association with videotex and transactions may be grouped into three classes: those that relate to uncertainties and possible conflicts in the organization and delivery of electronic transaction services, those that relate to possible impacts on the consumer of the service, and those that are broad societal/economic/industry structure issues. The issues in these three classes are as follows:

Conflicts in the Organization and Delivery of Electronic Transaction Services

- Who has what responsibility for transactions (e.g., service provider, system operator, network provider, credit provider, end users)?
- Are actions needed to prevent undue concentration of power in electronic transactions (e.g., AT&T, cable companies)?
- Is some guarantee of access to videotex systems needed to protect competition among commercial institutions?
- Are there specific standards implications of transaction services for videotex information and system operators?
- Will local and regional merchants need protection against competition from national teleshopping or telebanking services?

Impacts on the Consumer

- Are new laws or mechanisms needed to protect privacy and security of electronic transactions?
- Are new measures needed to clarify rights and obligations of consumers in the electronic marketplace (impulse buying, resolution of disputes and rights of redress, disclosure of terms)?
- Are new measures needed to protect the integrity of electronic transaction systems (electronic fraud and theft)?
- Should equity of access to videotex transaction systems be guaranteed for consumers? What happens to those without credit or access in an electronic world?
- Are current credit card laws regarding risk in the event of theft or loss of card sufficient?

Societal/Economic/Industry Structure Issues

- Is the national financial structure involving banks, savings and loans, finance brokers, etc., equitable? Are new measures needed to support or regulate some sectors to equalize opportunities?
- Are the national and state banking laws adequate for a national videotex electronic funds transfer system?
- Are controls needed to prevent government intrusion into private transactions?

- Teleshopping allows for the possibility of production on demand. Are actions needed to protect the existing distribution infrastructure (e.g., warehousing, transportation)?
- Are there any new job protection or retraining implications of widespread penetration of videotex systems for home banking and home shopping? Is there likely to be a job loss at the national level due to elimination of low-skill positions?
- Are changes in regulations needed to allow banks (and other major financial institutions) to be involved in retailing and other commercial activities?

These issues were then ranked in terms of their likely effect on the development of videotex systems for financial transactions and home-based shopping. The five highest-ranking issues identified for further analysis were:

1 Should equity of access to videotex transaction systems be guaranteed for consumers?

2 Who has what responsibility for transactions (e.g., service provider, system operator, network provider, credit provider, end user)?

3 Are actions needed to prevent undue concentration of power in electronic transactions (e.g., AT&T, cable companies)?

4 Are new laws or mechanisms needed to protect privacy and security of electronic transactions?

5 Is the national financial structure involving banks, savings and loans, finance brokers, etc., equitable? Are new measures needed to support or regulate some sectors to equalize opportunities?

POLICY OPTIONS FOR THE KEY ISSUES

The second step in the process was to generate a number of policy options for each of the five key issues. The aim was to search for options that were feasible although not necessarily involving formal participation by federal or state governments. An industry code of ethics, for example, is one option for regulating services and products.

1 Guarantee of Consumer Access to Electronic Transaction Systems

This issue can be divided into two components: (1) access to information about transactions, and (2) access to transactions per se.

The first component arises in the short term because of a lack of "computer literacy" and of publicly available terminals. Possible options are:

- To educate consumers in terminal literacy (the government is involved in the education process).
- To locate terminals in public institutions and provide information (free of charge) through these sources.

In the longer term, the issue is likely to be less important because publicly available, easily accessible coin-operated terminals provided by both public and private sources will be commonplace.

The second component only arises in the longer term if nonelectronic transactions are used to cross-subsidize electronic systems, thus forcing people without access (probably the more economically disadvantaged) to subsidize those with access. Possible options are:

- To do nothing about the problem, it being one of a number of inequities that exist within society.
- For government to offer subsidies to particular user groups to maximize the possibility of access.

There are two additional options for those individuals deemed "not credit worthy" in such a system:

- To offer services on a pay-as-you-go basis.
- To classify individuals according to risk (based on credit worthiness) and to charge high-risk groups some form of premium.

2 Responsibility for Videotex Transactions

There are four broad functional tasks in the provision of transactional services: communication networking, operating the videotex system, packaging transaction services for distribution through a videotex system, and the provision of the transaction service itself. The broad policy options are:

- To specify the functional responsibility for each actor and allocate risks accordingly; e.g., interindustry agreement.
- To apply existing regulation and marketplace mechanisms for banks, common carriers, etc., to those organizations performing particular tasks, there being no homogeneous solution over all types of transaction services.
- To rely on the courts to resolve disputes on a case-by-case basis and to determine precedents.
- To allow responsibility to be shaped by the above (especially case law) and ultimately documented by government bodies in public utility statutes.

3 Organizational Concentration of Power in Videotex Transaction Services

The key issue is the extent to which organizations, in particular network providers, could become vertically integrated in offering an array of functions in a videotex transaction service. The policy options are:

- To allow vertically integrated structures to emerge in the marketplace.
- To limit the range of functions that any communications company can perform to tasks of communication (common carrier approach for telephone, cable, and broadcast companies offering network services, including content-carriage separation).
- To give greater regulatory discretionary powers to the FCC and FTC for handling eligibility to participate in national systems.

- To forbid cross-ownership of videotex services at the local level.
- To allow (by regulation or lack of regulation) new roles to be played by existing actors such as banks, financial organizations, and retailers (common carriers being an exception).
- To create a publicly funded communication infrastructure (highway to support transaction services). This might involve the leasing of some percent of channels on cable systems to prevent concentration of power and to ensure competition and variety of service.

4 Privacy and Security

There are three areas where protections may be necessary: the home terminal (illegal access), the transmission network (monitoring type of transactions), and the system operator (monitoring transactions and user payment behavior).

The policy options are:

- To require user approval for use of any personalized data for commercial purposes (such as compilation of unsolicited mailing lists).
- To implement procedures to guarantee the electronic erasure of data after it has been used. (It is recognized that financial institutions must keep records. In this case the issue becomes one of accessibility to the records, especially consumer-sensitive records such as psychiatric services, blue movie viewing habits, etc.)
- To prohibit totally the sales of lists of consumers and/or their transaction patterns.
- To clarify the government (e.g., Internal Revenue Service and Social Security) rights of access to data or transactions held by videotex service organizations or those of others involved in the packaging and distribution of transaction services.
- To introduce no new regulations, recognizing that people voluntarily give away (or sell) personal information.
- To strengthen existing privacy and security laws.

5 Competitiveness of the Videotex Transaction Environment

This issue is of fundamental importance for the electronic delivery of financial and transaction services in a competitive market environment. There is currently a plethora of state laws and regulations that are inconsistent with an efficient national or international flow of financial information. Furthermore, most current regulations in this area have been introduced to protect the position of the financial institutions; any new regulatory thrust is likely to be toward the creation of a more competitive environment. However, representatives of the banking industry argue that restrictions remaining on their institutions will constrain them from competing fully with non-banking organizations across the range of possible services.

Possible options for achieving a competitive market environment for the delivery of financial services are:

- To decontrol ''one-way'' restrictions on classes of organizations, i.e., to remove restrictions that currently prohibit one class of institution, from offering particular services, e.g., banks engaging in retail where the converse does not apply e.g., retailers offering banking services. Such decontrol could be anticompetitive in areas where the position of an existing monopoly would be strengthened, e.g., by way of cross-subsidization.

- To change the emphasis of regulation and control from special interest groups (banks, savings and loan, retail) to the *classes of services provided* (for example, local daily transactions, national transactions).
- To strengthen the federal control of banks and financial institutions involved in providing national services, and, concurrently, to reduce the extent of state laws and regulations. (The thrust of this option is simple: the move to nationally coordinated electronic transactions is incompatible within a dual federal-state banking system unless the state laws are consistent between states as well as with federal regulations).
- To allow industry the right to do business in this environment without government intervention.

POLICY OPTION PROFILES

There are many ways that the public policy debate for videotex transactions could unfold. Following the framework developed in Chapter 11, two broad public policy profiles are outlined to indicate the key differences between a deregulatory policy environment and a proregulatory environment.

Numerous other factors will influence and be influenced by the development of any policy profile. These include the economic state of the nation, the rate of penetration (and hence the perceived impact) of the technology, and the communication networks (telephone, cable, or a hybrid of broadcast of cable and telephone) used to deliver teleshopping and telebanking. A comparison of two feasible policy profiles—under deregulation and proregulation—is contained in Table 13.2.

IMPACTS OF WIDESPREAD PENETRATION

There are a number of aspects of widespread penetration of videotex for transactions that are specifically influenced by the policy profile. These include:

- *The role of government in videotex transaction systems.* The possible roles, dependent on the prevailing politics, range from formulation and implementation of a consistent policy across media and videotex systems generally to total nonintervention the industry during its infancy (as was the case in the online computer timesharing industry).
- *Determination of technical standards.* This has been discussed at length in Chapter 11, and, obviously, its relevance carries over to transactions.
- *Consumer protection.* The policy profiles suggest sharp differences in the ways consumer issues such as guarantee of consumer access, equity of access, protection of privacy, and security are addressed. One profile, the market-determined approach, would result in litigation as a primary tool to correct wrongs. The other profile, which attempts to regulate in advance, will more clearly define the respective roles but at the

TABLE 13.2   POLICY PROFILES—VIDEOTEX TRANSACTION ISSUES

Policy issue	Deregulation	Proregulation
Guarantee of consumer access	Videotex is one way of obtaining transaction services. Those who choose to use it pay the going rate. Government does not intervene in the way the service is provided nor does it attempt to influence its use. "User-pay" principles adopted.	Government introduces terminal subsidies for lower-income groups so that access is not prevented on the basis of cost. Government provides opportunities to educate consumers in terminal literacy, database access, use of hardware, and subsidizes technology.
Responsibility	Market-determined roles and responsibilities. Industry associations assume responsibility only where necessary. No role for government.	Existing regulations carry over to industries involved in providing videotex transaction services. Responsibilities ultimately end up in a public utility and other such statutes. The use of broadcast for videotex transactions raises a number of key responsibility issues with respect to carrier ownership such as ownership of —the VBI on broadcast —the VBI on cable and whether or not MDS channels will be opened up for transactions.
Concentration of power	Market-determined response. Trade and industry associations monitor organizational arrangements. Government concerned with content of videotex but not the general industry structures providing the service. Common carrier regulations, for example, still apply.	Content separated from transmission or carriage. Mandated leased channels on all cable systems for commercial use by organizations other than the cable company. Range of functions that telephone companies can perform is limited. Government concern is with undue concentration in the *delivery* of services but not *content*, which is widely diversified. Cross-ownership and cross-subsidy forbidden at the local and national levels.
Privacy and security	Existing laws strengthened, e.g., Privacy Act and Omnibus Crimes Control Act with respect to government access. No new legislation introduced. Guidelines evolve primarily from industry initiatives.	User approval required before any personalized data can be sold; rights of redress clearly defined; erasure clauses built into most electronic transactions; and government rights to access data are clarified and limited.
Equalized opportunities to compete	With the exception of common carrier regulation, no constraints are imposed on industries wishing to provide various videotex transaction services.	Federal controls on major financial institutions involved in videotex to determine roles and to minimize antitrust and noncompetitive practices. In particular, banks have their scope limited and roles carefully defined. The extent to which bank activities are curtailed depends on the general health of the economy and the competitiveness of banks vis-à-vis foreign competitors.

price of implementing a complex framework of laws that may hinder the growth and economic viability of the industry.

- *Type of service.* The policy environment is only one factor that helps shape the technology, the marketplace, and the range of applications. One final impact of the particular policy profile is on the characteristics of the general videotex transaction service itself that is offered to the user. Such characteristics include the quality of the service, the compatibility with other services, and the reliability, comprehensiveness, and cost of the service (Table 13.3).

Finally, there are a number of general impacts of widespread penetration of videotex transaction systems, impacts that arise in either regulatory context. They include:

TABLE 13-3   CONTRAST OF POLICY PROFILES FOR USER IMPACT CHARACTERISTICS OF TELEBANKING AND TELESHOPPING

User impact characteristic	Deregulatory vs. proregulatory profiles
Quality of service	Difference is marginal if both the marketplace and the regulated environment have a series of checks and balances that are implemented. The major influence on quality is likely to be the competitiveness of the service; something that could be eroded if proregulation is associated with bureaucratic rather than market decisions.
Cost of services	In general, the market-determined or deregulatory environment should produce lower prices. There may still be a desire by government to redistribute, e.g., subsidize particular income groups. The absolute cost to the consumer will be determined by the extent to which the service penetrates the total market and the degree of competition among providers.
Timeliness of introduction of services	There would appear to be more likelihood of a service that was offered in an unregulated market environment being available before the comparable regulated service. This principle has been true in most modern industries.
Comprehensiveness of services	In the deregulatory environment those services that are economically marginal, even if socially valuable, would tend not to be offered.
Reliability of the service	Market pressure should guarantee a more reliable service than one developed under a regulatory umbrella. On the other hand, the existence of formal regulatory mechanisms defines a legal minimum acceptable level of service without the consumer having to suffer.
Compatibility of service	The multiple standards that could prevail under a deregulatory environment would result in numerous competing noncompatible videotex systems. This may limit penetration, slow down growth, and increase price.
Flexibility of service	The market-determined world opens greater flexibility and choice for the user in terms of transaction possibilities. If there is no guarantee of consumer access, this "flexibility" is of no concern to those excluded or unable to subscribe to the service.

- *Production on demand*. A videotex transactions system offers an incentive for manufacturers to gear production to electronic requests for goods. It may alter the "middleman" infrastructure of American retailing: the "product middleman" being replaced by an "information broker."
- *Reduction of inventories* and hence a change in the warehousing patterns and the vulnerabilities to shortages, stoppages, and delays.
- *Altered distribution (and transportation) networks*. Since production may be triggered electronically, geographic location becomes irrelevant at the time of making the choice of purchase (unless the transportation costs and time delays for receipt of product are significant). Not only is the structure of the distribution network susceptible to change but outlets to that network and how it functions, including delivery time, may alter. New demands will be placed on the efficiency and timeliness of the delivery system, a system that when juxtaposed with electronic ordering appears archaic.
- *Cost of production*. Videotex transaction systems could alter the cost of doing business from production through distribution. It may be that this change is primarily to do with *where* the cost occurs on the cycle rather than the absolute cost itself.
- *Closure of surburban banking branches*. The removal of the ceiling on interest rates on savings and checking accounts (1986) with the already competitive environment for funds and the emergence of at-home banking will put additional stress on the banking infrastructure to utilize capital efficiently. This puts considerable pressure on the increasingly uneconomic local branches.
- *Reduction of tellers*. The restructuring of the local branches offers productivity increases by replacing front-office staff with public and private terminals.
- *Decline in effectiveness of state barriers*. The current banking structure is organized at the state level. Telephone debit and credit arrangements across state lines make intrastate regulations difficult to enforce.
- *Change in the money flow*. It seems debatable whether or not a move toward an electronic system of payment will increase the velocity of money or influence inflation. It is acknowledged that the "float" may become a privilege, not a right, and that it may even be sold or offered to certain classes of customers.

CONCLUSION

The videotex public policy debate will unfold at two levels. The developmental issues described in Chapter 11 will be the starting points for most general debates. As videotex systems emerge, however, they will also raise a number of issues that are related specifically to the applications that are offered on the various systems. Financial transactions and home-based shopping are key videotex applications that highlight the potential policy impact of this new technology. Widespread penetration of videotex will heighten competition among both financial and retail institutions and is likely to have broad social and economic impacts. Eventually, videotex systems may help create a new electronic marketplace through which a substantial portion of consumer transactions are conducted. The challenge for policymakers will be to foster as much competition as possible while ensuring that the rights and interests of individuals are protected.

CHAPTER 14

TRANSFORMATIVE EFFECTS AND SOCIETAL IMPACTS

Any technology that achieves widespread penetration will bring with it direct and indirect changes in the ways people relate to one another, to institutions, and to the technology. The direct changes usually show up quickly; in general, they are extrapolations of the existing social order. Indirect changes, on the other hand, tend to appear more slowly. They are often not extrapolations of the past but second-order consequences of direct effects. They challenge some basic foundations of organizations and institutions.

Second-order consequences are also somewhat less predictable than direct, extrapolative changes. There is an element of the unintended and unanticipated in these effects. They often reveal themselves only after the technology has been adopted and adapted by the users. In short, they may be considered *transformative effects*. Television, for example, was developed to provide entertainment for mass audiences but the extent of its social and psychological side effects on children and adults was never planned for. The mass-produced automobile has impacted city design, allocation of recreation time, environmental policy, and the design of hospital emergency room facilities.

So far, we have focused on direct policy issues likely to arise as a result of introducing videotex and teletext (developmental policy issues), together with the extrapolative policy issues that may also arise as a result of widespread penetration (consequential issues). In this chapter, our aim is to consider the indirect long-term transformative effects of this technology. In some sense we are now hypothesizing not only widespread penetration but societal adoption and integration of the technology. The question being addressed is: what happens to the present organization of society?

The more direct developmental and consequential effects can be targeted with the help of the types of methodologies used in this study. Transformative effects are less easily identified. They are also less easily tied to direct public policy intervention. Yet they are important for policymakers to understand because they provide a backdrop for the more direct policy choices; they will influence how such choices will interleave with other social changes in the coming decades. It is also important to look at transformative effects to devise, where possible, early warning signs that call attention to likely serious negative side effects of the new technology.

To identify transformative effects of teletext and videotex we have taken a building block approach. Within a general scenario of widespread penetration of teletext/videotex, we identify four key areas of societal impact. Each key area is then described against a backdrop of general societal trends and indicators, and a number of general themes comprising societal impacts and transformative effects are discussed. A workshop setting, with participants who are familiar with the long-term effects of technological innovation—especially communication or information technology—was used to generate much of the following information.

TELETEXT/VIDEOTEX SCENARIO: THE TECHNOLOGY

In considering transformative effects, we first of all need to make some basic assumptions for a "high-penetration" scenario. Whether or not the technology will develop in exact accord with the scenario is not important; what is important is *to identify the major changes that might occur, given the assumption that teletext/videotex has become almost as common as television or the telephone*. Assume the following scenario for 1998:

> . . . It (teletext/videotex) has achieved relatively widespread penetration. It may not be in every home, but it is probably in a neighbor's home, and you might be considering getting the service yourself. It is used for all five classes of applications: information retrieval, messaging, transactions, computing, and telemonitoring. How it developed is not important for considering transformative effects. What is important is that the technology did develop and that it is being widely used. The service is paid for by information service providers, users, advertisers, and nonprofit/public organizations.
>
> Networking capabilities are widely developed. Through packet-switched networks and satellite and terrestrial communication links, data and broadband networks have vastly increased in numbers and in services offered. Low-cost terminals are accepted and available much as cheap pocket calculators and digital watches were in the early 1980s.
>
> Teletext and videotex exist in a world where digital communications have markedly altered the ways in which all electronic media—stand-alone or electronically linked—operate. Digital sound recording has not only upgraded the quality of sound reproduction but has also created an on-demand music delivery service for music and sound recordings. Video technology is comparable. Users are able to request particular programs from service suppliers when they want them. The dominant medium of the '60s and '70s—television—has become modular, with a variety of purposes, including program watching, terminal display, home movie projector, and more. The television set has also become interactive for both interconnect services (through telephone and cable) and for stand-alone uses (interactive videodiscs).
>
> Microprocessors have become commonplace, not only in communications technology but in scores of other home and business products as well. Security and protection services,

for example, have built-in microprocessors, allowing for multiple home, security, and emergency functions to be monitored simultaneously. Similarly, it has become more economical for utility companies to monitor and read home and office energy and water usage remotely.

For our analysis, then, the key features of the 1998 videotex and teletext environment are:

- widespread penetration
- growth of electronic networks
- interactive capabilities
- old media take new forms
- prevalence of microprocessors

KEY AREAS OF SOCIETAL IMPACT

The above scenario describes the society from a technological perspective. But as emphasized in this study, technological capabilities account for only half the pattern of technological development. User needs account for the other half. For two broad groups of users—consumers and business—it is possible to identify four key areas where teletext and videotex technologies may make an impact. These are:

- the home and family life
- the consumer marketplace
- the business office
- the political arena

For each area we now examine some of the longer-term implications of teletext and videotex.

As stated in the scenario, these are not the only electronic technologies that will transform society. The effects we discuss, of necessity, include some of the impacts of associated technologies.

THE HOME AND FAMILY LIFE

Context   By 1998, the post-Second World War baby-boom age group that has dominated society is between 45 and 55 years old. Lower death rates, higher income, and longer labor-force participation are prevalent for the elderly.

Rapid population growth continues in the south and the west, with most of the growth in ''small big cities''—cities located in semirural areas with pleasant climates and willing workers. The rising cost of housing and house financing results in more shared living space—consumer units (the number of unrelated people living in the same household) continue to grow at a greater rate than household formation. The upturn in the birthrate that began in the mid-1980s puts further pressure on housing and begins to increase household size. There are still, however, in excess of 20 percent of the population living in single households (Institute for the Future, 1981).

In addition to TV, low-cost modular videotex terminals have given households another communication resource for everything from store-and-forward messaging to directory assistance and instant news updates. Middle- and upper-income families have several terminals available for different family members. New networks of interest and professional groups are linked via videotex systems. Much of the routine shopping is done from a home terminal by browsing through electronic catalogs. Not only does the videotex-equipped home have access to vast amounts of information, but the providers of information have some knowledge of the home and can tailor their information to the needs and interests of that home. Many videotex and teletext options are available, so consumers can choose from a range that extends from no-frills teletext to super-equipped videotex.

Transformative Effects All of the following effects were evoked by asking the question ''What's going on in the American home in 1998?'' They portend some big changes in the American home in the coming decades.

Electronic Home Family life is not limited to meals, weekend outings, and once-a-year vacations. Instead of being the glue that holds things together so that family members can do all those other things they're expected to do—such as work, school, and community gatherings—the family is the unit *that does those other things*, and the home is the place where they get done. Like the term ''cottage industry,'' this view might seem to reflect a previous era when family trades were passed down from generation to generation and children were apprenticed to their parents. In the ''electronic cottage,'' however, one electronic ''tool kit'' can support many information production trades.

Interdependence In the electronic home familial ties of interdependence may be based on new skills. Spouses may be drawn to each other as much for their ability to manipulate databases as for their ability to prepare gourmet meals or to play racquetball. But even more significant, particularly for the transition period, will be the skills of children. While even computer-literate adults are somewhat awed by digital technology, children are not, and they are learning to make good use of electronic information systems. Are child labor laws enforceable in this environment?

Men, Women, and Child Rearing The electronic cottage is the perfect technological solution to the problem of how women can have traditional careers and rear children at the same time, e.g., use the electronic technology to participate in information-related tasks from secretarial services to real estate management. It is also possible to use electronic communication to coordinate home-based tasks ranging from meal planning to electronic component production. The problem with this solution is that the problem is not simply a technological one. Nevertheless, knowledge workers in the 1990s may be more productive because they are working at home, and perhaps this increased productivity will allow both parents to be part-time parents and part-time workers, which in teletext/videotex math adds up to one full-time parent and two full-time incomes.

The Electronically Extended Family If widespread use of in-home information and communication systems can recreate cottage industries, they might also recreate the extended family. In-home knowledge-related work seems a most suitable form of employment for the elderly. This possibility might conjure a view of self-sufficiency, independence, and isolation for the elderly. However, if the elderly are gainfully employed and fully occupied with productive work, it may be more desirable to have them around. This opens the whole question of government payment of retirement benefits and social security payments to a population that is employable. Mechanisms for enforcing retirement, or at least identifying bonafide retirees, are sought. Without IRS scrutiny of all electronic transactions, moonlighting and tax avoidance may become serious problems. We suspect that desirability may not be the key issue with respect to the elderly, but capital might be, as the next theme suggests.

Buying the Electronic Home The electronic home is capital intensive, and someone has to capitalize it. But who? and how? One possibility is a kind of localized travel-communication trade-off: the family's second car might be traded for elaborate videotex and other electronic equipment. Alternatively the government might issue ''information stamps,'' just as they now issue food stamps, if it became obvious that a lack of access to electronic information systems was creating a class of disadvantaged. That takes care of the cost of the electronics, but what about the home? It is costly and has been so for two decades. Consumer unit formation now significantly exceeds household formations. The elderly may hold a trump card here. It may be easier to add on rooms to existing homes and extend the family. Of course, the extended family need not be blood relatives. The house-holding elderly could ''adopt'' a younger family. Whether they are blood relatives or not, though, this extended family may wish to incorporate. Since a lot of work is going on in this home, there may be real tax advantages to this strategy. But what would be the impact on the tax base of communities and of the country as a whole? Are residential property taxes replaced by commercial taxes for home industries? What happens to zoning laws in such an environment? And finally, if families are corporations, are family relationships more subject to public scrutiny and legal sanctions than they are now?

The changes in work styles and family relationships described in these relations suggest other changes not directly related to work. Changes in the quality of interpersonal interaction in the family might include:

More Conflict The family has often been seen as the institution that insulates family members from the competitive, conflict-prone world outside. But if the home is also the workplace, and family members are sharing business resources and swapping ''knowledge'' skills, the struggles with the outside world may increasingly become struggles in the home. This scenario raises questions about the stability of the already somewhat shaky family. It suggests that some other institution will need to evolve to fill the safe-haven role. It also foretells a steady stream of clients for marriage and family counselors who may grow to become a privileged—and powerful—profession comparable to today's physicians.

Peers Redefined In contemporary society, peer groups are usually defined by age, thanks to an educational system that segregates students by age. Electronic home information systems such as teletext and videotex are blind to age. But they create classes of people based on interests, skills, and even specialized languages. As it becomes easier for individuals to link with various others of these classes, to establish relationships with members of these classes, to *identify* with them, ties with traditional peer group members may break down. And as these ties break down, so may the rules for appropriate behavior of one age group or another. Although the traditional rules for societal stratification are changing, it is not clear whether the new alliances result in a more or less socially stratified society.

The redefinition of peer relationships may *begin* with the establishment of personal electronic networks, but it probably won't get far so long as the educational system remains more or less intact. Some possible changes in the area of education are:

Retraining the Trainers   In the information- and data-rich environment of 1998, there are numerous electronic resources available to the student and teacher for the learning process. As a result, the traditional role of the teacher has been transformed. Rote learning is greatly facilitated by numerous interactive systems accessible at school, home, and work. The teacher's role in this environment is to help young learners synthesize data and develop creative problem-solving skills. The best teachers have always seen themselves in this role, but the "best" are far from "all." Retraining is a possibility, but the bureaucracy of education is big enough and powerful enough that many may resist. This resistance is likely to be politicized. In fact, the back-to-basics movement may be the first stage of this resistance.

Let There Be Generalists   Just as the pocket calculator has changed the way students approach mathematics, videotex will change the way they approach education. In 1998, information technology is so widespread that job requirements have undergone a major shift. Rather than increasing specialization, there is, at the end of the century, a much greater demand for generalists with multispecialties. Since so much of information technology is devoted to relieving tasks that have, in the past, required highly specialized skills, it becomes advantageous to hire a knowledgeable generalist—in other words, a person who knows enough about the job area to function in it and who knows how to use information tools. However, generalists have traditionally been free-lancers; there are not a lot of corporate generalists.

THE CONSUMER MARKETPLACE

Context   By 1998, consumer markets are characterized by a growing influence of upper-income buyers. The share of consumer spending accounted for by these upper-income households, which constitute only 30 percent of all households, exceeds 50 percent. Continuing inflation, however, influences consumption patterns by requiring a greater proportion of income to be spent on necessities such as housing, home heating, and transportation (Institute for the Future, 1981).

Today, the shopping mall and the large retail chains are the symbols of the American marketplace. Multiunit retail establishments, for example, increased their share of total sales from 36 percent in 1963 to 47 percent in 1977 (U.S. Bureau of the Census, 1980). By maintaining a little less than this growth rate, this share reaches 60 percent by the late 1990s. Shopping-related uses of teletext create a scenario in which electronic purchasing and electronic transactions have created a new electronic market-place. Bills, bank accounts, and credit card accounts are all serviceable from home terminals. National marketing services using electronic publishing networks are able to offer significant discounts for major brand items. Some products are produced only according to consumer demand, which now can be determined almost instantly. Market research is a much more precise tool since more is known, electronically, about the tastes, interests, and lifestyle of the consumer.

Transformative Effects   A variety of conflicting effects are identified:

Preprogrammed Purchases   One of the characteristics of the electronic house-hold is that a lot of information about the household's basic lifestyle patterns can be collected and stored automatically—by the family itself, by the organizations with which it interacts through the system, and by the people who offer the videotex service. This potential to maintain an in-depth profile of family habits has many implications. One of them is the potential for preprogrammed or automatic purchases. In its simplest form, this approach to shopping might simply involve a standing list prepared by the family of goods that they purchase on a regular basis, together with information about how much and how often it should be purchased and from whom. The in-home computer automatically orders two rolls of paper towels every week, a 25-pound bag of potatoes every other week, and vacuum cleaner bags every three months. But with a little more intelligence, the same computer might also remember that the family, given a choice between inexpensive paper towels with no brand name and more expensive brand name towels, always chooses the brand name. The com-puter then makes itself a little rule about criteria for paper towels or any other product and "shops around" to get the best deal for that family. Of course, if the computer is using such criteria to make purchases, information about these criteria is also avail-able from the distributor. Two strategies are possible: First, the distributors might make just basic standardized information available, in which case the family will need to specify its choices in terms of this information only; as a result, the buying patterns might become much more rationalized, much less subjective, and therefore less susceptible to the common advertising appeals of the television era. The second is that people will never rationalize their buying patterns, and if the videotex system is going to be the "electronic shopping mall" of the 1990s, a lot more information about products will need to be available so that different households can use different criteria for purchases and the ordering computer will be able to locate the necessary information to make the decision. In either case, there's going to be a lot of attention given to the question about the kind of information that should be available about pro-ducts and the requirements for truth in advertising.

Automated Individual Tailoring   Almost the polar opposite of the previous effort is the idea that videotex or teletext systems, together with computer-aided manufac-ture, might actually support the individual tailoring of items—not just standardized routine products. In this case, the purchaser might interact more or less directly with an industrial robot that makes, say, blue jeans. The purchaser could specify exact measurements, how many pockets in front, how many pockets in back, whose designer label to attach at the end. This possibility demands a lot of user interface design. The people who are now designing video games for home entertainment may be the "most-wanted" employees in the manufacturing world of tomorrow. The kinds of things that are tailor-made now are the kinds of things that need to be purchased in person, where people can touch the fabrics and imagine themselves in a particular design. Enhanced videotex plus computer-aided manufacture may assist in individualized mass production. If videotex purchases are boring, there will still be a lot of enthusiasm for traditional shopping malls. But if video purchasing takes on the qualities of a video game there will probably be the ultimate merging of TV advertising, home entertainment, and consumption.

Aside from the question of individually tailored, special-order purchases compared to routine standardized purchases, the idea of electronic shopping introduces an information flow problem in the realm of accounting:

Cash-Flow Complexities The consumers of the closing years of this century may spend all of their time trying to straighten out their personal bank accounts. If some purchases are preprogrammed and use up personal funds like a time-release capsule, while others are made on the spur of the moment with a videotex shopping service, while still others are made in more conventional at-the-store shopping trips, the potential for omissions and confusions is high. If we add multiple family members to the scene, it becomes even more confused. A lot of this problem falls into the realm of electronic funds transfer systems, and one approach is to imagine that the design of such systems will be sufficiently clever to provide a technological solution to this problem. Eventually, systems will provide on-line assistance in sorting out personal cash-flow problems. Of course, there is a more global impact. Cash flow at the macro level, at the societal level, is currently more or less invisible. However, as more of the flow is programmed, recorded, and condensed in time, the flow may actually become visible. New criteria may be introduced for credit rating, for solvency, for transaction. The credit card float may be a thing of the past.

Reorganizing Product Distribution A generation is growing up that doesn't know what home delivery means except in relation to the evening paper. And they don't know that home delivery went out of business because it became uneconomical. But electronic shopping implies home delivery of physical products. With home delivery comes reorganized distribution networks at the national, regional, and local levels. The mail service is in contact with every home and could become a delivery vehicle for both information and physical products at local and regional levels.

When we consider alternative forms of distribution—forms other than the local marketplace—we begin to suspect that the market or shopping center is more than just a display and an efficient means of distribution. By addressing the question: "What roles do shopping centers currently fill?" we are led to the following projections:

Remodeling the Shopping Center The shopping center is not going to disappear as a result of electronic shopping, but it will probably change a lot. Automated production, assembly, and distribution processes—the McDonald's hamburger effect—create an environment where the key interface is one of information exchange. One possibility includes a mix of convenience stores for picking up items that would normally be ordered routinely through the system, plus stores that provide an intense sensory experience, and perhaps electronic ordering centers; centers where a few items were displayed along with many terminals for shoppers to place their own orders for later delivery, or for pick up at a drive-through warehouse. Discount stores and large magnet stores could well disappear. Another possibility is a mix of shopping with other services—entertainment and recreation facilities, as well as hotels, might all find themselves sharing space in the shopping mall. The prototypes of these malls already exist, and American consumers can probably expect to see more of them, particularly if electronic shopping becomes a norm.

The needs that such shopping centers fill are twofold: the need for sensory stimulation plus the need for human interaction. A straight-line extrapolation from current models of shopping malls might suggest that these needs will be met by the most garish of sensory experiences and the most shallow of human interactions. However, it is equally possible to imagine a scenario in which, as consumption becomes apparently more or less divorced from the shopping center, the sensory and interaction components become subject to greater scrutiny and higher-quality demands. One thinks, for example, of the growth of bookstores that are also coffeehouses with readings and an alternative to the bar as a place for people to meet one another. This is not to suggest that everyone will prefer bookstores to bars but that consumers may expect any store to be a place to meet people with similar interests. In this sense, the stores and their clientele may mirror the kinds of interest-group networks that evolve through the use of in-home information systems. The end result may be just the opposite of the initial impulse: consumption, rather than being divorced from sensory experience and human interaction, will become completely integrated within it.

Electronic Spontaneous Purchase   One of the main foundations for the large department or discount store is spontaneous purchase. Eye-level counter space next to checkout counters rents for a premium. People spend more money when they are in a crowd, and crowds gather in big stores with lots of different "good buys" for lots of people with lots of different needs and interests. Someone will probably discover the electronic equivalent of spontaneous purchase and exploit it, for example, using the extensive records that will be available on family spending patterns and very carefully targeted sales efforts, directed at the individual in his or her home and appealing to known likes and dislikes may be possible. This approach is intrusive and is bound to raise regulatory issues of some kind as well as introduce protective legislation, such as the three-day "cooling off" period that now applies to sales in the home.

Whatever specific regulatory and protective measures evolve, they are going to feed into the electronic cash flow mentioned earlier. If someone can instantly transfer funds, what happens if they later "cool off"? One possibility is the evolution of a kind of consumer escrow in which a neutral company becomes the intermediate holder of all funds transferred and guarantees satisfaction to both consumer and seller, mediating disputes that may arise. This kind of insurance may be the natural evolution of today's bank courtesy cards that guarantee checks up to a certain amount, or today's bank credit cards that essentially insure purchases from the seller's point of view. The difference would be that the consumer might also gain protection—at a fee, of course—and that it might become impossible to make an electronic purchase without the insurance. This company may also be the employer of those people who help consumers manage their personal accounts on line. This is yet another source of detailed financial information on the individual and the household.

Homogenized World Views   The consumer marketplace reinforces selective perceptions of the world. Malls in different neighborhoods sell different products, display their products in different ways, have different styles. These styles portray different visions of what the world is all about, of how people should and actually do live. People who live a particular lifestyle go to the market that supports that lifestyle and are therefore reinforced in their belief that their lifestyle is the right one, and perhaps the commonest one. If people shop electronically, will world views become homogenized? Will already reduced style differences among classes and regions and age groups be effaced? The argument for homogenization comes from the analogy with television—television presents a monolithic view of the ideal American life. The analogy is not too convincing, though, since television is a broadcast medium, and home electronic shopping services need not be broadcast. In fact, they have the potential for being highly responsive to individual tastes and lifestyles, so rather than encouraging a single concept of the world, they are likely to promote even more diversity. This diversity will be quiet diversity most of the time, since people simply won't see what others are doing electronically; so conflicts that arise out of diversity are likely to be more surprising and less easy to categorize—and so less easy to address politically.

THE BUSINESS OFFICE

Context   During the 1980s and early 1990s, labor force growth slowed substantially to about 1 percent per year. Even so white-collar and service workers outnumbered blue-collar, manufacturing, and agricultural workers by nearly four to one. Entry-level labor and skilled labor were in short supply. Education enrollments, which began declining in 1975, continued to decline steadily throughout the 1980s, although they began picking up in the late 1990s as a result of the rising birthrate in the 1980s. Labor shortages were acute in the fast-growing, high-technology areas. This included engineers, who were in short supply throughout the 1980s, and programmers, of which there was a chronic shortage for two decades. Beyond specific skill shortages, the declining performance of high school seniors since 1968 on nationally administered tests such as the Scholastic Aptitude Test (SAT) continued (Institute for the Future, 1981).

Within the work force, satisfaction with work, which peaked in 1969 with over 90 percent satisfied (Public Opinion, 1981*b*), generally declined. Reasons for this decline included lack of pride, poor performance, lack of challenge, and financial rewards. This decline in satisfaction was highest among blue-collar workers and those white-collar workers being replaced by technology, such as clerical workers.

The office itself took on a new character, brought about by information and communication technologies. Many of the electronic innovations were integrated such that word processing, document storage, copying, and messaging were linked together. Videotex was part of this chain, as were other online data resources. The electronic office now permitted greater flexibility in location of office work: home or regional work stations are common. There is much less reliance on paper and more on electronic memos. Most office workers, including high-level executives, have a terminal on their desks. There is routine use of store-and-forward communication. Thus, it is relatively easy and natural for an office worker to use a terminal at home for office matters as well as to use an office terminal at work for personal services and enjoyment.

Transformative Effects

Blurred Boundaries between Work and Home—Telecommuting   In 1998, much more work is being done in the home than today. This work is primarily related to information production—and in particular to "creative knowledge" activities— programmers, software writers, data analysts, information brokers, database managers, forecasters, planners, stockbrokers, designers, modelers, architects, journalists, authors, real estate information service agents, travel agents, and so on. The office is integrated around electronic communications—electronic meetings, text composition and editing, electronic daily appointment calendars, computer problem solving, and electronic file cabinets. Home-based, or at least decentralized, location of employees is widespread.

In the "electronic cottage" there will be considerable competition for the available electronic tool kits. This flexible work environment will change the concept of overtime and the higher rates of pay usually associated with work outside of "normal hours." Information tasks rather than single organizational affiliations may become a way of life for people in these industries. This raises issues of accountability, for example, defining what is output of a home-based "creative knowledge" worker, payment for work done (traditionally, but not always, linked to time spent), compensation for illness or accidents, health insurance, vacation arrangements, and retirement plans. A new set of control and security problems arises for management. Can the performance and productivity of employees be maintained remotely? Will those who originate information also dictate how it is distributed? Who has the keys to the electronic file? Can the confidentiality of communications be protected and the integrity of software preserved? Can the organization prevent unauthorized access to information? This provides a very different role for the manager and senior executive.

Horizontalization   Since differential access to information creates hierarchies, the widespread availability of information in a teletext/videotex future could mean a collapse of these hierarchies. Automatic copying of electronic memos to everyone in a department and computer-based conferencing are examples of increased horizontal communication and, hence, a weakening of established hierarchy. However, while there may be fewer levels in the hierarchy, gaps between them may be wider. The medium might even define the structure of the hierarchy. Written communication and electronic information accessible through teletext/videotex and electronic mail systems could be at the lowest level. More personalized communication such as telephone communication and audio and video conferencing would be at a higher level and face-to-face communication would be at the highest level. Access to these different media could be used to maintain hierarchies. If this view is correct, how will these media-based hierarchies compare to existing organizational hierarchies? Will the same personalities populate the different levels or will a new power elite emerge?

New Leaders   The forecast of new leadership is based on observations that different "stars" emerge when new technology, such as the computer, is used. This phenomenon has been demonstrated in computerized conferences, research environments, and even senior management settings. The problem lies in determining just how global this stardom is. If computer conferences are where important organizational decisions are negotiated, then those who do well in this medium may indeed become the new leaders. But if, as was suggested previously, power is still linked to face-to-face communication, computer stars may be leaders only in very restricted circles. Still, managing the complicated communication in networks and between office and home may require very different styles than current managers exhibit, and those who find themselves in that elite network of face-to-face communicators may be those who can best understand and conceptualize the *process* of using information systems.

Redesigning Systems Design   The new kinds of management issues that arise in the environment of a mass information storage and retrieval system will lead to new approaches to the whole problem of system design. Already, military and office uses of systems are abandoning a strictly rational, task-oriented approach to systems design and are stressing instead human factors issues. This approach is aided by the growing base of knowledge in two areas: human behavior and artificial intelligence. The office will begin to be designed around information and communications technologies. This affects the physical design and layout as well as the location.

Externalization of the Organization   Teletext and videotex are not just media for internal communication within an organization. In addition to being used to manage the kind of day-to-day internal information needs along with other electronic communication media, they will also be used to take the pulse of the environment in which the organization is operating. The vastness of this external information environment, together with the accessibility it acquires with the introduction of a system such as teletext or videotex, means that the members of an organization are probably going to be spending an increasing portion of their time in external communications. Just as the new systems will blur the boundaries between home life and work life, so they will blur the boundaries between the internal affairs of an organization and its external affairs. This externalization could mean some redefinition of "corporate social responsibility," which has already been subject to considerable redefinition in recent years. At the same time, teletext and videotex are sometimes not as direct as the telephone or television for communicating with the external world. Organizations will, however, be able to use the technologies in assessing their public image and even "how public" they are—that is, how visible they are. Social responsibility will be difficult to assess for those outside the organization.

Finally, it was suggested that it would be a mistake to underestimate the ability of existing office structures and organizations to adapt to the technology, and that the efforts to adapt could actually redefine the technology itself!

THE POLITICAL ARENA

Context   Just as the physical marketplace has spawned an electronic offshoot, so, too, has the political process developed an interactive electronic political arena. The percent of the voting-age population voting in presidential elections has remained about 50 percent. The long-term disenchantment with government has continued, criticisms being that government wastes a lot of tax dollars, government is run for a few big interests, the federal government can't regularly be trusted to do what is right, and governments are run by people who don't know what they are doing (Public Opinion, 1981*a*). On the other hand, strong concensus has remained that it is the government's role to see that everyone who wants to work has a job (Public Opinion, 1981*c*). Interactive terminals in the home have led to the emergence of new forms of citizen-government communication. Before this technology, there was a clear distinction between political polling or opinion sampling and formal voting. Now, with interactive systems in the home and more frequent and extensive feedback, the line is less clear-cut. Where there is near universal penetration of videotex terminals—for example, in new housing developments—electronic referenda are regularly used for deciding routine problems. Political issue groups are more likely to find others who are interested in their cause because more is known about the opinions and preferences of individual household members. More information is available about government service, and citizen views, expressed through electronic media, are used more extensively in the planning and execution of those services. Just as government agencies had had to open their decision-making processes to greater public scrutiny in the 1960s and '70s, so have they had to incorporate uses of these new technologies in the political system in the 1990s. Thus, when a government transportation agency proposes changes in a mass-transit schedule, it conducted a public hearing. But in this new electronic environment, it seeks citizen opinion not only through traditional means (letters, telephone calls, etc.) but also through the new media. A basic need these systems meet in the political process is for a more informed and involved electorate.

Transformative Effects   With these uses of videotex and teletext as a starting point, there emerge a set of effects that pose contradictions, an indication of the extent of the uncertainty of the impact of the new technology:

Political Representation   Electronic voting provides profiles of candidates in terms of their platform and record and allows voters to cast their ballots remotely. Authenticity of remote voting has become a key concern. The interaction with elected representatives has many new aspects. A computer clearinghouse (electronic ombudsman) receives and investigates citizen complaints against the government and its agencies. Individuals have an electronic means of contacting and possibly influencing decision makers, who in turn can respond on a personal basis. Public opinion polls are much more extensive and timely and the results are available at many levels: by household, age group, income group, type of employment. Access, although ostensibly universal, becomes available to the privileged or those prepared to pay. Citizens who lodge complaints may find that their electronic profile of past political involvement is exposed. The protection of citizens' rights and protection of privacy are key concerns.

Image Candidate   One of the basic assumptions about politics in the theme just described is that extensive political information is available. That is, the politicians know more about the concerns and preferences of citizens, and the citizens know more about the issues and politicians' behavior with regard to these issues. In this kind of environment, there are two possibilities. Either politicians will become more accountable and will deal much more with data and empirical facts, or, with all of this information available, the candidate who is best able to manipulate public attention to focus on some facts and not on others will be more successful in presenting an "image" that matches public concerns. Similarly, those candidates who are most able to electronically manipulate their own image to appear to match, at any given moment, the mainstream concern of Americans will have the edge in the political process.

Diversity of Information   In 1998, there will be a great increase in information that is available about citizen concerns and preferences, about their lifestyles and values. This information will allow very rapid, very effective targeting of cohesive, special-interest groups who can become voting constituents for a variety of special-interest candidates running on single-issue platforms. This ability could lead to much greater diversity in the American power structure. So could the ability of citizens to write their own electronic slates and find their own candidates. Videotex might mean the end of the two-party system as networks of voters band together to support a variety of slates—maybe hundreds of them. At the same time, videotex, as a political system, could have as great an impact at the national level as the local level. In light of the emergence of the "image" candidate, the demise of the special-interest candidate is possible. Candidates in the 1990s may be measured by their overall image for the range of issues that confronts Americans. Those who advocate special interests may be seen as spoilers.

"Political" videotex allows governments and other groups to take electronic straw votes and display results to the voters in their homes—so that it is apparent where viewers fall in the overall distribution of votes. This may lead to the suppression of extreme viewpoints and to false consensus (or enhanced consensus) in elections. This would result not in diversity but in a kind of single-mindedness in American politics. If we try to imagine the resolution of these two opposing tendencies, we can envision a society that presents a facade of unity with an underlying foundation of unrecognized diversity. Videotex also raises the possibility for greater use of regional political advertising. Rather than producing national political ads, the same candidate might target different regions with different messages; the candidate might do the same with different partisan networks. The result, when the candidate is elected, will be the appearance of support and consensus, while the underlying reality is that different groups might be rallying behind very different issues and even stands on these issues.

The Power of Time   Political systems are concerned with power, and for several years now, people have been saying that information is power. Teletext and videotex are systems that channel information, and the way they channel that information will certainly shape the struggle for both personal and political power in a society. But the alternative ways for channeling information-as-power raise a new currency for power: time. Teletext and videotex alter the use of time. They extend the trend of the telephone and the television to make information more immediate, to make events happen faster, to make people respond to events more quickly. They also reorganize the use of time. The person who works at home via an electronic system is not bound by the eight-to-five schedule that binds the person who must work with people or machines at a common site. Home information systems will redefine societal rules for the use of time, and the social organization of time is a powerful means of political control. Conflicts in the use of time within a household may lead to "electronic home-time management counselors." Interactive time will become a highly valued commodity in the videotex society. And those who seek power will seek to control interactive time, not in months, weeks, or days, but in *minutes*. Time will be power.

New Era for Market Research   Whatever the complexion of American candidates and the political process in the 1990s, videotex and its use as a political instrument will usher in a new era of sophisticated market research. These techniques will emerge from (1) the need to make sense out of the vast amount of new data that is available about individuals through their use of electronic information systems, and (2) the need to understand the nature of a "network" market compared to a mass market. These needs will prompt a lot of interest in developing new types of forecasting tools and methods, many of which may involve online communication and the use of artificial intelligence. These areas will grow as priority areas for research in the coming two decades. At the same time, the development of these new tools and methods will reshape the research industry, including traditional scientific inquiry. Thus, a change in political practice has implications for scientific practice. The link between these two areas has certainly existed all along. But the link is going to become more apparent and is also going to become the focus of societal tensions. Science in the 1980s and 1990s is going to become increasingly politicized.

This chapter, by its very nature, is speculative. Our aim was to move to a world in which there is widespread penetration of teletext/videotex and attempt to identify how the technology would change the way we live our lives.

CHAPTER 15

CONCLUSION
AND FINDINGS

Teletext and videotex are driving forces in the evolution toward electronic delivery of information services into the home. The technologies themselves are undergoing rapid change, and it is likely that the systems that exist at the end of this century will only remotely resemble the current prototypes. Nevertheless, the present systems are already bringing together new alignments of technologies and services—all assembled within the living room.

The most prominent feature of teletext and videotex systems is their inclusiveness with respect to applications. Although conceived of initially as providing households and businesses with more information, they are now emerging as vehicles for home-based shopping and banking, electronic messaging, games, computer and financial management, and fire and security monitoring. While information retrieval will continue to be an important application of videotex, other services may prove to be as important or more important in the long run.

Several dozen teletext and videotex trials and services are under way in the United States. They utilize a number of hierarchical and limited keyword access database arrangements, several different communication networks, and a wide variety of user terminals. The one-way teletext trials are being conducted using the vertical blanking interval of broadcast and cable television channels as well as full channel cable. Teletext services may also be offered using low-power television, MDS, FM radio, and direct satellite broadcast. Two-way videotex trials have utilized the telephone network and two-way cable systems. In addition, a hybrid system, comprised of one-way cable with a telephone, appears to be a promising approach to initiating videotex services with currently available technology.

The user terminals being tested range from remote numeric keypads for use with modified television receivers to full alphanumeric keyboards and CRT computer display terminals. This, however, is only the beginning. Touch-sensitive screens and voice-activated and voice-response systems will provide substitutes for the user-unfriendliness of keyboards. In addition, basic videotex terminals will be enhanced by such devices as inexpensive hard-copy printers and floppy discs for local storage. Add to this the fact that videotex systems utilize an intelligent terminal, i.e., one with a central processing unit, and it is clear that videotex technology is moving toward a general purpose home information system that incorporates the power of a personal computer while not requiring the user to be proficient in computing or even typing!

The potential size of the U.S. market for these technologies has been widely speculated on. Teletext services are more widespread than videotex and are growing at a much greater rate. At the beginning of 1982, for example, there were more than 350,000 teletext sets in use in Britain, compared to a total of 15,000 videotex (Prestel) subscribers. Throughout Europe the number of teletext-equipped households was nearing 1 million in early 1982. If American consumers adopt teletext as readily as their European counterparts and if teletext services are as easily accessible as they are on European national television systems, it is reasonable to forecast a market approaching 40 percent of households by the early 1990s. Such a forecast assumes that cable operators will also be providing basic one-way teletext information services as part of their tiered package of offerings.

Teletext systems will offer three key services. First, the limited bandwidth broadcast systems are ideal for rapidly changing general information such as news, weather, sports, and entertainment guides. Second, the full-channel or wider-bandwidth services offer the potential for broadcasters or other information providers to offer subscription services such as electronic magazines and newsletters for distribution to those with addressable converters and scrambling devices. Third, the wider bandwidth systems also allow the down-loading of software to intelligent terminals for commercial and home use, for example, electronic games.

We identified five generic classes of applications for videotex: information retrieval, transactions, messaging, computing, and telemonitoring. The specific consumer markets that could potentially be served by videotex include some of the fastest growing markets in the nation, including direct nonstore sales, video games, home computers, and telemonitoring services, as well as other substantial print-media, nonprint-media, and bill-paying transaction markets.

The videotex market, even given the wide range of potential services, is expected to grow more slowly than that of teletext. Videotex is not emerging in response to any pent-up consumer demand but is being driven by industries that realize the potential of this technology. Banks are already heavily subsidizing consumers for checking accounts, and electronic banking presents a cost-effective alternative. Similarly, retailers are involved in exploring the potential of teleshopping. With industry-driven attempts to stimulate demand—telephone companies offering telephone terminals with electronic directory facilities for lease or purchase, banks encouraging customers by way of financial incentives to undertake home-based electronic transactions, or cable companies making information services available as part of a basic user cable package—we estimate that about 40 percent of households will have videotex services by the end of the century. The historical growth of other media in the United States tends to support such a projection in that television took 16 years from the time a commercial service was introduced to penetrate 90 percent of households, and color television has taken 27 years to reach 75 percent of households.

A key driving force for both teletext and videotex will be the extent to which advertisers are persuaded to adopt the new medium, since advertising revenues will subsidize the cost of the service, helping to reduce subscription charges to users. Teletext services, with the possible exception of public television, will be advertiser-supported. Videotex, with its interactive potential, allows for very specific targeting for advertisers. In addition, the ability to purchase goods directly and to arrange for payment through an electronic funds transfer in a single transaction makes videotex a highly attractive medium for direct marketing. The business market for videotex is very much oriented toward the provision of private videotex systems or closed-user group facilities on public systems. In the former case, videotex resembles the mini-computer market, while the latter will be one additional stimulus to the public videotex systems.

The public policy issues that are likely to arise from the emergence of teletext and videotex may be arranged into three broad classes: those that shape the technology and marketplace (*developmental issues*), those that are associated with widespread penetration of the technology (*consequential issues*), and those that transform or affect the general society given the service is adopted (*transformative effects*). How the issues will be resolved will depend in large part upon the underlying regulatory climate of the day. Videotex brings together technologies that have, traditionally, existed under different regulatory mechanisms (broadcast, telephone, cable, and print) and information services with different regulatory requirements (banking, mail, and computing). If the regulation of videotex is not to be fragmented, there is a very strong need to develop public policy from the point of view of the system rather than from that of the distribution medium.

The most significant developmental issues that arise as teletext and videotex evolve are standards, guarantee of access, competition, content regulation, and copyright. Teletext and videotex are currently on divergent paths with respect to standards. In 1981, the Federal Communications Commission decided against setting a single United States standard for teletext. Field tests of all three major teletext systems (Ceefax, Antiope, Telidon) are currently under way. Teletext systems are becoming commercial ventures before videotex, as the basic distribution and transmission systems are in place at the broadcast station.

Videotex standards, on the other hand, are being determined in national and international forums. The European nations have prepared a standard embracing the English Prestel system and the French Antiope system for presentation to the International Telegraph and Telephone Consultative Committee (CCITT). AT&T has proposed a standard derived from the Canadian Telidon system that is almost, but not exactly, compatible with the European standard. Neither standard is compatible with teletext, nor are these predominantly telephone-based standards compatible with various cable-based systems under development in the United States. Thus, there is a very real possibility of dual American and European videotex standards.

Another important developmental issue involves ensuring competition in the emerging teletext/videotex market. Newspapers, cable companies, banks, retailers, hardware manufacturers, finance houses, and broadcasters are creating organizational conglomerates to deliver videotex services. The organizational structure of this new industry is not clear. It requires careful delineation of the responsibilities of major actors with respect to content, and ongoing monitoring to ensure that small, local, and regional organizations still have access to videotex systems. The joint sponsorship of various trials is creating a new potential for anticompetitive practice because of horizontal controls by the major actors. The issues of copyright raised by videotex are not unique. But they take on special importance with videotex due to the ease of downloading databases and storing and then reformatting the information.

Consequential issues fall into two categories: those that impact the user (privacy and security, equity of access, and consumer protection) and those related to the broad economic infrastructure of the society (industry structure, employment, and international trade and communication). Just as television brought the salesperson into the household along with entertainment and information, so widespread penetration of teletext/videotex will bring the home into the marketplace. The technology will also bring with it a number of major societal changes. Videotex offers householders the opportunity to work, educate themselves (do-it-yourself) or be retrained, be informed, be entertained, shop, conduct transactions, monitor personal health, track household energy consumption, make appointments, be reminded of activities, transmit and receive messages, and vote and contribute to community affairs, all from a home-based terminal.

Thus, videotex brings a wide array of resources into the household, but it can take out information about every aspect of the household and ultimately reduce privacy and choice. For example, by monitoring the actual videotex pages that are read, it is possible to custom design new bulletins limited to only those topics the average householder reads or is interested in. No longer is there any need to report news on a variety of subjects or to present information in an objective fashion. Current regulations on privacy do not prevent a videotex system operator from compiling and making available profiles of household and possibly even individual activities such as expenditure patterns, voting patterns, viewing habits, and reading habits. Such profiles, unlike A. C. Nielsen ratings, would be available instantly, would be much more focused and comprehensive in their coverage, and would be more accurate. At present, the householder's only redress is to plead the protection of the Fourth Amendment. Telemonitoring also raises the possibility of selective household pricing for energy services and for fire, security, and health insurance.

The consequential impacts on the economic infrastructure are also likely to be important. Videotex services blur the line between previously separate businesses, raising the possibility of competition between hitherto separate industry segments. For example, videotex advertising shares characteristics with both newspaper classified advertising and yellow pages advertising and has already provoked conflict between the telephone and newspaper industries. In the area of financial services, videotex makes it possible for transaction-type services previously offered only by banks to be provided by other kinds of institutions as well. New industries will enter fields previously controlled by monopolies, such as messaging, and information service industries will emerge to support electronic shopping and financial management. Major changes will come in the customer service and distribution sides of the financial services industry. As consumers can order on demand, the distribution of goods and services will become even more interdependent and will probably add a link from the regional city to the household. There will be considerable displacement of "middlemen" in retail and financial industries. Customer service personnel, such as bank tellers, will be replaced by an electronic equivalent. Videotex requires new skills such as graphic design, page layout, color mixing, database design, electronic information "brokerage" services, and electronic quality control/audit functions. These are not yet well integrated in the curricula of teaching institutions.

The international trade and communication issue poses a genuine dilemma for U.S. industry. The creation of an international standard, while providing American industry with greater economies of scale for production, also opens up the United States to coordinated foreign competition for the hardware market such as terminals, switching, and microcomputers.

Finally, the widespread penetration of technology will bring with it a number of major societal changes:

- The dwelling unit will also become a place of employment. This not only affects the type of structure, in terms of architecture and layout, but also the geographic location.

- The electronic household will facilitate a new home-based cottage industry in electronic products.

- Home-based shopping, along with computer-aided production, will allow consumers to control the manufacturing process. Consumers will be able to order exactly what they want for "production on demand."

- The family will determine the electronic schooling (or, more correctly, the education) required for children and for retraining adults. There will be a shift away from the traditional school and work socialization processes to ones in which peer groups and alliances are electronically determined.

- New skills and career paths associated with the management of information will emerge. These will range from information brokers who provide the "best" deal on a used car to gatekeepers who monitor politicians and corporate activities and selectively release this information to interested parties.

- There will be an increase in opportunities to participate in educational, social, and political areas as the interactive capacity of the new media allows greater involvement.

These societal impacts must eventually be included in the social cost-benefit analysis for evaluating this new technology. The trade-offs that society makes, either implicitly or explicitly, are not inconsequential. In the long run, in fact, it may be these unintended or unexpected effects, rather than the more focused concerns and issues discussed in the previous chapters, that represent the most significant aspects of this technology.

One fundamental trade-off that arises again and again as we consider the impact of videotex concerns the question of *control*: this trade-off can be found on the level of individuals, of the videotex industry, and of social policy. Because they are interactive, videotex systems create opportunities for individuals to exercise much greater choice over the information available to them. Individuals may be able to use videotex systems to create their own newspapers, design their own curricula, compile their own consumer guides. On the other hand, because of the complexity and sophistication of these systems, they create new dangers of manipulation or social engineering, either for political or economic gain. Similarly, at the same time that these systems will bring a greatly increased flow of information and services into the home, they will also carry a stream of information *out* of the home about the preferences and behavior of its occupants.

This same trade-off is reflected in the choices to be made about the structure of the videotex industry: on one hand, the technology holds the promise of creating a new kind of marketplace in which many voices will be able to compete for attention. The technology is sufficiently versatile that it can accommodate many sources of information—in fact, that is one of its most salient characteristics. At the same time, there is the potential that videotex will be dominated by the same entities that dominate existing media. The economies of scale inherent in the technology may simply enable the largest organizations in publishing, banking, retailing, etc., to grow still larger.

Finally, the issue arises in terms of government regulation and control. On the one hand, videotex in the United States seems to offer an opportunity for a new medium to develop in the marketplace with a minimum of government intervention. However, because videotex encompasses a wide range of applications as well as several different distribution networks, each of which raises different concerns and entails a different legal/regulatory framework, the regulation of the technology may prove fragmentary and inconsistent in the absence of a process to develop new policies specifically designed for videotex.

Teletext and videotex clearly have the potential to influence many aspects of our lives. Though few of the issues it raises are novel or unique, they still provide a difficult challenge for policymakers. It is only through continued monitoring and anticipation of its consequences that the technology will be shaped to maximize its benefits and minimize its threats to society.

APPENDIX **ONE**

EXAMPLES OF
CHARACTER SETS

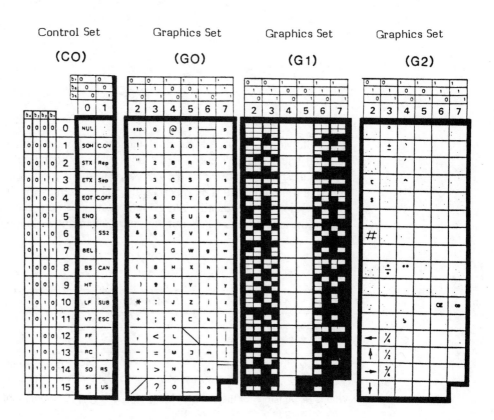

Control Set (CO)   Graphics Set (GO)   Graphics Set (G1)   Graphics Set (G2)

TELETEXT AND VIDEOTEX ASSOCIATED POLICY ISSUES

The following is a list of potential teletext and videotex issues associated with selected teletext and videotex applications. The left-hand column lists the specific policy questions identified at the project workshops discussed in Chapter 10. The right-hand yolumn indicates general policy issue themes for the specific questions. Our intention was to identify a large number of specific policy concerns and then to narrow the analysis to a number of recurring, important general policy themes.

I Information retrieval	
Application area: **1** Electronic yellow pages	
Policy question	**General policy**
Who is responsible for the content of the advertising (e.g., massage parlors)?	Content regulation
What happens if the transmission medium and the database are divorced?	Competition
What are the implications of adopting the newspaper or broadcasting analogues?	Content regulation

---

I Information retrieval

Application area:
1 Electronic yellow pages (cont.)

Policy question	General policy
What happens if the service follows other communication tendencies? (The most likely providers will be the telephone companies, newspapers, and perhaps cable or timesharing companies.)	Competition
What happens if access for all information providers is not guaranteed?	Guarantee of access
If one buys into the yellow pages, can one control who may advertise?	Guarantee of access
What happens to the national industrial structure if, with the sale of brand name items, price becomes the ultimate criterion and economies of scale benefit national and regional retailers at the expense of local merchants?	Industry structure
What are the implications of increasing the demand on the transportation system? (With the movement of more goods from a central distribution outlet there is an increased fragility of the distribution system).	Industry structure
Will there be protection against impulse buying if a system includes transactions?	Consumer protection
What are the effects of telephone and cable companies offering this EYP service on the advertising revenues and associated viability of newspapers as an alternative source of information?	Industry structure
An "enhanced" yellow pages (product promotion, purchasing) creates a structure for teleshopping that affects current retail and wholesale distribution of products. What are the implications for current products and service marketing?	Industry structure

---

Application area:
2 Education/retraining—upgrading skills, acquiring new skills

---

Is there to be a minimum standard for information literacy?	Equity of access

---

I   Information retrieval

Application area:
2. Education/retraining—upgrading skills, acquiring new skills (cont.)

Policy question	General policy
If equity of access is to be guaranteed, who chooses who gets terminals? Will public library or public school functions be expanded to provide such access? Will the poor receive "information" stamps?	Equity of access
What are the implications of greater stratification in society between the information-rich and information-poor?	Equity of access
What are the implications of a massive demand for retraining?	Employment
Are standards and accreditation of courses to be required? Who will provide the gatekeeper function for such services?	Content regulation
What are the implications of concentration of educational providers?	Industry structure
Are software owners to be protected under copyright statutes?	Copyright
What safeguards are to exist so that present or future employers might not know about an employee's skill upgrading?	Privacy and security
What are the implications of competition from national courses offered by prestigious schools on local area instruction (e.g., job losses, focus of cultural events, industry attraction)?	Industry structure

Application area:
3 "Storefront medicine" (expert systems accessed by professionals or consumers with details on diagnosis, treatment, pharmaceuticals, health risk, cost, and efficacy of alternatives)

Policy question	General policy
Who is responsible for the reliability of the information? Publishers, carriers, information providers? i.e., who would be sued for malpractice?	Content regulation
What are the implications of governments using the system to monitor health costs?	Privacy and security
Would there be a requirement for pharmaceutical companies, for example, to describe side and harmful effects of drugs?	Consumer protection

I Information retrieval

Application area:

3 "Storefront medicine" (expert systems accessed by professionals or consumers with details on diagnosis, treatment, pharmaceuticals, health risk, cost, and efficacy of alternatives) (cont.)

Policy question	General policy
Could such information systems contravene basic public decency codes (e.g., graphics of bodily functions)?	Consumer protection
What happens if there is censorship of material (e.g., from children, drug addicts, alcoholics)? Who would enforce community standards (e.g., an implicit censor as in library or individual electronic censors)?	Consumer protection
Does increased access to information create additional concerns for families and special-interest groups?	Consumer protection
How are good taste and ethical aspects guaranteed? Right to information vs. professional societies protectionism.	Consumer protection
What are the implications of restricting access to specified individuals?	Equity of access
Who has access to all or some of the medical records? Could employers check on disease/illness patterns of potential employees?	Privacy and security
Would drug abusers and infectious-disease carriers be immune to identification?	Privacy and security
What are the implications for existing health care industries?	Industry structure
Would there be serious professional shortages/imbalances due to occupational protectionism?	Employment
Nationally coordinated systems may allow the user access to the best medical practitioner and viability of existing local professional structure. What happens to the community professional support?	Industry structure
What are the market/ethical implications of professional advertising? Can clear distinctions be drawn between information (e.g., on a drug's healing powers) and advertising?	Content regulation

---

I Information retrieval

Application area:

4 Multiple listing services (e.g., as real estate or travel packages)

Policy question	General policy
Are new legal protections needed against fraud and piracy of data in these high-value-added services?	Privacy and security
What are the protections or redresses against fake and misleading electronic advertising?	Consumer protection
Are there safeguards over user profiles and use statistics?	Privacy and security
Are there cancellation mechanisms for erroneous input or changes of mind?	Consumer protection
What is the impact on professional employment? Will the size of the "middleman" class be reduced? Will there be widespread do-it-yourself endeavors?	Employment
Is the "little person" still guaranteed equal access to advertise and sell or will the new skills and resources required create financial penalties?	Guarantee of access
Who gets the benefits of efficiency of production, that is, the resources currently consumed by the middle retail industries—the consumer, the carrier, the information provider, or the owner of the product?	Industry structure

Application area:

5 Real-time job market (identifying positions, matching skills to vacant positions)

What are the implications of numerous competing employment agencies?	Industry structure
What are the effects on equity of access for minority groups? What happens to fairness in employment opportunities, in the spirit of EEO?	Equity of access
The widespread market in information services raises the concept of property rights for the product (information) being traded. What happens if these databases are readily accessible in the home?	Copyright
Individual privacy is a two-edged sword. What are the implications of allowing employees, governments, etc., access to individual data? Does complete	Privacy and security

---

I Information retrieval

Application area:
5. Real-time job market (identifying positions, matching skills to vacant positions) (cont.)

Policy question	General policy
privacy, if enforceable, mitigate against appeals or disputes for unfair treatment?	
Can the market be used to discriminate against certain applicants, as through selective access?	Equity of access
What protection does an employer have against discrimination appeals?	Consumer protection
A national system of real-time job matching implies a mobile labor force. Will people move to take the job openings? Will the increase in worker movement create societal and local tensions? What are the implications for support industries and services— housing, commerce—and professions—teachers, medics—and social support?	Industry structure/Employment

---

Application area:
6 Consumer information for comparative pricing and performance

---

The ready availability and easy changing of price information by product suppliers may lead to a convergence of product prices and other competitive features. What are the implications of the elimination of price and present competitive features?	Industry structure
What are the implications for smaller firms or suppliers with smaller economies of scale?	Industry structure
What are the effects of market or government enforcement of quality standards or measures for electronically described products?	Consumer protection
What is the analogue of malpractice for electronic information?	Consumer protection
Will goods have to be sold at the price as "marked" or "quality as indicated"? Can caveats that the information may not be current or that the assurances are not scientifically verified alter this? What are the implications of having a floating electronic price for all products?	Consumer protection

I Information retrieval	

Application area:
7 Electronic library—a universal book information delivery service

Policy question	General policy
What are the implications of increased demand for the loaning of physical books by the public library system? What are the impacts on library budgets and the delays on access?	Industry structure
Will "universal access" be required?	Equity of access
Does the electronic down-loading of loaned books violate copyright statutes?	Copyright
What are the implications of maintaining and enforcing copyright statutes?	Copyright
What are the implications of private information providers siphoning off the more lucrative areas of public library services?	Industry structure

II Electronic Messaging	

Application area:
8 Electronic mail

How does one assure the consumer that his or her communication is secure (e.g., the analogy of licking the envelope flap)?	Privacy and security
What are the implications in the use of various coding/decoding security systems (the public/private key encryption algorithm developed at Stanford University has been side-tracked under intense pressure from the National Security Agency)?	Privacy and security
What happens if vendors are allowed to nominate levels or degrees of privacy and to change them accordingly?	Privacy and security
Who is allowed to put what things in an electronic mailbox?	Content regulation
Is the carrier required to provide a means for preventing objectionable material from entering the mailbox (e.g., obscene material)?	Content regulation
What happens if the anonymity of the sender cannot be preserved?	Privacy and security
Is it possible to develop individually tailored censors, i.e., to electronically censor obscene or unwanted mail?	Consumer protection

II Electronic Messaging	

Application area:
8 Electronic mail (cont.)

Policy question	General policy
With 85 percent of all mail computer-prepared, the demand for the U.S. Postal Service will be significantly reduced. What is the role for the post office?	Industry structure
What are employment retraining implications for postal union members?	Employment
The proliferation of private systems may lead to compatibility problems. What happens if the telephone analogy is chosen (leading to a regulated industry) or the FAX machine analogy is selected (which converts to an ITT protocol as standard)? In the latter case, how, if at all, is the national (or international) standard arrived at and when?	Standards/International trade and communication
Who are the "gatekeepers" for electronic mail? What tasks do they perform and what responsibilities do they have?	Content regulation
What happens if there is to be "universal" access? How are rates and costs to be determined?	Equity of access
For the less advantaged, how will terminals and interconnection be provided? Will a two-class system of electronic mail vs. more expensive paper-based mail lead to a widening between the information-rich and information-poor? Will there be a "mail stamp" program income-based policy?	Equity of access
What is the effect of allowing teletext and/or videotex electronic mail services to be implemented as national or international monopoly services?	Market structure/International trade and communication
Who owns the electronic message during transmission or storage?	Content regulation
Can the service provider restrict the types of messages transmitted or is this a violation of the First Amendment (e.g., obscenity)?	Content regulation
Who has access to electronic directories?	Guarantee of access

---

III Transactions

Application area:

9 Electronic funds transfer—home-based billing, banking, and transactions

---

Policy question	General policy
What mechanisms prevent access to accounts? Who monitors? Single or multiple personal codes? Encryption devices?	Privacy and security
Who is responsible for the transaction?	Content regulation
What guarantees are there for confidential transactions? Who ensures?	Privacy and security
What is the legality of a non-paper-based computer record in a court of law? For example, current legal requirements for record keeping (type, length of time) are not derived from an electronic transaction-based society.	Consumer protection
Is the current statute of limitations with respect to redress still appropriate? What happens if it is not?	Consumer protection
Are there rights for minors within families regarding parental access to details of transactions?	Consumer protection
Who ensures that what is purchased matches what was requested? How are disputes resolved?	Consumer protection
How can defective or unsatisfactory information products be returned (e.g., not wishing to watch or pay for a movie)? What are the terms for a return of goods?	Consumer protection
What protection does the purchaser and supplier have against unauthorized electronic purchase (e.g., children ordering without parental approval)?	Consumer protection
What are the implications for widespread impulse buying?	Consumer protection
What are the implications of an increased velocity of money in circulation for inflation and economic cycles?	Industry structure
Can products be sold (electronically) prior to being produced? Does this affect the structure of the support industries (e.g., warehousing, transportation)?	Industry structure

---

III Transactions

Application area:

9 Electronic funds transfer—home-based billing, banking, and transactions (cont.)

Policy question	General policy
What are the effects of "producing on demand"? With no local inventory slack in the system, is there increased vulnerability to transportation and warehousing delays?	Industry structure
Will easier access to consumer behavior such as ordering and canceling or "not showing" result in pricing tiers, payment policies, and service options based on an individual's past record?	Consumer protection
What new regional or national implications are there from local disruptions (e.g., strikes) given the nature of the production and distribution cycles?	Industry structure
What happens if there are many noncompatible systems offering partial services? This issue is relevant to every application area.	Standards
What are the implications of no "soft edges" in a system whose nature is precision (e.g., extension of payment schedules in periods of layoffs)?	Consumer protection

---

IV Computing

Application area:

10 Video games and interactive gambling (e.g., off-track betting on horses)

What happens if signals to consumers on quotations are intercepted?	Privacy and security
Who guarantees that participants receive their rightful winnings (e.g., what mechanics can be developed to prevent electronic theft)?	Consumer protection
Who is responsible for the conduct of the activity (e.g., to ensure games are "above board")? What are the implications of many government or private controllers?	Content regulation
What happens if a new, broader credit base for all classes in society is introduced? Does it increase the velocity of money and hence inflation?	Societal impact

---

IV Computing

Application area:
10. Video games and interactive gambling (e.g., off-track betting on horses) (cont.)

Policy question	General policy
Are there implications for widespread "indebtedness," bankruptcy, and financial irresponsibility?	Societal impact
What happens if the telephone or cable networks are unable to handle popular interactive services such as an interactive lottery?	Industry structure/Standards
What are the implications for economic productivity, the work ethic, family life, and societal interactions?	Societal impact
Would participants be guaranteed anonymity against government, employers, colleagues, and organizations?	Privacy and security

---

V Telemonitoring

11 Electronic mother (A cradle-to-grave information "assistance" system and database. This is an extreme example introduced to demonstrate the positive and negative effects of electronic monitoring and surveillance.)

Will service have a deleterious effect on adult/child relationships?	Societal impact
Will children become even more passive than they are now (e.g., with television)?	Societal impact
Will it be a threat to family structure by eroding the authority of parents?	Societal impact
Will it challenge societal values by making explicit individual attitudes, wants, and needs over a long period, as contrasted with society's norms?	Societal impact
What will be the effect on interpersonal relationships? On human contact and loneliness?	Societal impact
Will the service offer the user a behavioral and social profile that can be manipulated by an external party or harnessed by the individual either for good or evil?	Privacy and security
What are the safeguards for unauthorized access to user attitudes and beliefs encoded in the database?	Privacy and security
What are the implications of external monitoring of extreme, antisocial, or unacceptable behavior patterns?	Privacy and security

BIBLIOGRAPHY

Adler, Richard, Michael Nyhan, and John Tydeman: *Report: Workshop on Electronic Directories/Electronic Yellow Pages*, paper P-119, Institute for the Future, June 1981.

Ahlhauser, John W.: *The Electronic Newspaper: U.S. Editor's Reactions to Teletext*, Research Report No. 9, School of Journalism, Indiana University, Bloomington, Ind., 1979.

Allan, Roger: "New Applications Open Up for Silicon Sensor: A Special Report," *Electronics*, November 6, 1980.

Allen, John: "Special Report: Cassette and Cartridge Tape Drives," *Computer Business News*, July 7, 1980, pp. 1–13.

Allen, John: "Speech Recognition Gear Pronounced Immature," *Computer Business News*, February 23, 1981(*a*), p. 6.

Allen, John: "Sophisticated Chips Enhance Speech I/O," *Computer Business News*, February 2, 1981(*b*), pp. 1, 18.

Allen, John: "N-Channel MOS RAM Growth Booms," *Computer Business News*, April 6, 1981(*c*), pp. 1, 6.

Allen, John: "Winchester Drive Paces Dynamic, Chaotic Mart," *Computer Business News*, October 12, 1981(*d*), pp. 1, 11.

Alternate Media Center: *Access Time and Reception Quality in the Field Trial in Washington, D.C.*, Research on Broadcast Teletext, working paper no. 1, September 1981.

Anderson, James C.: "An Extremely Low-Cost Computer Voice Response System," *Byte*, February 1981, pp. 36–43.

Applebaum, Simon: "Videotex: A New Bandwagon for Cable?" *Cablevision*, November 10, 1980, pp. 24–30.

Aranda, R. R.: "The Home Office: Videotex and Personal Computers for Office Automation," *Telecommunications*, September 1981, pp. 43ff.

Arlen, Gary: "Everyman's On-Line Information Resource," *Watch*, May 1980, pp. 28–32.

Arlen, Gary: "Getting Ready for Videotex, Teletext and the World Beyond," *Cablevision*, June 1, 1981, pp. 233ff.

Auerbach, Lewis: "The Distinction Between Carriage and Content," *Telecommunications Policy*, vol. 5, no. 1, 1981, pp. 3–11.

Background Papers, Workshop on Teletext and Videotex in the U.S.: Institute for the Future, Menlo Park, Calif., June 1979.

Baker, Jeri: "The Low-Power Story: FCC's Grass Roots Ploy Sprouts Confusion Over Television's Future," *CableVision*, April 20, 1981a, pp. 60–67.

Baker Jeri: "Cable Stakes Claim to Unmined Markets," *Cablevision*, June 1, 1981b, pp. 154–206.

Baldwin, Thomas R., Thomas A. Muth, and Judith Saxton: "Public Policy in Two-Way Cable," *Telecommunications Policy*, vol. 3, no. 2, 1979, pp. 126–133.

Ball, A. J. S., G. V. Bochmann, and Jan Gessel: "Videotex Networks," *IEEE Transactions*, December 1980, pp. 8–14.

Bankers Desk Reference: Warren, Gorham, and Lamont, Boston, 1978.

Baran, Paul: *Potential Market Demand for Two-Way Information Services to the Home, 1970–1990*, Report R-26, Institute for the Future, Menlo Park, Calif., 1971.

Bassak, Gil: "Low Data Rate Yields Lifelike Voice," *Electronics*, March 24, 1981, pp. 48–57.

Beere, Max P.: "The Power of Telecommunications: We've Unleashed It. Can We Control It?" *Computerworld Extra!*, vol. XV, no. 11a, 1981, pp. 5–12.

Bell, Daniel: "The Information Society," in Tom Forester (ed.), *The Microelectronic Revolution*, M.I.T. Press, 1980.

Bernstein, Peter W.: "Television's Expanding World," *Fortune*, July 2, 1979, pp. 64–69.

Bernstein, Peter W.: "Atari and the Video-Game Explosion," *Fortune*, July 27, 1981, pp. 40–46.

Bigi, F., G. Bonaventure, and S. Senmoto: "Optical Fibre Studies in the CCITT," *Telecommunication Journal*, vol. 46, VIII, 1979, pp. 487–491.

Blank, John: "System and Hardware Considerations of Home Terminals with Telephone Computer Access," *IEEE Transactions on Consumer Electronics*, vol. CE-25, no. 3, 1979, pp. 311–317.

Bloom, L. R., A. G. Hanson, R. F. Linfield, and D. R. Wortendyke: *Videotex Systems and Services*, National Telecommunications and Information Administration Report 80-50, U.S. Department of Commerce, Washington, D.C., October 1980.

Bolt, Beranek and Newman Inc.: *Communications Technology Forecast*, report no. 4037, prepared for The National Library of Medicine, Bethesda, Md., January 1979.

Bown, H. G.: "The Canadian Videotex System—Telidon," *IEEE, ICC '79 Conference Record*, 1979, pp. 56.3.1–56.3.4.

Bown, H. G., C. D. O'Brien, W. Sawchuk, and J. R. Storey: *A General Description of Telidon: A Canadian Proposal for Videotex Systems*, Department of Communications (Canada), Ottawa, 1978.

Brandon, George: "FCC May Exempt Teletext from Fairness Rules," *Editor & Publisher*, December 12, 1981, p. 10.

Branscomb, Lewis M.: "Information: The Ultimate Frontier," *Science*, vol. 203, January 12, 1979, pp. 143–147.

Branscomb, Lewis M.: "Computer Technology and the Evolution of World Communications," *Telecommunications Journal*, vol. 47, no. IV, 1980, pp. 206–210.

Braunstein, Yale M.: "The Functioning of Information Markets," *Issues in Information Policy*, Chap. IV, U.S. Department of Commerce, National Telecommunications and Information Administration, Washington, D.C., February 1981.

Braunstein, Yale M., and Laurence J. White: *Setting Technical Compatibility Standards: An Economic Analysis*, Brandeis Economics Research Center, Department of Economics, April 1981.

Broad, William J.: "Upstart Television: Postponing a Threat," *Science*, November 1980, pp. 611–615.

Broadcasting: "Teletext: TV Gets Married to the Printed Word," August 20, 1979, pp. 30–35.

Broadcasting: "The Curtain's Going Up on DBS: Television's next frontier," September 15, 1980(*a*), pp. 36–46.

Broadcasting: "MDS: Growing Older, Wiser and a Little Concerned," October 13, 1980(*b*), pp. 60–62.

Broadcasting: "DBS Down to Business," December 8, 1980(*c*), p. 7.

Brown, Merrill: "Information Industry's Future Is at Stake," *Washington Post*, March 5, 1981.

Brown, James A., Jr., and Kenneth Gordon: *Economics and Telecommunications Policy: A Framework for Analysis,* Federal Communications Commission, Office of Plans and Policy, December 1980.

Bulkeley, W. M.: "Competition to Equip the Paperless Office Heats Up as Potential for Devices Expands," *Wall Street Journal*, May 6, 1981.

Business Week: "Videodiscs: A Three-Way Race for a Billion Dollar Jackpot," July 7, 1980(*a*).

Business Week: "America's New Financial Structure," November 17, 1980(*b*).

Business Week: "TV: A Growth Industry Again," February 23, 1981(*a*), pp. 88–91.

Business Week: "The Home Information Revolution," June 29, 1981(*b*), pp. 74–83.

Business Week: "Why the Market Burst for Bubble Memories," September 7, 1981(*c*).

Business Week: "Electronic Banking," January 18, 1982, pp. 70–80.

Bygrave, Mike: "Writing on an Empty Screen," *InterMedia*, vol. 7, no. 3, 1979, pp. 26–28.

Cablevision: "TOCOM Stresses Marketing, Mass Production for Two-Way System," November 3, 1980, pp. 31–33.

Cablevision: "Japan Develops Camera for High-Definition TV," March 16, 1981(*a*).

Cablevision: "Cable Gets the Word; Getting Ready for Videotex, Teletext and the Interactive Worlds Beyond," June 1, 1981(*b*), pp. 233ff.

Cablevision: "Order via Videotex: French Retailer Adopts Internal Videotex System," December 7, 1981(*c*).

Campbell, James: "Videotex in Business Applications," paper presented at Socioscope Conference on Telidon and Its Applications, Ottawa, Ontario, September 29, 1980.

Campbell, James A., and Hilary B. Thomas: "The Videotex Marketplace: A Theory of Evolution," *Telecommunications Policy*, vol. 5, no. 2, June 1981, pp. 111–120.

Carey, John: *Teletext and Public Broadcasting*, Alternate Media Center, New York University, New York, 1980.

Carey, M. L.: "Occupational Employment Growth Through 1990," *Monthly Labor Review*, August 1981.

Carpenter, Ted: *Electronic Information Systems*, Office of Consumers' Education, U.S. Department of Education, Washington, D.C., 1980.

CBS: "Petition for Rulemaking," submitted to FCC re Amendment of Part 73, Subpart E of the Rules Governing Television Broadcast Stations to Authorize Teletext, July 29, 1980.

CCITT Recommendation S.100: "International Information Exchange for Interactive Videotex," Geneva, November 1980.

CCITT Recommendation F.300: "Videotex Service," Geneva, November 1980.

CEPT, Subworking Group CD/SE, Recommendation no. T/CD 6−1: "European Interactive Videotex Service Display Aspects and Transmission Coding," June 1981.

Champness, Brian: "Social Uses of Videotex and Teletext in UK," *Videotex '81*, Online Conferences Ltd., and Infomart, 1981.

Champness, Brian, and Marco DiAlberdi: *Measuring Subjective Reactions to Teletext Page Design*, Alternate Media Center, New York University/School of the Arts, New York, September 1981.

Chapman, T.: "Videotex Applications: An International Review of Current and Potential Market Penetration," *Viewdata '81*, Online Conferences Ltd., 1981.

Chase, Susan: "Makers Bet Millions That Big TV Screens Will Be the Next Rage in Home Entertainment," *Wall Street Journal*, March 9, 1981.

Chen, Kan, Kenan Jarboe, and Janet Wolfe: "Long-Range Scenario Construction for Technology Assessment," *Technological Forecasting & Social Change*, vol. 20, no. 1, August 1981, pp. 27−40.

Cherry, Susan Spaeth: "Teleconference: The New TV Information Systems," *American Libraries*, February 1980, pp. 94−110.

Chitnis, A. M., and J. M. Costs: "Videotex Services: Network and Terminal Alternatives," *IEEE Transactions on Consumer Electronics*, vol. CE-25, no. 3, 1979, pp. 269−277.

Christakis, Alexander, Samuel Globe, and Kazuhiko Kawamura: "Choosing Topics for Technology Assessment," *IEEE Transactions on Systems, Man, and Cybernetics*, vol. SMC-10, no. 1, January 1980, pp. 1−7.

Ciciora, Walter S.: "Twenty-Four Rows of Videotex in 525 Scan Lines," *Viewdata '81*, Online Conferences Ltd., 1981.

Ciciora, Walter, Gary Sprignoli, and William Thomas: "An Introduction to Teletext and Viewdata with Comments on Compatibility," *IEEE Transactions on Consumer Electronics*, vol. CE-25, no. 3, July 1979, pp. 235−245.

Codd, E. F.: "SQL/DS: What It Means," *Computerworld*, February 16, 1981, pp. 27−30.

Cohen, Tedd A.: "For Whom Does the Bell Toll?" *Forbes*, January 19, 1981, p. 40.

Collins, Hugh: "Forecasting the Use of Innovative Telecommunications Service," *Futures*, April 1980, pp. 106−112.

Colton, K. W., and K. L. Kraemer: *Electronic Fund Transfers (EFT) and Public Policy: Formulating an Agenda for Research*, University of California, Irvine, August 1978.

Communication Studies and Planning, Ltd.: "Teletext Standards in the USA: The Electronic Industries Association Approaches Decision Time," *In Context*, report 3, May 1980.

Computer Business News: "Graphics Terminals Seen Heading for '80s Success," May 14, 1979, p. 11.

Computer Business News: "Special Report: Rigid Disk Drives," February 4, 1980(*a*), pp. 9−16.

Computer Business News: "Low Cost Printer Sales Forecast at $300 Million," December 22, 1980(*b*), p. 18.

Computer Business News: "Nonimpact Market Growth Forecast at 23 Percent/Year," January 26, 1981, p. 26.

Computerworld: "Drops in Disk Drive Prices Likely to Continue," February 12, 1979, pp. 79−82.

Computerworld: "The Hard Disk Industry: A Strategic Analysis," February 18, 1980, p. 66.

Computerworld: "Report Forecasts 23% Annual Growth in Nonimpact Printers Through 1985," February 16, 1981(*a*), p. 67.

Computerworld: "Canada's Videotex Providing Government Data," October 19, 1981(*b*), p. 18.

Cooney, John E.: "With Video Shopping Services, Goods You See on the Screen Can Be Delivered to Your Door," *Wall Street Journal*, July 14, 1981.

Cooper, Robert B.: "Explosive Growth of Low-Cost Satellite Television Technology," *IEEE News*, January 1980, p. 5.

Cooperative Extension Service, University of Kentucky: "Green Thumb: A Test of an Information Delivery System for Farmers," Lexington, Ky., (unpublished), 1980.

Cotton, Ira W.: "Making Machines Talk: Simulated Speech, Part Two," *Data Communications*, February 1981, pp. 101–106.

Criner, Kathleen M.: "U.S. Videotex Activities and Policy Concerns," *Telecommunications Policy*, vol. 4, no. 1, 1980, pp. 3–8.

Criner, Kathleen M.: "Videotex: Implications for Cable TV," in Mary Louise Hollowell (ed.), *The Cable/Broadband Communications Book 1980–81*, vol. 2, Communications Press, Washington, D.C., 1980.

Crowther, G. O.: "Teletext and Viewdata Systems and Their Possible Extension to Europe and USA," *IEEE Transactions on Consumer Electronics*, vol. CE-25, no. 3, July 1979, pp. 288–294.

Data Communications: "Users Turning to Packet Radio for Future Communications Needs," August 1980, pp. 32–34.

Datamation: "Data Communication Carriers," August 1980, p. 107ff.

Date, C. J.: *An Introduction to Database Systems*, 3d ed., Addison-Wesley, Reading, Mass., 1981.

Dawson, Fred: "Wire Wars—A Cable Fable," *Cablevision*, December 15, 1980, pp. 72–118.

Dawson, Fred: "The Dilemma on Cable's Two-Way Street: Is Now the Right Time to Go with Interactive, Addressable Technology?" *Cablevision*, March 2, 1981, pp. 26–29.

De Peyster, Deborah: "Tape Streams in for Winchester Backup," *Computer Business News*, September 17, 1979, pp. 1, 5.

Doran, H. Michael: "The Old Order Changes," *Telecommunications: Trends and Directions*, E.I.A., 1980, pp. 59–67.

Dordick, Herbert S., Helen G. Bradley, Burt Nanus, and Thomas H. Martin: "Network Information Services: The Emergence of an Industry," *Telecommunications Policy*, vol. 3, no. 2, 1979, pp. 217–234.

Easton, Anthony T.: "Viewdata—A Product in Search of a Market?" *Telecommunications Policy*, September 1980, pp. 221–224.

The Economist: "Generation Game: 32-bit Microprocessors," February 28, 1981, pp. 84–85.

Eger, John M.: "The Internation Information War," *Computerworld*, March 18, 1981, pp. 103–109.

Electronic Industries Association: *Telecommunications: Trends and Directions*, Electronic Industries Association, Washington, D.C., 1980.

Electronic Publisher: "Franchising Process Reveals Clues to Text's Future on Cable," no. 11, February 24, 1982.

Elton, Martin: *Access Time and Reception Quality in the Field Trial in Washington, D.C.*, Research on Broadcast Teletext, working paper no. 1, Alternate Media Center, New York University/School of the Arts, New York, September 1981.

Emmy Magazine: "Television in the Eighties," Winter 1980, pp. 20–28.

FCC: "Statistics of Communications Common Carriers: Year Ended December 31, 1979," Government Printing Office, 1980.

Fedida, Sam: "Viewdata: An Interactive Information Service for the General Public," *European Computing Conference on Communications Networks*, 1975, pp. 261–282.

Fedida, Sam: "Viewdata," *Wireless Word*, part 1: vol. 83, February 1977, pp. 32–36; part 3: April 1977, pp. 65–69.

Fedida, Sam, and Rex Malik: *The Viewdata Revolution*, Wiley, New York, 1979.

Fentiss, R. H.: "The Impact of Satellite Communications on Distributed Data Processing," *Proceedings of the Fifth International Conference on Computer Communication*, North-Holland Publishing Co., Amsterdam, 1980, pp. 171–175.

Ferrarini, Elizabeth: "Videotex: The Race to Plug In," *Computerworld*, March 18, 1981, pp. 64–84.

Flohrev: "Electronic Mail: Communication of Tomorrow," *Telephony*, March 12, 1979, pp. 89–92.

Ford, Michael L.: "Prestel—The British Post Office Viewdata Service," *IEEE ICC '79 Conference Record*, 1979(*a*), pp. 56.5.1–56.5.4.

Ford, Michael L.: "The Search for International Standards," *Intermedia*, vol. 7, no. 3, 1979(*b*), pp. 48–50.

Frank, Howard: "Telecommunications: 1970–1990," *Computerworld Extra!*, vol. XV, no. 11A, 1981, pp. 85–88.

Fred Cohen and Associates: *Public Broadcasting and Videotex/Teletext: The Present and Future*, Fred Cohen and Associates, Washington, D.C., 1979.

Futures Research Institute: *The Multiple Perspective Concept with Applications to Technology Assessment and Other Decision Areas,* report 81–1A, Portland State University, Portland, Ore., September 1981.

Gelder, M.: "Local Measured Service—A Regulation's Viewpoint," *Telephone Engineer and Management*, January 1, 1980, pp. 91–95.

Gillmor, Donald, and Jerome Barron: *Mass Communication Law, Cases and Comment*, 2d ed., West, St. Paul, Minnesota, 1974.

Goddard, J.B.: "Technology Forecasting in a Spatial Context," *Futures*, April 1980, pp. 90–105.

Godfrey, D., and D. Parkhill (eds.): *Gutenberg 2: The New Electronics and Social Change*, 2d ed., Press Porcepic, Toronto, 1980.

Goldman, Ronald J.: "Demand for Telecommunications Services in the Home," *Telecommunications Policy*, vol. 4, no. 1, 1980, pp. 25–30.

Gollin, Albert E.: "Who Will Pay? Consumers and Advertisers in the Electronic Marketplace," paper read at conference, titled "Electronic Home News Delivery: Journalistic and Public Policy Implications," Indiana University, November 7, 1980.

Gotlieb, C. C., and Z. P. Zeman: *Towards a National Computer and Communications Policy: Seven National Approaches*, Institute for Research on Public Policy, Toronto, 1980.

Green, Roger: "Post Office Gives Viewdata a Wrong Number," *New Scientist*, October 30, 1980, pp. 300–303.

Greenhouse, Lee R.: "Communications and Networking Invade the Home Front," *Data Communications*, February 1981, pp. 56–59.

Grundfest, Joseph, and Walter Baer: *Regulatory Barriers to Home Information Services*, P-6106, The Rand Corporation, 1978.

Grundfest, Joseph, and Stuart Brotman: *Teletext and Viewdata: The Issues of Policy, Service, and Technology*, Aspen Institute Workshop Report, New York, 1979.

Gussman, T.K.: *Consumer Issues to Be Addressed in the Implementation of Electronic Fund Transfer Systems*, Task Force on Consumer Credit, Vancouver, British Columbia, March 1980.

Hamilton, L. Clark: "Copyright Laws—Will They Be Effective in an Information-Oriented Society?" *Information Utilities '81*, New York, March 1981.

Hamilton, P., and T. Manuel: "Relational Databases Do It More Easily," *Electronics*, March 24, 1981, pp. 102–103.

Hanson, Göte, and Riccard Montén: "Getting to See: Political Information on TV," Sveriges Radio, Stockholm, 1980.

Harris, A. B.: *Guide to Videotex Presentation-Level Standards*, Prestel, London, 1981.

Head, Sydney W.: *Broadcasting in America*, 2d ed., Houghton Mifflin, Boston, 1972.

Hirsch, Phil: "GTE Unveils Videotex Terminal for Office Use," *Computerworld*, February 16, 1981, p. 10.

Hirsch, Phil: "Videotex in U.S.: The Portland Project," *Computerworld Extra!*, vol. XV, no. 11*a*, 1981, p. 64.

Hooper, Richard: "An IP's View: Six Propositions," *InterMedia*, vol. 7, no. 3, 1979, pp. 17–21.

Hopengarten, Fred: "Direct Satellite TV Reception," *Telecommunications: Trends and Directions*, EIA, 1980, pp. 99–106.

Hutin, Francois Regis: "Informatics Is a Political Issue," *InterMedia*, vol. 9, no. 1, 1981, pp. 17–19.

Hutt, G.: "BL's In-House Viewdata Service," *Viewdata '81*, Online Conferences Ltd., Northwood, U.K., 1981.

IEEE Transactions on Consumer Electronics: "Special Issue: Consumer Text Display Systems (Teletext and Viewdata)," vol. CE-25, no. 3., July 1979.

Information Industry Association: "The 'Invisible' Information Industry Turns Out to Be a $9.4 Billion Industry," press release, September 30, 1980.

Institute for the Future: *Ten-Year Forecast*, Institute for the Future, Menlo Park, Calif., 1981.

Intelmatique press release: October 1981

Interim Report of the Teletext Subcommittee of the Electronic Industries Association, Broadcast Television Systems Committee, Electronics Industries Association, Washington, D.C., December 1, 1981.

Intermedia: "Special Survey: Videotex," vol. 7, no. 3, 1979, pp. 6ff.

International Institute of Communications: "France: La Telematique; The Technological Bubble," *InterMedia*, vol. 9, no. 1, 1981, pp. 12–16.

International Resource Development, Inc.: "Growing Keyboard Market," *Technology Forecasts*, November 1980.

International Resource Development, Inc.: *On-Line Data-Base Services*, International Resource Development, Norwalk, Conn., 1981.

International Videotex Teletext News: February 1981(*a*).

International Videotex Teletext News: November 1981(*b*).

International Videotex Teletext News: February 1982.

Irwin, Manley R., and John D. Ela: "U.S. Telecommunications Regulation: Will Technology Decide?" *Telecommunications Policy*, vol. 5, no. 1, 1981, pp. 24–32.

Jackson, Robert S.: "Industries in Conflict," *Computerworld*, April 16, 1979.

Johansen, Robert: "The Common Sense of Trying Out Videotex," *Viewdata '80*, Online Conferences Ltd., 1980.

Johansen, Robert, Michael Nyhan, and Robert Plummer: "Issues and Insights for the U.S.A.: Report of a Workshop," *Telecommunication Policy*, vol. 4, no. 1, 1980, pp. 31–41.

Kalba, K. K.: "The Impact of the New Media Technologies on Consumer Access to Market Information," *Proceedings of the Fifth International Conference on Computer Communication*, North-Holland Publishing Co., Amsterdam, 1980, pp. 605–610.

Kar, Saroj K.: "Closing the Gap: Compatibility with SNA," *Computerworld Extra!*, vol. XV, no. 11*a*, 1981, pp. 89—91.

Kawamura, Kazuhiko: *Technology Assessment or Technology Hindrance?*, occasional paper no. 17, Battelle Memorial Institute, Columbus Division, February 1981.

Kimmel, Hans: "Germany: A Battle Between Broadcasters and the Press," *InterMedia*, vol. 7, no. 3, 1979, pp. 39—40.

Kumamoto, Takao, and Tadashi Kitamura: "Captain System: Videotex in Japan," *IEEE 1980 International Conference on Communications*, June 1980, pp. 3.7.1—3.7.5.

Lachenbruch, David: "Your Complete Guide to VCRs," *Panorama*, February 1980, pp. 68—73.

Larratt, Richard (ed.): *Inside Videotex: The Future . . . Now*, Proceedings of seminar held March 13—14, 1980, Infomart, Toronto, 1980.

Leduc, Nicole F.: "Teletext and Videotex in North America: The Canadian Perspective," *Telecommunications Policy*, vol. 4, no. 1, 1980, pp. 9—16.

Lemasters, John N.: "Satellite Communications," *Telecommunications: Trends and Directions*, EIA, 1980, pp. 41—46.

Lerner, Eric J.: "Microcomputer Standards: Weighting the Pros and Cons," *IEEE Spectrum*, May 1981, pp. 47—50.

Lindsay-Smith, I.: "Newspapers and the Electronic Challenge," *Electronics and Power*, November/December 1979, pp. 784—787.

Lipinski, Hubert, John Tydeman, and Laurence Zwimpfer: *Teletext and Videotex Standards for the United States: Report of a Policy Workshop*, paper P-122, Institute for the Future, Menlo Park, Calif., June 1981.

Little, Arthur D.: *The Consequences of Electronic Funds Transfer*, Report to National Science Foundation NSF-C844, June 1975.

Logue, T. J.: "Teletext: Towards an Information Utility?," *Journal of Communication*, vol. 29, no. 4, 1979, pp. 58—65.

Loveless, William, and Gary Robinson: *KSL-TV Test Results and NTSC Broadcast Teletext Standards*, Salt Lake City, 1980 (unpublished).

Machalara, Daniel: "More Publishers Beam Electronic Newspapers to Home Video Sets," *Wall Street Journal*, January 2, 1981, p. 1.

Madden, John: *Videotex in Canada*, Department of Communications, Canada, 1979.

Mahn, Terry G.: "Rules Impede Teletext," *Information World*, October 1979, pp. 19—20.

Mahony, Sheila, Nick Demartino, and Robert Stengel: *Keeping Pace with the New Technologies*, VNU Books, New York, 1980.

Manildi, A. Bruce: "Tape Storage for Minis and Micros—A Technology That Has Come of Age," *Computer Business News*, July 7, 1980, pp. 8—9.

Marchand, Donald: "Privacy, Confidentiality, and Computers," *Telecommunications Policy*, September 1979, pp. 192—208.

Marcus, Stanley: "Keeping Pace with the Marketing Revolution," *Broadcasting*, April 13, 1981, p. 24.

Marsh, Donald J.: "Optical Communications—An Update," *Telecommunications Trends and Directions*, EIA, 1980, pp. 47—57.

Marshall, Martin: "Computer, Video Disk Link to Form System," *Electronics*, March 24, 1981, pp. 42—44.

Marti, Bernard: "France: A Total Approach to All Systems," *Intermedia*, vol. 7, no. 3, 1979, pp. 33—35.

Martin, J.: *Telecommunications and the Computer*, Prentice-Hall, Englewood Cliffs, N.J., 1969.

Martin, J.: *The Wired Society: A Challenge for Tomorrow,* Prentice-Hall, Englewood Cliffs, N.J., 1978.

Martin, T. H.: "Regulatory Constraints upon the Development of Consumer Information Services in the United States," *Teleinformatics '79,* IFIP, North-Holland Publishing Co., Amsterdam, 1979, pp. 275−278.

Marvin, Carolyn: "Delivering the News of the Future," *Journal of Communication,* vol. 30, no. 1, 1980, pp. 10−20.

Maurer, H. A., W. Rauch, and I. Sebestyen: *On Alphabetic Searching in Videotex Systems,* working paper 81−111, International Institute for Applied Systems Analysis, Laxenburg, Austria, August 1981.

Maurer, H. A., W. Rauch, and I. Sebestyen: *Videotex Message Service Systems,* working paper 81-113, International Institute for Applied Systems Analysis, Laxenburg, Austria, August 1981.

McCann-Erickson, Inc.: *Total Annual Advertising Expenditures,* Marketing Services Research Department, McCann-Erickson, Inc., 1981.

McCrum, William A., and Michael G. Ryan: "Risks and Benefits of New Communications Services," *Telecommunications Policy,* vol. 5, no. 1, 1981, pp. 33−39.

McKee, Edward A.: "Local Distribution Networks: Safe from Competition?" *Telephony,* April 20, 1981, pp. 22−24.

Mecca, Raymond G.: "Newspapers and Home Video Information Systems: The Present, the Promise and the Peril," *Videodisc/Teletext* (now *Videodisc/Videotex*), vol. 1, no. 1, Winter 1981, pp. 18−29.

Meltale, John: *The Changing Information Environment,* Westview Press, Boulder, Colo., 1976.

Menkes, J.: "Epistemological Issues of Technology Assessment," *Technological Forecasting and Social Change,* vol. 15, 1979.

Metcalf, R. M., and D. R. Boggs: "Ethernet: Distributed Packet Switching for Local Computer Networks," *ACM Communications,* July 1976, pp. 395−404.

Miami Herald: "Tune in to Tomorrow," August 24, 1980, p. 16.

Mier, Edwin: "Bank Data Networks: Moving Millions Electronically," *Data Communications,* April 1981, pp. 69−97.

Miller, Darby: "Where is Videotex going?," *Data Communications,* September 1981, pp. 97−105.

Miller, Robert B.: "Response Time and Man-Computer Conversational Transactions," *Proceedings Spring Joint Computer Conference,* vol. 33, AFIPS Press, 1968, pp. 267−277.

Mitchell, B. M.: "Optimal Pricing of Local Telephone Service," *The American Economic Review,* vol. 68, no. 4, September 1978, pp. 517−537.

Mokhoff, Nicholas: "A Computer Center for the Homeowner," *IEEE Spectrum,* September 1980, pp. 73−77.

Mosher, Lawrence: "The Approaching Boom or the Tube—The Regulatory Boxes No Longer Fit," *National Journal,* February 23, 1980, pp. 304−310.

Müller, Jürgen: "Potential for Competition and the Role of PTTs," *Telecommunications Policy,* vol. 5, no. 1, 1981, pp. 18−23.

Nash, D. Collingwood, and J. B. Smith: *Interactive Home Media and Privacy,* prepared for Office of Policy Planning, Federal Trade Commission, January 1981.

National Commission on Electronic Fund Transfers: *EFT and the Public Interest,* Washington, D.C., February 1977.

National Telecommunications and Information Administration: *The Foundations of United*

States Information Policy, U.S. Department of Commerce, June 1980.

National Telecommunications and Information Administration: *Issues in Information Policy*, U.S. Department of Commerce, February 1981.

Neustadt, Richard, Gregg Skall, and Michael Hammer: "The Regulation of Electronic Publishing," *Federal Communications Law Journal*, vol. 33, no. 3, 1981, pp. 331–417.

New Media: Broadcast Teletext, Videotex: The Commission on New Information Technology, Stockholm, 1981.

New Views: Computers and New Media—Anxiety and Hopes: The Commission on New Information Technology, Stockholm, 1979.

New York Times: "The Computer as Retailer," January 9, 1981.

Nielsen, A. C.: *1981 Nielsen Report on Television*, A. C. Nielsen Co., Northbrook, Ill., 1981.

Nilles, Jack M., F. Roy Carlson, Jr., Paul Gray, et al.: *A Technology Assessment of Personal Computers*, Vol. 1: *Summary*; Vol. 2: *Personal Computer Technology, Users, and Uses*; Vol. 3: *Personal Computer Impacts and Policy Issues*, Office of Interdisciplinary Programs, USC, Los Angeles, 1981.

Nisenholtz, Martin: *Early Uses of Graphics in the Alternate Media Center/WETA Teletext Trial*, Alternate Media Center, New York University/School of the Arts, New York, October 1981.

Noll, A. Michael: "Service and System Implications," *Telecommunications Policy*, vol. 4, no. 1, March 1980, pp. 17–24.

Noll, A. Michael: "Teletext and Videotex in North America: Service and System Implications," *Telecommunications Policy*, vol. 4, no. 1, 1980, pp. 17–24.

Nora, Simon, and Alain Minc: *The Computerization of Society*, MIT Press, Cambridge, Mass., 1980 (originally published in French in 1978).

Nyborg, Philip: "Computer Technology and U.S. Communications Law," *Telecommunications Policy*, no. 5, December 1977, pp. 374–380.

Nyborg, Philip: "Regulatory Inhibitions on the Development of Electronic Message Systems in the USA," *Telecommunications Policy*, vol. 2, no. 4, 1978, pp. 316–326.

Nyhan, Michael J.: "The Videotex Invention: Marvel or Myth?" in Katherine S. Rutkowski (ed.), *Videotex Services*, Institute for the Future paper P-98, National Cable Television Association, Washington, D.C., 1980.

Nyhan, Michael J., Robert Johansen, and Robert Plummer: "Home Information Systems: Some Thoughts on the Role of Public Broadcasting," *Public Telecommunications Review*, May 1979, pp. 20–25.

Nyhan, Michael J., Robert Johansen, and Robert Plummer: "Videotex and Teletext in the U.S.: Prospects for the 1980s," *Telecommunications Journal*, vol. 47, no. 6, 1980, pp. 396–400. Institute for the Future paper P-81.

O'Brien, C. D., H. G. Bown, J. C. Smirle, Y. F. Lum, and J. Z. Kukulka: *Telidon: Videotex Presentation Level Protocol: Augmented Picture Description Instructions*, Communications Research Centre, Department of Communications, Ottawa, Canada, February 1982.

OCLC: "Channel 2000 Project Report," Columbus, Ohio, 1981.

O'Connor, Robert A.: "Teletext Field Tests," *IEEE Transactions on Consumer Electronics*, vol. CE-25, no. 3, 1979, pp. 304–310.

Oettinger, Anthony G., Paul J. Berman, and William H. Read: *High and Low Politics: Information Resources for the 80s*, Ballinger Publishing Co., Cambridge, Mass., 1977.

Omnico brochure: Distributed at West Coast Computer Fair, San Francisco, April 3–5, 1981.

Pacific Telephone: *Openline*, March 1981.

Parker, Edwin B.: "New Approaches to Development/An Information-Based Hypothesis," *Journal of Communication*, Winter 1978, pp. 81–83.

Paul, Lois: "Joint Venture Tests Home Banking Waters," *Computerworld*, March 16, 1981, p. 34.

Pergler, P.: *The Automated Citizen: Social and Political Impact of Interactive Broadcasting*, Institute for Research on Public Policy, occasional paper no. 14, Montreal, 1980.

Peyton, David: "Videotex Policy Issues in the United States," *Electronic Publishing Review*, vol. 1, no. 4, 1981, pp. 251–262.

Plowright, Teresa: *Social Aspects of Videotex Services: Proposed Research Directions*, Social and New Services Policy Branch, Broadcasting and Social Policy Branch, Department of Communications, Ottawa, Canada, 1980.

Plummer, Robert: "The Source: A Pioneer Home Information System," *In Context*, report 2, Communication Studies und Planning, Ltd., March 1980.

Plummer, Robert, Robert Johansen, Michael J. Nyhan, and P. G. Holmlov: "4004 Futures for Teletext and Videotex in the USA," *IEEE Transactions on Consumer Electronics*, CE-25, vol. 3, 1979, pp. 317–325.

Porat, M.: *The Information Economy*, U.S. Government Printing Office, 1977.

Posa, John G.: "Memories," *Electronics*, vol. 53, no. 23, 1980, pp. 132–145.

Powers, Ron: "They Surely Won't Throw This Newspaper on Your Doorstep," *Panorama*, November 1980, pp. 52–55.

Public Opinion: "Opinion Roundup: The Bad News is . . . ," June/July 1981(*a*), p. 34.

Public Opinion: "Opinion Roundup: Satisfaction with Work Remains High," August/ September 1981(*b*), p. 29.

Public Opinion: "Opinion Roundup: Government in the Workplace," August/September 1981(*c*), p. 34.

Pye, Roger: "The Birth of a Videotex Industry," *InterMedia*, vol. 7, no. 3, 1979, pp. 41–47.

Raggett, R. J.: "Informatic Systems: Technology in Search of a Market?" *Telephony*, May 26, 1980, pp. 70–78.

Ragland, John D.: "Project Green Thumb Pilot: Information Harvest for Farmers," *Telephony*, May 11, 1981, pp. 20–24.

Reid, Alex: "Flying Fast But Safe," *InterMedia*, vol. 7, no. 3, 1979, pp. 22–25.

Reid, Alex (ed.): *Prestel 1980*, Post Office Telecommunications, London, 1980.

Risher, Carol: "Electronic Media and the Publishers, Part I: Teletext," *Videodisc/Videotex*, vol. 1, no. 3, Summer 1981, pp. 162–167.

Robinson, Gary, and William Loveless: "'Touch-Tone' Teletext: A Combined Teletext-Viewdata System," *IEEE Transactions on Consumer Electronics*, vol. CE-25, no. 3, 1979, pp. 298–303.

Robinson, Glen O.: *Communication for Tomorrow: Policy Perspectives for the 1980s*, Praeger, New York, 1978.

Rochell, Carlton C.: *An Information Agent for the 1980s*, American Library Association, Chicago, 1981.

Roger W. Hough and Associates, Ltd.: *A Study to Forecast the Demand for Telidon Services over the Next Ten Years*, Ottawa, Canada, December 1980.

Roizen, Joe: "The Technology of Teletext and Viewdata," in E. Sigel (ed.), *Videotext: The Coming Revolution in Home/Office Information Retrieval*, Knowledge Industry Publications, 1980.

Rosch, Gary D.: "Viewdata and Teletext: What the Europeans Are Doing," *Executive Perspective*, second quarter 1979.

Rosch, Gary D.: "Videotex and the Publishing Industry," *Vital Speeches*, SLVIII, no. 4, 1980, pp. 98–100.

Rothman, John: *Video Information Systems*, paper presented at Aspen Institute Conference on

Teletext and Viewdata, 1979.

Rothman, John, and John Werner: *Video Information Systems*, The New York Times Company, New York, June 1979.

Rutkowski, Katherine S.: *Videotex Services*, National Cable Television Association, Washington, D.C., 1980.

San Francisco Chronicle: "FCC OKs Use of New Mobile Phones," April 10, 1981, p. 4.

Sawyer, Robert: "The Shape of Things to Come," *Advertising Age*, January 18, 1982.

Scannell, T.: "Demand Soaring for Low-Cost Color Printers," *Computerworld*, February 9, 1981, p. 11.

Schuster, Stewart A.: "Relational Database Management for On-Line Transaction Processing," *Computerworld*, 1981.

Selwyn, Lee L., and William P. Montgomery: "Deregulation, Competition, and Regulatory Response in the Telecommunications Industry," *Public Utilities Fortnightly*, November 22, 1979, pp. 13−22.

Shaw, Louise G.: "Why Touch Sensing," *Datamation*, August 1980, pp. 138−141.

Sifton, John, and David Wright: "Canada: Major Support for Telidon," *InterMedia*, vol. 7, no. 3, 1979, pp. 30−32.

Sigel, Efrem (ed.): *Videotext: The Coming Revolution in Home/Office Information Retrieval*, Knowledge Industry Publications, White Plains, New York, 1980.

Sirbu, Marvin A., Jr.: "Innovation Strategies in the Electronic Mail Marketplace," *Telecommunications Policy*, September 1978, pp. 198−210.

Smirle, J. C., and H. G. Bown: "New System Concepts and Their Implications for the User," *Teletext and Viewdata in the U.S.: A Workshop on Emerging Issues, Background Papers*, Institute for the Future, Menlo Park, Calif., 1979.

Smirle, J. C., Y. F. Lum, and H. G. Bown: *International Videotex Standardization: A Canadian View of Progress Towards the Wired World*, Canadian Federal Department of Communications.

Smith, Anthony: "Will Consumerism Inhibit the New Media?" *InterMedia*, vol. 7, no. 3, 1979, pp. 51−53.

Smith, Anthony: *Goodbye Gutenberg: The Newspaper Revolution of the 1980s*, Oxford University Press, New York, 1980.

Smith, Kevin: "Prestel to Transmit Color Photographs," *Electronics*, October 23, 1980, pp. 78−80.

Snowden, Richard: "Home Information Systems: Problems of Scale," MIT Research Program on Communications Policy, Cambridge, Mass., March 19, 1981.

Standards Policy Workshop Background Papers: working paper WP-34, Institute for the Future, Menlo Park, Calif., June 1981.

Sterling, C. H., and T. R. Haight: *The Mass Media: Aspen Institute Guide to Communication Industry Trends*, Praeger, New York, 1978.

Sullivan, D.: *Information Utilities: Myth or Reality*. Unpublished keynote address at Information Utilities Conference '81, New York, March 1981.

Swerdlow, Joel: "Why Is Everyone Afraid of Ma Bell?" *Channels*, October/November 1981, pp. 29−33.

Takasaki, N., and K. Mitamura: "Social Acceptability of New Video Systems in Japan," *Proceedings of the Fifth International Conference on Computer Communications*, North-Holland Publishing Co., Amsterdam, 1980, pp. 611−616.

Tanton, N. E.: "UK Teletext—Evolution and Potential," *IEEE Transactions on Consumer Electronics*, vol. CE-25, no. 3, July 1979, pp. 246−250.

Taylor, John P.: "Direct Satellite-to-Home?: Just Around the Corner?" *Television/Radio Age*, October 22, 1979(*a*).

Taylor, John P.: "Direct Satellite-to-Home Broadcasting: How Comsat's Pay-TV Plan Figures In," *Television/Radio Age*, November 19, 1979(*b*).

Taylor, John P.: "Direct Satellite-to-Home Broadcasting: How Will WARC '79 Decisions Affect It?" *Television/Radio Age*, January 28, 1980(*a*).

Taylor, John P.: "Direct Satellite-to-Home Broadcasting: Other Countries Ahead of the U.S.," *Television/Radio Age*, March 24, 1980(*b*).

Technology Forecasts: "Bubble Memory Outlook Bullish," January 1981, p. 12.

Tedesco, Albert S., and F. Anthony Bushman: "Teletext in the Public Interest: The Case of New Consumer Information Services," *Teletext and Viewdata in the U.S.: A Workshop on Emerging Issues, Background Papers*, Institute for the Future, Menlo Park, Calif., 1979.

Telecommunications Reports: "STC Files Application for First Phase of Direct Broadcast Satellite Service with FCC; Charyk Calls 'Far-reaching Development' a 'Logical Extension' of COMSAT's Endeavors," December 22, 1980, p. 35.

Telephony: "British Telecom Tells Yellow Pages Changes," February 23, 1981(*a*), pp. 41−42.

Telephony: "Int'l Telecom Spending Continues Its Climb," February 23, 1981(*b*), pp. 52−61.

Teletext: no. 5, Marketing Solutions Ltd., London, February 1982.

Telidon Reports: no. 8, Department of Communications, Canada, January 1982.

Thomas, H. B.: "Tree Structure: The Root of Videotex?" *In Context*, report 2, Communication Studies and Planning, Ltd., March 1980.

Thomas, H. B., and M. Tyler: "Videotex and Teletext: Computing and Communication for the Mass Market," *Proceedings of the Fifth International Conference on Computer Communication*, North-Holland Publishing Co., Amsterdam, 1980, pp. 490−496.

Thompson, G. B.: *Memo from Mercury: Information Technology Is Different*, Institute for Research on Public Policy, Montreal, 1979.

Thompson, G. B.: "Some Potential Socio-Economic Implications of Videotex Systems," *Proceedings of the Fifth International Conference on Computer Communications*, North-Holland Publishing Co., Amsterdam, 1980, pp. 617−620.

Tomita, Tetsuro: "Japan: The Search for a Personal Information Medium," *InterMedia*, vol. 7, no. 3, 1979, pp. 36−38.

Trak, Ayse, and Michael MacKenzie: "Appropriate Technology Assessment: A Note on Policy Considerations," *Technological Forecasting and Social Change*, vol. 17, 1980, pp. 329−338.

TransData Corporation: *National Study of Home Banking Services in the USA*, 1981.

Trauth, Eileen M., and Denise M. Trauth: "The Policy Implications of the Use of Cable Television for Data Transmission," in Alan R. Benenfeld and Edward John Kazlauskas (eds.), *Proceedings of the 43rd Annual ASIS Meeting*, vol. 17, American Association of Information Science, White Plains, N.Y., 1980, pp. 106−108.

Tydeman, John: *Forecasting a New Hybrid Technology—Teletext and Videotex*, Proceedings of the Fifteenth International Symposium Mini and Microcomputers, ISMM, Mexico, April 1981.

Tydeman, John: *Teletext and Videotex: Tomorrow's Technology Today*, paper P-120, Institute for the Future, Menlo Park, Calif., August 1981.

Tydeman, John, Robert Johansen, Hubert Lipinski, and Michael J. Nyhan: "Videotex: A Dozen Public Policy Concerns and a Design to Understand Them," *Proceedings of the Fifth International Conference on Computer Communication*, North-Holland Publishing Co., Amsterdam, 1980, pp. 621−628.

Tydeman, John, and Hubert Lipinski: *Futures Workshops*, working paper WP-32, Institute for the Future, Menlo Park, Calif., December 1980.

Tydeman, John, and Laurence Zwimpfer: *Videotex in the United States—Toward Information*

Diversity, Videotex '81, Online Conferences Ltd., Infomart, 1981.

Tyler, Michael: "Electronic Publishing: A Sketch of the European Experience," *Teletext and Viewdata in the U.S.: A Workshop on Emerging Issues, Background Papers*, Institute for the Future, Menlo Park, Calif., 1979(*a*).

Tyler, Michael: "Videotex, Prestel, and Teletext—The Economics and Politics of Some Electronic Publishing Media," *Telecommunications Policy*, vol. 3, no. 1, 1979(*b*), pp. 37–51.

Uhlig, Ronald P., David J. Farber, and James H. Bair: *The Office of the Future: Communications and Computers*, vol. 1, Monograph Series of the ICCC, North-Holland Publishing Co., Amsterdam, 1976.

University of Michigan, Institute for Social Research: *Nonhousehold Mailstream Study*, Ann Arbor, Mich., 1980.

U.S. Bureau of the Census: *Statistical Abstracts of the United States: 1980*, Washington, D.C., 1980.

Vallee, Jacques: "Questioning Some Assumptions: An Information Scientist Looks at Teletext/Videotex," *Teletext and Viewdata in the U.S.: A Workshop on Emerging Issues, Background Papers*, Institute for the Future, Menlo Park, Calif., 1979.

Venture Development Corporation: "The Hard Disk Industry: A Strategic Analysis," *Computerworld*, February 18, 1980, p. 66.

Venture Development Corporation: "Bubble Outlook Bullish," *Technology Forecasts*, January 1981, p. 12.

Vermilyea, David L.: *Teletext in the Year 2000: A Delphi Forecast*, unpublished thesis, San Diego State University, 1980.

VideoPrint: vol. 1, no. 13, International Resource Development, Inc., September 22, 1980.

Video Systems: "Teletext Systems (Electronic Newspapers) Introduced in United States," July 1978, p. 4.

Videotex '81 (Proceedings): Online Conferences Ltd., Northwood Hills, Middlesex, U.K., 1981.

Videotex Services (National Cable Television Association Executive Seminar Series): The National Cable Television Association, Washington, D.C., 1980.

Videotex Standard: Presentation Level Protocol: American Telephone and Telegraph Company, Parsippany, N.J., May 1981.

Viewdata '80 (Proceedings): Online Conferences Ltd., Northwood Hills, Middlesex, U.K., 1980.

Viewdata '81 (Proceedings): Online Conferences Ltd., Northwood Hills, Middlesex, U.K., 1981.

Viewdata Code of Practice 1980: The Association of Viewdata Providers Ltd., London, 1980.

Vigilante, Frank S.: "The Digital Network Becomes Reality," *Telecommunications Trends and Directions*, 1980, pp. 69–71.

Waller, Larry: "Memories," *Electronics*, vol. 53, no. 12, 1980, pp. 41–42.

Wenk, Edward Jr.: *Technology Assessment: Concepts, Practice and Experience in the United States*, document 5, Workshop on Technology Assessment, Government of Australia, Department of Science, Canberra, July 3, 1978.

Wessler, B. D.: "United States Public Packet Networks: An Update," *Telecommunications Journal*, vol. 47, no. VI, 1980, pp. 373–374.

White, C.E.: "New Services for Subscribers," *Telecommunications*, January 1981, pp. 17–22.

White, James A.: "AT&T, in Antitrust Settlement, Is Putting Its Bets on the Future, While It Writes Off the Past," *Wall Street Journal*, January 12, 1981, p. 1.

White, Wade, and Morris Holmes: "The Future of Commercial Satellite Telecommunications," *Datamation*, July 1978, pp. 94−102.

Williams, Gerald W.: "Focus on Data Terminal Equipment," *Computerworld Extra!*, vol. 15, no. 11*a*, 1981, pp. 29−32.

Wilson, Larry G.: *Planning a Videotex Trial*, paper presented at 2d CCITT Interdisciplinary Colloquium on Teleinformatics, Montreal, June 10, 1980.

Winsbury, Rex: *The Electronic Bookstall: Push-Button Publishing on Videotex*, International Institute of Communications, London, 1979.

Winsbury, Rex: *Viewdata in Action: A Comparative Study of Prestel*, McGraw-Hill, London, 1981.

Winsbury, Rex, and Martin Lane: "Prestel Is the First to Start," *InterMedia*, vol. 7, no. 3, 1979, pp. 10−16.

Wirth, Timothy E.: "Exploring Government Policy and the Telecom Revolution," *Telephony*, September 14, 1981, pp. 120ff.

Woolfe, Roger: *Videotex: The New Television/Telephone Information Services*, Heyden & Sons, Ltd., London, 1980.

Yao, Margaret: "Two-Way Cable TV Disappoints Viewers in Columbus, Ohio, as Programming Lags," *Wall Street Journal*, September 30, 1981.

Yasuda, Kimikazu: "Conception of CAPTAIN System—Background, Experiment and Future Plans," *Viewdata '80*, Online Conferences Ltd., 1980, pp. 107−111.

Young, Ian, and Ian Gray: *The Cultural Application and Implications of Videotex Services in the United Kingdom*, Council of Europe, Strasbourg, 1980.

Yurow, Jane H., Robert Aldrich, Robert Belair, et al.: "Managing Information," *Issues in Information Policy*, Chap. VI, U.S. Department of Commerce, National Telecommunications and Information Administration, Washington, D.C., February 1981.

Zientara, Marguerite: "Publishers Decry Bell Home System," *Computerworld*, December 15, 1980, p. 1.

Zientara, Marguerite: "Companies Experiment with Telecommuting," *Computerworld*, November 30, 1981, p. 23.

Zwimpfer, Laurence: *The Chip at Work: Emerging Home Information Services*, paper P-121, Institute for the Future, Menlo Park, Calif., August 1981.

INDEX

ABC (American Broadcasting
    Corporation), 41
A. C. Nielsen (firm), 80, 271
A. H. Belo (firm), 43
Access:
    database and, 24, 25, 89–92, 95–98
    in foreign systems: cost of, 18, 19
        database for, 24, 25
        equity of access in, 37
        methods of, 32–34
        responsibility for, 23
    input devices and, 51
        (*See also* Input devices)
    private-port, 119
    random-access memory, 30, 137
    (*See also* Access time; Equity of
        access; Guarantee of access)
Access time, 5
    billing based on, 34
    defined, 4
    elements affecting, 96
    of existing systems, 155
    storage and, 139–143
    trade-off between service
        characteristics and, 97
    with two-way transmission, 5–6,
        96, 98
Adams-Russell (firm), 239
Addressable converters, 112–113
Advanced Communication System
    (ACS), 98, 118
Advanced Research Project Agency
    (ARPA), 119–120
Advertising, 41, 53–54, 62, 67, 270
    consumer protection and, 221, 223
    in EYP, 55
    in foreign systems, 18, 38–39

Advertising (*Cont.*):
    market potential for, 67, 72–73
    regulation of, 13, 173, 176–178, 199,
        204–206
Agriculture, Department of, 43
Allen, 127, 131, 135, 138
Alphageometric graphic codes, 29,
    30, 155, 194, 197
Alphametric keyboards, 92
Alphamosaic graphic codes, 29–30,
    155, 194, 197
Alphanumeric characters, 4
Alphanumeric keyboards, 5, 27, 29,
    51, 52, 122, 126, 132, 136, 149,
    268
Alphaphotographic graphic codes,
    29, 30, 32, 155, 194–195
Alternate Media Center (New York
    University), 41
AM radio, 83–84
American Express, 224, 239
American National Standards Institute
    (ANSI), 191, 192, 194, 197, 198
American Newspaper Publishers
    Association (ANPA), 226
Amplifiers, 108, 110
Analog data, transforming, into
    digital data, 26
    (*See also* Digital data)
Analog models of human voice, 130
Analog signals, 5, 101, 105
Analog-switching problems, 101, 106,
    115, 118, 119, 121
Analog-video equipment, 105
Anderson, 130
Antelope (system), 14
Antennas, 4, 5, 102, 106–108, 110

Antiope, 1, 16, 18, 116, 144, 155,
    158, 209
    standards for, 197, 270
    structure of, 21, 22, 24, 25, 28
    turnkey systems using, 90
    U.S. trials with, 41, 44
Antitrust legislation, 13, 174–175,
    201, 211, 213
Apple computers (I; II), 20, 134–135,
    144, 224
Application layer of standards, 194
Application standards, 195–196
Applications:
    based on current trials, 60–65
    most popular, 85, 86
    policy and, 204–208, 210, 212, 213
    (*See also* Future applications *and
        specific types of services*)
Arthur D. Little (firm), 239
ASCII coding, 116, 118, 130, 144, 148
Associated Press, 43, 53, 224
Asynchronous (variable-format) signal
    transmission, 28, 29, 38, 196
Atari, 77, 224
ATARI CX70 (light pen), 134
ATMs (automated-teller machines),
    55, 74, 89
AT&T (American Telephone & Tele-
    graph), 6, 13, 15, 98, 116–118,
    144, 174–176, 185, 224
    antitrust legislation and, 174, 175
    common-carrier regulations and,
        175–176
    content-carriage issue and, 202, 203
    current trials with, 43, 47, 55
    divestiture decision of, 117, 175,
        176, 179

AT&T (*Cont.*):
  expenditure study conducted by, 81
  industry structure and, 226
  standards setting and, 191, 192, 197,
    198, 270
  state regulation and, 179
Attributes:
  of basic systems, 150–152, 154
  of communication technologies, 104
  of display devices, 123, 124
  of enhanced systems, 158, 159, 163,
    164, 166, 168
  of foreign systems, 23, 28
  of input devices, 132, 133
  serial and parallel, 28, 195, 196
  storage, 139, 141
  (*See also* Technological attributes)
Audio cassettes and recorders,
    140–143, 178
Audio information:
  display of, 123
  storage of, 139–142, 144
Australian telecommunications, 16
Austrian telecommunications, 15, 16,
    18
Authentification of messages, 164
Automated Data, 242
Automated individual tailoring
    (individualization) effect, 35–36,
    259–260, 273
Automated-teller machines (ATMs),
    55, 74, 89
Average waiting time (*see* Waiting
    time)

Baker, 105
Ball, 90
Banc One, 43, 55, 224, 242
Bandwidths, 2, 15, 102, 104, 105, 209
  for cable television-based networks,
    110, 113–114
    for cellular radio-based systems,
      114
  for DBS-based systems, 109
  display format and, 51
  for enhanced systems, 160, 161,
    163, 167–169
  for FM radio-based systems, 109
  for MDS-based systems, 106–108
  for one-way transmission, 120
  for optical-fiber cable-based
    systems, 111
  for packet-switched network-
    based systems, 118
  regulation and scarcity of, 36
  for switched-telephone network-
    based systems, 115, 117
  (*See also* Full-channel systems)
Bank of America, 44, 224
Bank Holding Act (1956), 240
Banks and banking industry:
  consumer protection and, 222, 226
    implications and deregulation for,
      238–239
  regulatory legislation for, 13, 177
  (*See also* Telebanking services)
Barriers to entry, 172, 213, 225

Barron, 199
Barron's (newspaper), 43, 54
Basic service, Federal
    Communications Commission
    definition of, 176
Basic systems, 12, 147–158
  attributes of, 150–152, 154
  defined, 150–155
  existing systems compared with,
    155–158
BBC (British Broadcasting
    Commission), 14, 16, 17, 23, 34,
    36, 37, 39
Belgian telecommunications, 15, 16
Bell Canada, 15, 16, 22, 127
Bell System, 114, 115, 175, 179,
    180, 203
  (*See also* AT&T)
Benton & Bowles, 61, 78
Betamax case, 178
Bigi, 111
Bildschirmtext, 24, 26, 37, 89, 98
Bill-paying services (*see* Telebanking
    services)
Billing, 1, 3, 5, 45, 195, 196
  external database and, 92
  foreign methods of, 23, 25, 34, 36
  identification for purposes of, 96
BISON, 43, 48, 50, 51
Bits and bit rate, 50, 96–98
  access time and, 4, 34
  with basic services, 148, 150
  with cable television-based
    systems, 113
  defined, 6n.
  display and, 123–126, 131
  with enhanced services, 160, 161,
    163
  with existing services, 157
  with FM radio-based services, 109
  with packet-switched networks, 118
  with Prestel, 6
  and 7-bit code, 26
  with switched-telephone networks,
    30, 115
  VBI and, 102
Black-and-white displays, 15, 49
  for basic services, 148–150, 156
  European, 27
  for graphics, 126, 127
  printers for, 128
Bloom, 98
Boggs, 113
Bolt, Deranek and Newman, Inc., 118
Bonneville International, 40
Brandon, 205
Branscomb, 137
Braunstein, 201, 212
British Leyland (BL), 82
British Post Office, 225
British telecommunications, 209
  broadcast television-based, 102, 103
  development of, 14–17
  market for, 80–82, 84
  public policy and, 24–29, 206, 234
  services offered by, 17–18, 20
  structure of systems in, 21, 24, 25,
    33, 34

Broadcast-based systems (*see specific
    broadcast-based systems, for
    example:* Broadcast television-
    based systems; Direct-broadcast
    satellite-based systems)
Broadcast hybrids, 120, 122, 211, 268
  basic systems using, 153–156
  enhanced systems using, 169
  European, 18, 22
  FM radio and, 110
  policy issues related to, 186, 187
  policy options for, 209, 211
Broadcast radio (*see* Radio)
Broadcast television-based systems, 2,
    4, 14, 52, 102–109, 121, 212, 269
  advertising potential of, 72, 73
  attributes of, 50, 96
  basic, 148, 153–156
  employment with, 231
  foreign, 14, 22, 25
  frequency spectrum and number of
    channels for, 104
  in local distribution networks, 100
Broadcasting industry:
  regulation of, 35, 171, 173, 176–177,
    190, 199, 200, 202, 205
  standards of, as models, 193–194
BSE (satellite), 107
Buffer memory, 4, 96, 97
Bulkeley, 127
Burglary protection (*see* Tele-
    monitoring services)
Business information services,
    potential for, 81–83
Business market, 63, 77, 127, 144
  foreign, 17, 18, 27
  potential for, 72, 80–83, 86, 87
Business office:
  automation of, 80
  transformative effects on, 262–265
Business organization, transformative
    effects on, 264–265
Bytes:
  access time and, 96, 97
  defined, 6n.
  in LISA, 131
  storage and, 94, 131
  transmission speed and, 97
  waiting time dependent on, 96
  (*See also* Waiting time)
  (*See also* Bits and bit rate)

C56500 (desk-top calculator), 131
Cable (*see* Cable television-based systems;
    Copperwire-based cable; Optical-
    filter cable)
Cable television (CATV), 84, 101, 107,
    172, 173
  franchising of, 13, 112, 179–180,
    199, 201, 202, 204, 211
Cable television-based systems, 4, 5,
    50, 63, 80, 100–103, 105, 109–114,
    120, 121
  applications of, 53, 55, 56, 76
  (*See also specific services*)
  attributes of, 104

Cable television-based systems (*Cont.*):
bandwidths for, 110, 113–114
basic services using, 152, 154–155
DBS and, 109
employment with, 231
enhanced services using, 166, 169
guarantee of access to, 200
market for, 6
MDS-based systems vs., 106
policy issues related to, 172, 175, 186, 190
policy options on, 209, 212, 213
trials with, 42, 44, 45
(*See also specific cable television-based systems, for example:* Coaxial cable-based systems)
Cablecom, 22
Cabletext, 42, 47, 50
Calculators, 131
Cameras, digitizing video, 132, 135, 136
Canadian Communications Department (DOC), 22
Canadian telecommunications, 15–17, 21, 23, 27, 29, 30, 107, 197, 234
Captain (system), 15, 23, 26, 28
Captioning, 2, 7, 54, 61, 66, 194
closed, 41, 47, 50, 122
Carey, 233
Carriage-content separation issue, 13, 36, 190, 201–204, 211, 213, 270
Carterphone decision (1981), 172
Cartridge tapes and players, 140–143
Cash flow, transformative effects on, 260
Cassettes (*see* Audio cassettes and recorders; Digital cassettes and players; Videocassettes and recorders)
Catalog shopping (*see* Teleshopping services)
Cathode-ray tube (CRT) terminals, 36, 126–127, 134, 153, 268–269
CATV (*see* Cable television)
CBS (Columbia Broadcasting System), 41, 50, 176, 197, 224
CBS/AT&T trial, 44, 47, 48
Ceefax, 1, 14, 16–19, 21, 22, 25, 28, 34, 155, 209, 270
Cellular radio, 5, 114–115
Census, Bureau of the, 125, 258
Censorship (*see* Content regulation)
Center for the Study of Tele-communications and Television (CCETT), 14
Centigram (firm), 131
Central indexing, 23
Central processing unit (CPU), 136–137, 150–152, 194
Centralized database, 89–90, 148, 166
Channels (*see* Bandwidths)
Chapman, 81
Characters, 119
alphanumeric, 4
character generators, 28
codes for, 28–30
color for, 28

Characters (*Cont.*):
sets of, 26, 30, 155
storage requirements affected by number of, 94
Chase Manhattan Bank, 45, 125, 241
Chemical Bank, 44, 224, 241
Child rearing, transformative effects on, 256
Ciciora, 148
Citibank, 44, 224, 225, 241
Classified advertising (*see* Advertising)
Clayton Act (1914), 174, 201
Closed captioning, 41, 47, 50, 122
Closed-user groups (CUG), 3*n.*, 19–20, 33–34, 82–83, 90
Coaxial cable-based systems, 2, 98, 101, 104, 110–114, 121, 194
Coaxial Cable Information System (CCIS), 15
Codes:
alphageometric, alphamosaic and alphaphotographic, 29–30, 32, 155, 194–195, 197
ASCII, 116, 118, 130, 144, 148
attribute, 28
character, 28–30
for digital data transmission, 26, 196–197
7-bit, 26
(*See also* Decoders)
Color, 28
as factor affecting potential application, 72
Color displays, 27
with basic services, 149
European, 27, 28
for graphics, 126, 148
on printers, 128, 129
on television, 4, 6, 12, 29, 41, 49, 104, 125, 156
Colton, 221
Common carriers:
dominant (*see* AT&T)
guarantee of access and, 201
regulation of, 13, 173, 175–176
standards setting and, 197
Communication links:
cost of, 100–102, 104, 105, 108, 116, 117, 119, 121
types of, 45
Communication network providers, 6, 22, 45–48, 222, 224
competition issue and, 202–203
decoders leased from, 101
standards development and, 192
Communication networks (*see specific types of communication networks, for example:* Radio; Television)
Communications Act (1934), 13, 171–176, 215, 217
carriage-content separation issue and, 13, 36, 190, 201–204
Communications Technology Management, Inc., 243
Competing technologies, 62–63
for business market, 80

Competing technologies (*Cont.*):
effects of advertising potential on, 72–73
role of advertising in supporting, 72
(*See also specific types of technologies, for example:* Audio cassettes; Newspapers and magazines)
Competition, 270, 271
policy issues relating to, 182, 183, 185, 187, 189–190, 201–203
policy options on, 203–204, 210–212
in telebanking case study, 248–250
Comp-U-Card, 43, 55, 192, 224, 239
CompuServe, 7, 101, 144, 148, 156, 158, 192, 224
centralized database of, 90
standards for, 198
telebanking with, 240
U.S. trials with, 43, 44, 48, 50–56
Comp-U-Star, 43, 48, 50, 51, 240
Computer-aided instruction, 64, 71
Computer databases (*see* Databases)
Computer inquiries I and II, 175
Computers, 1, 20
as display terminals, 123, 124, 126–128
international market for, 236
public policy on, 171
(*See also* Terminals *and specific types of computers, for example:* Personal computers)
Computing services, 3, 10, 59, 182, 183, 187, 269
basic, 150, 151
consumer protection and, 222
described, 57–58
employment with, 233
enhanced, 164–166, 169
equity of access to, 220
industry structure and, 227, 228
popularity of, 85, 86
potential market for, 69, 76–77, 79
privacy issue and, 216
storage requirements for, 94
technologies competing for, 63
Comsat, 108
Conditioned (leased) circuits, 26, 115–116, 118, 119, 167
Confidentiality (*see* Privacy and security)
Congress, policy-making power of, 193, 201, 203, 206, 207, 236
Connection procedures, 95–96, 192
Consent Decree (1956), 175
Consequential policy issues, 184, 187, 270
defined, 13
(*See also* Consumer protection; Employment; Equity of access; Industry structure; Inter-national trade and communication; Privacy and security)
Consumer market (*see* Market)
Consumer protection, 13, 271
foreign policies on, 37

Consumer protection (*Cont.*):
  FTC and, 178
  policy issues related to, 182, 183, 185, 187, 220–222
  policy options on, 222–223
  in telebanking case study, 245, 249, 251
Content-carriage separation issue, 13, 36, 190, 201–204, 211, 213, 270
Content regulation, 39, 176, 199
  competition and, 202
  policy issues related to, 182, 183, 185, 187, 190, 204–206
  policy options on, 207, 210–212
  standards and, 199
Continental Telephone, 44, 48, 50, 224
Convenience criterion, 62, 75
Converters, addressable, 112–113
Cooper, 107
Copiers/printers, 80
Copperwire-based systems, 2, 101, 114, 118
Copyright, 13, 270, 271
Copyright Act (1976), 178, 207, 208
Copyright Royalty Tribunal, 208
  foreign policies on, 37–38
  policy issues related to, 178, 182, 185, 187, 190, 207–208
  policy options on, 208, 210–212
Corporation for Public Broadcasting, 41
Cost, 3
  of access, in foreign countries, 18, 19
  of addressable converters, 113
  of basic systems, 152, 156, 158
  of cable television-based systems, 108, 110
  of color television, 12
  communication links, 100–102, 104, 105, 108, 116, 117, 119, 121
  of computer terminals, 126–128
  of CPU, 137
  of database, 90
  of decoders, 102, 105, 106, 109, 116, 117
  of enhanced systems, 161, 162, 164–169
  as factor affecting application, 62, 72
  of input devices, 134, 135, 138, 139
  interface unit, 100–102, 121, 135, 143–144
  of microprocessors, 6
  of modems, 115–117, 126, 157
  of packet-switched network-based systems, 119–120
  of printers, 128, 129
  of satellite communication, 107
  of services, in U.S., 45
  of speech synthesizers, 130–132
  of storage, 139–145
  of switched-telephone network-based systems, 116–117

Cost (*Cont.*):
  of user terminals, 123, 125
  (*See also* Billing)
Cox Cable Communications, 44, 54, 55, 198, 200, 224, 239, 241
CPU (central processing unit), 136–137, 150–152, 194
Creative Strategies International (CSI), 129
Crowther, 29
CRT (cathode-ray tube) terminals, 36, 126–127, 134, 158, 268–269
CTS⁵/Hermes (satellite), 107
CUG (closed-user groups), 3*n*., 19–20, 33–34, 82–83, 90
Cycle and cycle pages, 4–6
  access time and, 96, 97
  of European systems, 19
  transmission speed and, 96, 97
  waiting time and, 5, 6

Dallas Morning News (newspaper), 43, 53
Danish telecommunications, 34–36, 38, 39
Data link layer, 195
Data processing (*see* Processing)
Databases, 18, 98–99
  access procedures and accessibility of , 89–92, 95–98
  for basic systems, 151, 154, 157
  for enhanced systems, 159
  foreign, 18, 19, 23–26, 36
    arrangements of, 23–25
    of Prestel, 20
    responsibility for quality of, 38
    service centers and, 25, 26
    structure of, 23
  international market for, 235, 237
  percentage of world, in U.S., 17
  structures of, 23, 89–94
  (*See also specific types of databases, for example:* Distributed database)
Dataspeed, Inc., 109
Datavision, 33
Date, 92
DBS (direct-broadcast satellite-based systems), 5, 102–104, 107–109, 120, 131, 156, 180, 268
Decoders, 4, 5, 15, 27, 34, 101, 106, 137
  ASCII, 116, 118
  for basic systems, 154
  for broadcast television-based systems, 102
  for cable television-based systems, 106, 112
  cost of, 102, 105, 106, 109, 116, 117
  for enhanced systems, 164, 165, 167, 169
  for FM radio-based systems, 109
  for foreign systems, 18, 22, 28, 29
  for program captioning, 54
  for satellite-based systems, 108

Decoders (*Cont.*):
  for switched-telephone network-based systems, 116–118
  U.S. trials with, 41, 42
Defense, Department of, 119
Demand and supply (*see* Market)
Depository Institution Deregulation and Monitoring Control Act (1980), 239
Deregulation, 190, 209–213
  impact of: on consequential issues, 217–218, 220, 222, 227, 228, 233–234, 237
  on developmental issues, 199, 201, 204, 207, 208–213
  of satellite communications, 107
  in telebanking study, 250, 251
  of telecommunications industry, 102
Deutsche Bundespot (firm), 36
Developmental policy issues, 184, 187, 189–213, 270
  defined, 13, 189
  policy options on, 198, 201, 203–204, 207–213
  (*See also* Competition; Content regulation; Copyright; Guarantee of access; Market, structure of; Standards)
Dial-It services, 203
Dial-up telephones, 118, 119
Dialogue standards, 195, 196
Didon (system), 18
Digital cassettes and players, 140–143
Digital code, 26, 196–197
Digital data, 4, 26
  in broadcast TV-based networks, 102
  conversion of, 129
  storage of, 139, 140, 142
  transmission of, 26, 196–197
Digital signals, 5, 101, 102, 105
Digital speech, 130, 131, 163
Digital-switching telephone networks, 115, 118, 121
Digital television receivers, 125
Digital video systems, 105
Digitizing video cameras, 132, 135, 136
Direct-broadcast satellite-based systems, 5, 102–104, 107–109, 120, 121, 156, 180, 268
Direct Mail Marketing Association, 75
Direct-mail shopping (*see* Teleshopping services)
Direct Vision, 41
Directories (*see* Electronic directories; Meta-service directories; Paper directories)
Discovision Associates, 144
Discs and drives (*see* Floppy discs and drives; Hard discs and drives; Videodiscs and players)
Dish antenna, 106–108
Display, 4, 122
  with basic systems, 148, 149, 151, 154, 157, 158
  display advertising, 53, 67

Display (*Cont.*):
with enhanced systems, 159, 160, 162–168
with European systems, 21, 27–32
with existing systems, 158
format of, in U.S. trials, 50–51
of graphics (*see* Graphics, display of)
options available for, 11
(*See also specific options, for example:* Cathode-ray tube terminals; Printers)
performance attributes relevant to, 49
standards of, 196
technological attributes and, 50–51
touch-sensitive, 132, 133, 135, 136, 269
transmission speed and quality of, 101
(*See also* Transmission speed)
with two-way systems, 98
(*See also* Black-and-white displays; Color displays; Television sets, display on)
Distributed database, 91–92
Dominant carrier (*see* AT&T)
Dow Jones, 7, 43, 48, 50, 51, 144, 192, 206, 223, 224
Down-converter earth terminals, 108
Down-loading, 30, 38, 150, 269
of data, 57
of PDIs, 30
policy issues and, 185, 186, 190, 208
of software, 58, 165
Downstream transmission (*see* Transmission)
DRCS (dynamically-redefinable character sets), 30, 155
Dutch telecommunications, 15–17
Dynamically-redefinable character sets (DRCS), 30, 155

Ease-of-use criterion, 62, 72, 95, 192
Echo unit, 130
Economic concentration:
antitrust legislation and, 13, 174–175, 201, 211, 213
in telebanking case study, 245–250
Editing, 46
with digital video equipment, 105
on European systems, 27
on keyboards, 4, 5
of mail messages, 149
multiple-page, 27
(*See also* Editorial content)
Editorial content:
content regulation and control of, 206
(*See also* Content regulation)
effects of government support on, 208
information advertising and, 54
Education:
computer-aided, 64, 71
transformative effects on, 258, 272
Education, Department of, 41

Educational information services, 61
policy issues related to, 182
potential market for, 66, 71–73
U.S. trials with, 43, 44, 54
Effective transmission speed, defined, 50*n*.
Electronic directories, 6, 15, 20, 27, 32, 43, 55, 61, 66
EYP, 55, 67, 80, 174, 182, 203, 226
Electronic Funds Transfer Act (1978), 216
Electronic funds transfers (*see* Telebanking services)
Electronic Information Service (EIS), 43, 48, 50, 52
Electronic magazines, 17, 41, 42, 75
Electronic mail, 52
potential market for, 76, 80
public policy and, 183, 220, 222, 232
U.S. trials with, 36, 42, 55
Electronic messaging (*see* Messaging services)
Electronic Money Council, 74
Electronic newspapers, 43, 52–53, 61, 65, 173
(*See also* News, sports, and weather information services)
Electronic yellow page (EYP), 55, 67, 80, 174, 182, 203, 226
Electronics Industries Association (EIA), 191, 192, 197
Employment, 271, 272
effects on European, 35
effects on telebanking on, 252
industry structure and, 227–234
policy issues related to, 182, 183, 186, 187
policy options on, 233–234
and telework services, 83
Encyclopedias, video, 43
Energy management (*see* Telemonitoring services)
Enhanced services, 3, 9, 12, 157–168
computing, 164–166, 169
Federal Communications Commission definition of, 176
information retrieval, 158–162
messaging, 158, 162–165
telemonitoring, 166–168
transactions, 161–164
Entertainment (*see* Games and entertainment)
Equal Credit Opportunity Act, 220
Equal Time Provision, 205
Equity of access, 225, 271
to European systems, 23, 27
policy issues related to, 182, 183, 185, 187
policy options on, 218–220
in telebanking case study, 245, 246
Error rate, 32, 63, 93, 110, 196
ESS No. 4 (switching system), 115
Ethernet (system), 113
European Conference of Postal Telecommunications Authorities (CEPT), 28, 38, 197, 198

Existing systems:
basic systems compared with, 155–158
U.S. trials with, 40–49
analysis of, 45–47
applications of, 52–58
technological attributes of, 47–52
Expenditures:
for advertising, 72
for business information, 81
for computing services, 77, 79
for education, 71–72
for information-retrieval services, 71–72, 79
for messaging services, 75–76, 79
summary of potential markets based on, 79
for telemonitoring services, 78, 79
for transaction services, 73–74, 79
usage and, 87–88
Express Information, 44, 48, 50–52
Extended family, electronically-created, 256–257
External database, 91, 92
EYP (electronic yellow page), 55, 67, 80, 174, 182, 203, 226

F.300 (recommended standard), 191, 197
Fair Credit Reporting Act, 178, 220
Fair Packaging and Labelling Act, 178
Fair-trade laws, 236
Fairness Doctrine, 174, 176, 178, 185, 199, 202, 205
Family and home life, transformative effects on, 255–258, 272
Federal banking laws, 13, 177
Federal Communications Commission (FCC), 13, 54, 97, 109, 114, 173–177
AT&T divestiture and, 175
broadcasting regulation and, 176–177
cable regulation and, 111, 112, 179–180
common-carrier regulation by, 173, 175–176
communication policy and, 202–203
content regulation and, 185, 205–207
DBS and, 107
economic concentration and, 247
market structure and, 201
MDS channels allowed by, 106–107
standards setting and, 176–177, 191, 192, 197–199, 270
Federal Deposit Insurance Corporation (FDIC), 177
Federal Reserve Bank, 177, 222, 239
Federal Savings and Loan Insurance Corporation (FSLIC), 177
Federal Trade Commission (FTC), 177–178, 204, 207, 220, 247
Fedida, Sam, 2, 3, 14
Ferrarini, 14
Field Electronic Publishing Inc., 41
Fifth Amendment (1790), 215

Financial information services, 3, 17, 18, 42, 43, 54, 61, 62
Financial institutions (see Banks and banking industry)
Financial Times (newspaper), 34
Finnish telecommunications, 15–17, 32, 34, 36–39
Fintel (firm), 33
Fire alarm and detection (see Telemonitoring services)
First Amendment (1790), 13, 173, 175, 199, 204, 205, 207
First Bank System, Minneapolis, 45, 224, 244
First Interstate Bank, Los Angeles, 44, 224
First Interstate Bank of California, 242
Fixed-format (synchronous) signal transmission, 28–29, 38, 196
Floppy discs and drives, 140, 141, 143, 162–164, 166–169, 269
FM radio-based systems, 102–104, 109–110, 120, 121, 156, 205, 268
432 micromainframe, 137
Fourth Amendment (1790), 215, 271
Frame and frame size:
    in basic services, 148
    consumer protection and, 221
    design of, 3
    display and, 29, 30, 123
    distinction between advertising and informational, 216, 223
    down-loading and modification of, 38
    privacy and types of, 216–217
    in U.S. trials, 44
    waiting time per average, 97
Frame video, 123, 124
Franchising, cable, 13, 112, 179–180, 199, 201, 202, 204, 211
Frank, 110
Freedom of the press (and of speech) (see Content regulation; First Amendment)
French Postal, Telegraph and Tele-communication Department (PTT), 6, 14, 15, 22, 25, 36, 37, 236
French telecommunications, 14–18, 21–25, 27, 29, 34, 35, 37–39, 107, 234
Frequency-shift keying techniques, 142
Front-end minicomputers, 20
Full-channel systems, 5, 103, 120, 125, 268, 269
    coaxial cable-based systems and, 104
    with enhanced systems, 160
    public policy on, 186, 187, 209
    (See also Direct-broadcast satellite-based systems; FM radio-based systems; Multi-point distribution systems)
Full-text searching, 34
Future applications, 10, 60–88
    based on teletext and videotex growth potential, 83–88

Future applications (*Cont.*):
    in business market, 72, 80–83, 86, 87
    insights about, based on current U.S. trials, 60–64

Game paddles (joystick), 71, 132–134, 136
Games and entertainment, 3, 17, 144, 269
    classification of, 10
    on European systems, 20
    popularity of, 85, 86
    potential market for, 67, 76–77
    public policy and, 183
    in U.S. trials, 42, 43, 54–56, 60, 61
Gateway design, 3, 92, 98, 236
Gelder, 117
General classes of information, 56–59, 152
    (See also Computing services; Information-retrieval services; Messaging services; Telemonitoring services; Transaction services)
General Electric Company (GE), 20, 224
General information services, basic systems for, 148
Gillmor, 199
Glass-Steagall Act (1933), 240
Government regulation (see Regulation)
Government support and subsidies, 46
    for creation of information, 208
    equity of access and, 219, 220
    foreign, 14, 15, 22, 37, 46
    guarantee of access and, 247
    international trade and, 236, 237
Graphics, 3, 125–128, 147
    alphageometric and alphamosaic codes for, 29–30, 32, 155, 194–195, 197
    character sets for, 26, 30
    devices for input of, 132–135
    display of, 29, 144
        on computer terminals, 126
        on European systems, 29–32
        on printers, 128
        standards of, 155, 184
        on television sets, 123, 125
    editing of, 4
    in enhanced systems, 159, 163, 167
    as factor affecting potential application, 63, 72
    quality of, as key difference among systems, 155
    standards for, 155, 184, 196
Graphics terminals (see Graphics, display of)
Green Thumb, 43, 48, 50–52, 54, 234
Great Britain (see British tele-communications)
Growth:
    of cable television, 84, 101
    of cable television-based systems, 110–113

Growth (*Cont.*):
    new technology development and, 273
    potential for, 80, 83–88
    of printers, 128, 129
GTE-Telenet, 120, 224
Guarantee of access, 270
    policy issues related to, 182, 183, 185, 187, 189, 200–201
    policy options on, 201, 210–213
    in telebanking case study, 246–247, 250
Gussman, 221

Hamilton, 94, 208
Hard copy, 123, 124, 159, 160, 230, 232
Hard (rigid) discs and drives, 140, 141, 143, 166
Hardware industry:
    employment with, 233
    international trade and, 237
    standards setting and, 192
HDLC (high-level data link control), 195
HDTV (high-definition television), 105–106, 125
Hearing impaired, services for, 7, 17, 19, 41, 54
Helsingen (firm), 22
Hierarchical database, 92–93, 148, 268
High-definition television (HDTV), 105–106, 125
High-level data link control (HDLC), 195
High-resolution displays, 64, 148, 149
HI-OVIS (Highly Interactive Optical Video Information System), 15
Home banking (see Telebanking services)
Home Box Office (HBO), 63, 106
Home computers (see Personal computers)
Home and family life, transformative effects on, 255–258, 272
Home market, 63–79, 144
    foreign, 14–18, 27, 36
    potential, 64–79, 86–87, 269
    technologies competing for, 63
    (See also specific services)
Home security (see Telemonitoring services)
Home shopping (see Telemonitoring services)
Home-work boundaries, blurring of, 263
HomeServ, 192
Hopengarten, 107, 111
Horizontalization, 263–264
Host computers, 6, 150
H&R Block, 43
Hutt, 82

IBM (International Business Machines), 20, 94, 191, 192, 224

Image candidates, 266
Impact printers, 127–128
Import quotas, 236
INDAX, 44, 48, 50–52, 54, 55, 95, 145, 158, 198, 200, 203, 239
Independent Broadcasting Authority (IBA), 14
Independent Television (ITV), 17, 34
Indexes and indexing, 1, 45, 55, 90, 95
  central, 23
  and keyword searching, 92
  as responsibility of informative providers, 45
  by service centers, 90
  uniformity of services and, 203
Individualization (automated individual tailoring) effect, 35–36, 259–260, 273
Industry:
  new technologies and growth of, 273
  structure of, 46–47, 271
    public policy and, 182, 183, 186, 187, 199, 200, 223–227
    in telebanking case study, 245–246
  (See also specific industries, for example: Banks and banking industry)
Infomercials, 54
Informat, 15, 22, 44
Information Industry Association, 81
Information pages (see Page and page format)
Information providers, 6, 22, 223, 224
  content regulation and, 206
  database structure and, 90, 91
  employment with, 186
  foreign, 25, 39
  guarantee of access and, 200, 201
  responsible for database quality, 38
  role and tasks of, 45, 46
  as service providers, content-carriage separation and, 13, 36, 190, 201–204, 211, 213
  standards setting and, 192, 194, 197
  technological attributes important to, 48
  terminals used by, 27
  in U.S. trials, 41, 47
Information-retrieval services, 2, 3, 10, 43, 182, 183, 187, 211, 213, 269
  basic, 147–149, 154, 157
  computer terminal displays for, 126
  consumer protection and, 221
  content regulation and, 204, 205
  database structure and, 92
  described, 56
  employment with, 228–231
  enhanced, 158–162
  equity of access to, 219
  foreign, 17, 20, 21, 34
  industry structure and, 225–226, 228
  for library information, 54
  popularity of, 85, 86
  potential market for, 64–66, 70–73, 79
  with Prestel, 61
  privacy and security and, 215–216

Information-retrieval services (Cont.):
  storage requirements for, 94
  technologies competing with, 63
  U.S. trials with, 43
Information services, 80
  centralized database for, 90
  foreign, 14, 15, 17–20
  generic classes of, 56–59, 152
    (See also Computing services; Information-retrieval services; Messaging services; Telebanking services; Telemonitoring services; Transaction services)
  most popular, 61–62
  regulation of, 15
Information sets, 18
In-house (private) systems, 20, 33–34, 194
Input, 122, 132–139
  in basic systems, 151, 153, 157
  computing and, 150
  in enhanced systems, 162
  in European systems, 23
  in existing systems, 158–160, 162–164, 166–169
  performance attributes relevant to, 49
Input devices, 27
  messaging capability and, 57
  technological attributes of user, 49, 51
  (See also specific types of input devices, for example: Keypads)
Institute of Electrical and Electronic Engineers (IEEE), 191, 192
Institute for the Future's Forecast of Consumer Expenditures, 70
Intel (firm), 137
Intellectual property (see Copyright)
Intelligent copier/printers, 80
Intelligent terminals, 3, 122, 136, 150
  computing with, 57, 58
  conclusion and findings on, 269
  for enhanced systems, 160
  page identification with, 104
Intellivision, 76, 77
Interface:
  with enhanced systems, 159, 160, 163, 164, 166, 168
  with existing and basic systems, 158
  performance attributes relevant to, 48, 49
Interface devices:
  for cable television-based systems, 110
  of European systems, 23
  function of, 101
  for packet-switched network-based systems, 119
  in U.S. trials, 48
  (See also specific types of interface devices, for example: Modems)
Interface unit cost, 100–102, 121, 135, 143–144
Internal Revenue Service (IRS), 224, 257
International communication (see International trade and communication)

International Computers Limited (ICL), 20
International Consultative Committee for Radio (CIR), 191, 192, 198
International Electrotechnical Commission (IEC), 191, 192
International Organization for Standards (ISO), 26, 191, 192, 194, 198
International Resource Development, Inc. (IRD), 81, 125, 126
International Telecommunication Union (ITU), 26
International Telegraph and Telephone Consultative Committee (CCITT), 2, 26, 191, 192, 195, 197, 198, 270
International trade and communication, 38, 271, 272
  policy issues related to, 183, 186, 187, 234–236
  policy options on, 213, 236–237
Intersystem standards, 194
Intrasystem standards, 194
Italian telecommunications, 15
ITFS (service), 107

J. C. Penney (firm), 224, 226
J. Walter Thompson (firm), 239
Japanese telecommunications, 15, 17, 22, 105–107
Japanese Telegraph and Telephone Public Corporation (PTT), 15, 22, 236
Joystick (game paddles), 77, 132–134, 136
Justice, Department of, 174, 175, 178, 203

KCET (station), 54
Keyboards, 23, 77
  alphametric, 92
  alphanumeric, 5, 27, 29, 51, 52, 122, 126, 132, 136, 149, 268
  for basic systems, 153
  editing on, 4, 5
  for enhanced systems, 162, 164, 165, 167
  input, 126
"Keyfax," 41
Keypads, 27, 77
  for basic systems, 153, 154, 156
  editing on, 4
  hand-held, 55
  and menu searches, 32
  messaging capability and, 57
  numeric, 5, 22, 52, 132, 133, 136, 148–150, 268
  price of, 105
Keyword searches, 32, 92, 98
Kitamura, 15
KMart, 226
Knight-Ridder Newspapers, 43, 47, 53, 223
KNXT/KCET/KNBC trial, 41, 47, 50, 54
KPIX (station), 47, 50

Kraemer, 221
KSL trial, 40, 47, 50, 51

Labor Statistics, Bureau of, 233
Laser videodisc players, 144
Layers of standards, 194–195
Leadership, emergence of new, 264
Leased telephone circuits (manually-conditioned lines), 26, 115–116, 118, 119, 167, 201
Legislation (*see* Deregulation; Proregulation; Regulation)
LeMasters, 107
Library information services, 43, 54–55, 66, 182
Licensing (*see* Franchising; Station licensing)
Light pen, 132–134, 136, 164, 165
Limited database, 268
Line-printer graphics, 126
LISA (logically-integrated speech annunciator), 131
LNA (low-noise amplifiers), 108
Local distribution networks, 100, 101
    (*See also specific local distribution networks*)
Local loop (local exchange)-based systems, 45, 100, 114–115, 117, 118, 120, 194
Log-on, 90, 95, 195, 196
Logically-integrated speech annunciator (LISA), 131
Long-distance telephone networks, 100, 115–117
Long-haul networks, 100
    (*See also specific types of long-haul networks*)
Low-noise amplifier (LNA), 108
Low-power television broadcast (LPTV), 5, 104, 105
Low-resolution alphamosaic system, 97
LPTV (low-power television broadcast), 5, 104, 105

Madden, 2, 200
Magazines (*see* Newspapers and magazines)
Magnetic bubble memory, 137–138
Mail, 75, 76
    expenditures for business, 81
    public policy on, 171
    (*See also* Electronic mail)
Mainframe computers, 20, 235
Malik, 2
Management:
    energy (*see* Telemonitoring services)
    of information, 45
Manuel, 94
Market, 6–7, 9–10
    European, 18–19
    impacts on, 258–262
    penetration of, 120, 121
        factors affecting widespread, 101–102

Market, penetration of (*Cont.*):
    growth potential and, 86
    of local distribution networks, 100–102
    of telebanking services, 249–250
    transformative effects of, 253–254
    TV, 125
    of user terminals, 123
standards development and, 191, 192
structure of: policy issues related to, 47, 198–203
    policy options on, 201, 203–204, 210, 212, 213
    (*See also* Competition; Guarantee of access)
target, 3
    (*See also* Business market; Home market)
    (*See also* Cost; Expenditures; Growth)
Market research, transformative effects on, 267
Martin, 114
Master/replicated database, 91–92
Matsushita (firm), 131
Mattel (firm), 77
MDS (multipoint distribution systems), 5, 102–104, 106–107, 120, 121, 156, 180, 268
Memory:
    for basic systems, 152
    computer terminal, 126
    decoder, 28
    input and, 137–139
    microprocessor memory cards, 37
    random-access and read-only, 30, 137
    semiconductor, 18, 137, 139
    silicon chip, 137, 138
Memory chips, 105, 137, 138
Menkes, 8
Menu (tree) searches, 32–33, 36, 51, 61, 92–93
Merrill Lynch, 224, 239
Messaging services, 3, 10, 36, 59, 182, 183, 187, 212, 213, 269, 271
    access procedures for, 96
    basic, 149, 151, 157
    computer terminal display for, 126
    consumer protection and, 222
    content regulation and, 205
    described, 56–57
    employment with, 231–232
    enhanced, 158, 162–165
    equity of access to, 219
    European, 20, 23
    industry structure and, 226–228
    popularity of, 86
    potential market for, 64, 68, 75–76, 79, 83
    privacy and, 216
    service centers and, 23
    storage requirements for, 94
    technologies competing for, 63
    U.S. trials with, 55, 60
    user-to-user, keyboards and, 52
Meta-service directories, 92
Metcalfe, 113

Meter reading (*see* Telemonitoring services)
Microband, 106
Microband Corporation of America, 106
Microcomputers, 89, 90, 235, 272
Microprocessor memory cards, 37
Microprocessors, 6, 136–137, 142, 144
Microtel, Inc., 243
Microwave frequency, 21, 104, 107, 109
Mier, 74
Miller, 99
Minc, Alain, 14
Minicomputers, 20, 89–90, 147, 235
Mission Cable System, 44
Mitchell, 117
Mobile telephones, 114
Modems (modulator-demodulator), 26, 101, 110, 125
    cost of, 115–117, 126, 157
    for enhanced systems, 164, 165, 169
    function of, 5, 26
    for switched-telephone network-based systems, 115–116
Mokhoff, 125
Monopoly:
    regulated, 180
    (*See also* Economic concentration)
Montgomery Ward (firm), 226
Mosher, 106
Multilingual audiences, services for, 54
Multiplexing:
    of data messages, 98
    with switched-telephone network-based systems, 114, 117
Multipoint distribution systems (MDS), 5, 102–104, 106–107, 120, 121, 156, 180, 268
Muzak, 109, 205

Narrowband transmission (*see* Bandwidths)
Nash, 217
National Association of Broadcasters (NAB), 176, 206
National Association of Regulatory Utility Commissioners (NARUC), 193
National associations of viewdata information providers (AVIP), 206
National Bureau of Standards/Institute for Computer Sciences and Technology (NBS/ICST), 191, 198
National Cable Television Association (NCTA), 78, 197
National Captioning Institute, 47, 54
National Commission on Electronic Funds Transfer, 216, 218
National Science Foundation, 41
National Telecommunication and Information Agency (NTIA), 41, 193, 215, 217–219, 223
National television networks, 100

NBC (National Broadcasting Corporation), 44, 224
NC1/closed captioning, 41, 47, 50, 51
Network layer of standards, 195
Networks (*see specific types of networks, for example:* Radio)
Neustadt, 215
New York Times Information Bank, 54
News, sports, and weather information services, 9, 60–62, 269
  access to, 96
  enhanced, 158
  U.S. trials with, 41–44, 52, 53
News Retrieval Service, 43
Newspapers and magazines, 9, 171, 173
  content-carriage separation issue and, 202
  effects of advertising potential on, 72, 73
  electronic, 17, 42, 52–53, 61, 65, 75, 173
  (*See also* News, sports, and weather information services)
  European, 17, 18, 35–37
  expenditures for, 70, 71
  regulation of, 13, 197, 199, 202, 205, 207
  in Europe, 35, 36
Nilles, 77, 80, 236
912 CRT terminal, 127
Nokia Electronics, 22
Nonimpact printers, 127–129
Nora, Simon, 14
Norpak (firm), 15, 20, 144
Northern Telecom, Ltd., 127
Norwegian telecommunications, 15
NTSC (system), 4
Numeric keypads, 5, 22, 52, 132, 133, 136, 148–150, 268

OCLC (On-Line Computer Library Center), 43, 48, 50–52, 54, 55, 206
Office (*see* Business office)
Official Airline Guide, 64
Omnibus Crime Control and Safe Streets Act (1968), 215
On-Line Computer Library Center (OCLC), 43, 48, 50–52, 54, 55, 206
One-way transmission, 2, 102, 111, 120, 268
  access time and, 96
  for basic systems, 151, 152, 156
  for computing services, 57, 58, 150
  future applications with, 64
  for general information services, 148
  terms describing, 1
  waiting time with, 96–97
Online database, 2, 80–82, 235
Optical fiber-cable, 2, 101, 114, 115, 118
Oracle, 1, 14, 21, 22, 25, 26, 28, 34, 97

Original equipment manufacturers (OEM), 117

Pacific Telephone, 114
Packet-switched network-based systems, 42, 90, 98, 100, 102, 103, 113, 118–121, 149
  database structure and, 92
  European, 24–26
  standards for, 195
  U.S. trials with, 42, 45
Page and page format, 37, 53, 96–98, 103, 155
  accessing capability for groups of, 32
  in basic general information services, 148
  billing on cost-per-page basis, 34
  chaining of successive, on broadcast television-based systems, 103
  databases and, 89–92
  display time of, 148
  error rate and, 155
  in European systems, 17–20, 37
  keeping record of accessed, 36
  Keyfax, 41
  means of increasing number of, 103–104
  selection of, 5, 27, 42
  in VBI-based systems, 64
  (*See also* Bits and bit rate; Bytes; Cycle and cycle pages)
Paired-copper cable (copperwire-based systems), 2, 101, 114, 118
PAL, 4
Paper-based computer terminals, 123, 124, 126
Paper directories, 6, 15, 19, 32–33
Parallel attributes, 29, 195, 196
Passwords, 23
Pay television, 80, 84, 105, 109, 111, 113
  (*See also* Cable television)
PBX (private branch exchange), 113
PDI (picture description instruction), 30, 32
Peer relations, redefinition of, 257–258
Penetration of market (*see* Market, penetration of)
Performance attributes, technological options and, 48–49
Personal attack rule, 202
Personal computers, 2, 17, 20, 76–77
  business use of, 80
  input devices for, 134–135
  market for, 236
  retrieval with, 43
  support services for, 56, 61, 76–77
  telebanking with, 55
  timesharing with, 42, 43
  U.S. trials with, 53, 54, 56
Personal storage files, 56, 60, 61, 69
Phoneme synthesizers, 130, 131
Physical layer of standards, 195
Picture description instruction (PDI), 30, 32

Pioneer (firm), 78
Pixels, 29, 30, 125
Playcable, 77
Plotters, 123, 124, 127–129, 164, 165
Plowright, 217
PLP (Presentation Level Protocol), 116, 117, 144
Plummer, 119
Point-of-scale terminals, 89
Policy issues, 1, 6, 7, 12–13, 47, 60, 171–188
  background to, 171–172
  conclusions and findings on, 270
  current, 173–180
  foreign, 34–39
  identifying, 180–183, 244–245
  and public policy defined, 171
  in telebanking case study, 245–249
  themes of, 184–186
  (*See also* Consequential policy issues; Developmental policy issues)
Policy options (policy profiles), 209–213
  (*See also* Deregulation; Proregulation)
Political arena, transformative effects on, 265–267, 272
Porat, 81, 228
Portable radio, FM, 109, 120
Portable terminals, 119
Portable voice communication, 120
Postal Service, U.S. (USPS), 76, 180, 218, 226, 232
Preformatted messages, 27, 55, 149
Presentation layer of standards, 194–195
Presentation Level Protocol (PLP), 116, 117, 144
Presentation standards, 196, 198
Prestel, 1, 3, 6, 14–16, 19–25, 35, 80, 82, 100, 198, 225, 236
  access procedure to, 95
  application layer and, 194
  content regulation for, 206
  database for, 89, 91, 93
  decoder for, 116
  enhanced, 161
  number of subscribers to, 84, 269
  services on, 60–61
  standards for, 197, 270
  structure of, 26, 27, 29, 30, 32, 33
  text and graphics display on, 144
Price, 105
Price (*see* Cost)
Prime Time Access Rule (PTAR), 173, 202, 205
Print media (*see* Newspapers and magazines)
Printed copies of messages, 164
Printers, 80, 123, 124, 126–129, 164, 165, 269
Privacy Act (1974), 215, 217, 218
Privacy Protection Study Commission, 218
Privacy and security, 149, 183, 211
  in European countries, 36, 37
  policy issues related to, 185, 214–217

Privacy and security (*Cont.*):
  policy options on, 217–218
  in telebanking case study, 245, 246, 248, 250
Private bank exchange (PBX), 113
Private-port access, 119
Private (in-house) systems, 20, 33–34, 194
Processing, 20, 34, 36, 81, 122, 135–139, 175
  with basic systems, 148–154, 157
  computing to enhance, 57
  CPU, 136–137, 150–152, 194
  with existing and basic systems, 158–160, 162–164, 166, 168
  and microprocessors, 6, 136–137, 142, 144
  performance attributes relevant to, 49
  terminals with local, 144
Product distribution, transformative effects on, 260–262
Program captioning (*see* Captioning)
Promotional advertising (*see* Advertising)
Proregulation, 190, 209–213
  defined, 209
  impact of: on consequential issues, 218, 220, 223, 227, 228, 234, 237
    on developmental issues, 201, 203–204, 207, 208
  in telebanking case study, 250, 251
Public Broadcasting Service (PBS), 41, 54
Public policy (*see* Policy issues; Policy options; Regulation)
Public services, 15, 17, 19, 20, 24, 25, 224
Public terminals, 220
Public utility commissions (PUCs), 179, 203
Publishing industry:
  consumer protection and, 222
  employment in, 229
  (*See also* Newspapers and magazines)
Purchases (*see* Teleshopping services)
Push-button telephones, 40

Qasar microwave oven, 131
QUBE, 1, 42, 50–52, 111, 200, 202, 203, 239

Radio, 173
  AM, 83–84
  cellular, 5, 114–115
  enhanced systems using, 165
  FM-radio based systems, 102, 104, 109–110, 120, 121, 156, 205, 268
  for packet-switched network-based systems, 119–120
  potential for advertising on, 72
Radio Shack, 43, 135, 144, 145
RAM (random-access memory), 30, 137

RAMTER 5911 (light pen), 134
Random-access memory (RAM), 30, 137
Rate regulation, 179, 199
Read (read-write) capability, storage and, 139, 141
Read-only memory (ROM), 30, 137
Reader's Digest (magazine), 42
Receivability of information, 45
Recorders (*see* Audio cassettes and recorders; Videocassettes and recorders)
Reference Model for Open Systems Interconnection-OSI, 194, 195
Regulated monopoly, 180
Regulation, 15, 171, 172, 273
  (*See also* Deregulation; Proregulation; Self-regulation)
Regulation E, 220, 223
Relational database, 93–94
Reliability criterion, 62
Remote computers, 58, 136, 149, 150, 152
Repeaters for packet-switched network-based systems, 119
Reservation services, 25, 43, 44, 62, 73–75
Resolution, 49
  of graphics terminals, 126
  high-resolution displays, 64, 148, 149
  low-resolution alphamosaic system, 97
  television-tube, 125
Response time (*see* Waiting time)
Retail industry:
  consumer protection and, 226
  (*See also* Teleshopping services)
Retrieval (*see* Information-retrieval services)
Reuters News Service, 42
RGB beam circuits, 4, 5, 28
Right to Financial Privacy Act, 216
Roizen, 4
ROM (read-only memory), 30, 137
Roman Associates International, 140
RS232-C (connector), 116

S.100 (recommended standard), 191, 197
Sales, regulation of, 177–178
Sanoma Oy (firm), 22
Satellite subscription television (SSTV), 108
Satellite Television Corporation, 108
Satellites, 2, 3, 45, 100, 173
  DBS-based systems, 5, 102, 104, 107–109, 120, 121, 156, 180
  dish antennas for, 107, 108
  transponders of, 43, 107
Savings and loan associations (*see* Banks and banking industry)
Sawyer, 238
Schuster, 93, 94
SDLC (synchronous data-link control), 195
Search:
  with basic systems, 151

Search (*Cont.*):
  database structures and, 92–94
  full-text, 34
  keyword, 32, 92, 98
Sears Roebuck, 41, 224, 226, 239
Security (*see* Privacy and security; Telemonitoring services)
Security Pacific Bank, 44
Self-regulation, 206, 207, 209, 217, 222, 223
Semiconductor chips, 122, 131
Semiconductor industry:
  and cost of technology, 18
  employment in, 233
  international market for, 236
Semiconductor memory, 18, 137, 189
Semiregulated monopoly, 180
Sensors (transducers), 9, 132, 133, 135, 136, 150, 153, 168
Serial attributes, 28, 195, 196
Service centers, 23–26, 45, 90
Services:
  selected organizations involved in, 224
  selection of, 95–96
  trade-off between access time and characteristics of, 97
  (*See also specific types of services*)
Session layer of standards, 195
7-bit code, 26
Sharp (firm), 131
Sherman Antitrust Act (1890), 174, 201
Shopping centers, remodeling of, 260–261
Signal-to-noise ratio, 117
Signals:
  analog, 5, 101, 105
  asynchronous and synchronous transmission of, 28–29, 38, 196
  converters of, 106
  DBS, 107
  digital, 5, 101, 102, 105
  FM, 109
  television (*see* Vertical blanking intervals, systems using)
  video, 118
Silicon chip memory, 137, 138
Smart cards, 135, 161
Smith, 217
Societal impacts, 1, 255–267
  on business office, 263–265
  conclusion and findings on, 270, 272–273
  in foreign countries, 35–36
  on home and family life, 256–258
  on market, 259–262
  in political arena, 265–267
Society of Motion Picture and Television Engineers (SMPTE), 191, 192
Software, 1, 20, 56, 165
  down-loading of, 58, 165
  for enhanced services, 165
Software industry:
  copyright laws and, 178, 185
  international trade and, 235
  standards setting and, 192

Source, The, 7, 144, 156, 159, 203
  application layer of standards and,
    194
  average usage of, 119
  centralized database with, 90
  services offered by, 54, 55, 61, 62,
    240
  U.S. trials with, 42, 48, 50–52,
    54–56
Southern Satellite Systems, 42
Southwestern Bell, 179
Southwestern Cable TV, 42
Spanish telecommunications, 16
Speak and Spell (learning aid), 131
Specific information services, 148
Specificity of retrieved information,
  147
Speech recognition devices, 131, 132,
  135, 136, 160–162, 164
Speech synthesizers, 123, 124,
  129–132, 160, 162
Spontaneous purchases, electronic,
  261–262
Sports information (see News, sports,
  and weather information
  services)
SRI International, 131
SSTV (satellite subscription
  television), 108
Stand-alone computers, 20
Standards, 13, 52, 64, 270
  AT&T and, 197
  coding, 144
  display, 125, 155, 184
  Federal Communications
    Commission and, 176–177,
    191, 192, 197–199, 270
  in foreign countries, 38
  graphics, 155, 184
  policy issues related to, 183, 184,
    187, 190–198
  policy options on, 210, 212
  for telebanking services, 249
State, Department of, 191, 192
State banking laws, 13, 177
State regulation, 13, 177–179, 199
State regulatory commissions, 193
  public utilities, 179, 203
Station licensing, 176
Storage, 57, 64, 122, 139–144
  with basic services, 150, 151, 153,
    154, 157
  database structure and, 92–94
  with European systems, 23
  with existing systems, 158–160,
    162–169
  with European systems, 23
  message, 149
  performance attributes relevant to,
    49
Store-and-forward communication,
  118, 149
Strategic Business Services, Inc., 129
Streets Electric Corporation (SEC),
  130
Structural regulation (see Industry,
  structure of)
Structured Query Language Data
  System, 94

Subinformation providers, 46
Subscription television (STV), 180,
  206
Subsidiary Carrier Authorization
  (SCA), 205
Sullivan, 6
Summagraphics (firm), 135
Support services for personal
  computers, 56, 61, 76–77
Support and subsidies (see
Government support and sub-
  sidies)
Sveriges Radio AB, 16, 19, 23, 35, 39
Swedish telecommunications, 15, 19,
  34–37, 39
Swiss telecommunications, 15–17, 34
Switched-telephone network-based
  systems, 102, 103, 114–119, 126,
  167–168, 272
Synchronous data link control
  (SDLC), 195
Synchronous (fixed-format) signal
  transmission, 28–29, 38, 196
Synthesizers, speech, 123, 124,
  129–132, 160, 162
System characteristics, 11–12
  (See also Display; Input; Storage)
System design, redesigning, 264
System operators, 6, 22, 26, 27, 224
  guarantee of access and, 200
  public policy and, 223, 224
  role and tasks of, 46
  service centers of, 23–26, 45, 90
  standards development and, 192
  technological attributes important
    to, 47
System organization:
  foreign, 22–23
  U.S., 45–46

Tablet/pen, 132–136, 165
Takasaki, 15
Tanton, 103
Tapes and players, cartridge, 140–143
Tariffs, 236, 237
Taylor, 111
Teachers (see Education)
Technological attributes, 47–52
  applications and, 63–64
  of foreign systems, 22–23
  needed for educational use, 72
  (See also specific technological
    attributes, for example:
    Transmission speed)
Technological options:
  performance attributes and, 48–49
  (See also specific technological
    options, for example:
    Decoders)
Technology path (see Standards)
Telebanking services, 61, 62, 73, 74,
  177, 183, 187, 201, 269
  case study of, 239–252
    context of, 239–240
    impact of penetration of services,
      249–252
    policy issues identification in,
      244–245

Telebanking services,
case study of (Cont.):
    policy options on key issues
      affecting, 246–249
    workshop structure in, 240–244
  consumer protection and, 220–222
  database structure for, 92, 93
  impacts of, 259, 260
  industry structure and, 226
  potential market for, 67, 73–74
  privacy and, 216
  U.S. trials with, 43–45, 55, 56
Telecom (firm), 37
Telecommunications Deregulation
  Bill, 172
Telediffusion de France (TDF), 14, 16
Telemail (service), 149
Telemonitoring services, 3, 10, 152,
  269, 271
  basic, 150, 151
  cable television and, 180
  consumer protection and, 222
  described, 58–59
  employment with, 233
  enhanced, 166–169
  equity of access to, 220
  industry structure and, 227, 228
  policy options on, 212
  popularity of, 86
  potential market for, 64, 69, 77–80
  privacy and, 216–217
  sensors used with, 9
  technologies competing for, 63
Telenet (firm), 42, 149, 202
TELENET (system), 118–120
Telephone-based systems, 5–6, 43, 268
  attributes of, 104
  basic, 156–157
  enhanced, 166, 169
  foreign, 15, 22, 26
  interface unit and communication
    link cost of, 101
  with local distribution networks,
    101
  market for, 6
  with MDS-based systems, 106
  PBX in, 113
  policy issues related to, 171, 173,
    178, 186, 190, 202–203, 209
  policy options on, 212, 213
  transmission speed of, 50
  two-way transmission with, 98
  U.S. trials with, 44, 45
  waiting time with, 98, 99
Telephone-bill paying (TBP) (see
  Telebanking services)
Telephones and telephone industry,
  112, 194
  dial-up telephones, 118, 119
  growth of industry, 84, 101
  push-button telephones, 40
  regulation of, 199
  role of, in teletext and videotext
    development, 22
  touch-tone telephones, 132, 133,
    136, 153, 154
Teleshopping services, 9, 61, 62, 226,
  272
  access procedures for, 96

Teleshopping services (*Cont.*):
  case study of (*see* Telebanking
    services, case study of)
  database structure and, 92, 93
  enhanced, 161
  impacts of, 258–262
  international, 236
  policy issues related to, 183
  potential market for, 73, 74
  U.S. trials with, 43, 44, 55, 56
Teletel, 1, 23, 27–29, 82, 89
Teletext magazine, 41, 42, 205
Teletext and videotex technology:
  assessed, 10–12
  conclusions and findings on,
    268–273
  definitions of, 2–3, 35
  overview of, 1–2
  workings of, 4–6
  (*See also specific aspects of teletext
    and videotex technology*)
Television (TV):
  advertising on, 72
  educational, 54
  European, 37, 38
  growth of, 83–84, 101, 270
  role of, in development of home
    market, 80
Television-based systems (*see*
  Broadcast television-based
  systems; Cable television-based
  systems)
Television Code, 176, 206
Television sets, displays on, 44, 45,
  51, 101, 123–125, 194
  of basic systems, 154
  decoders and, 105
  for enhanced services, 165
  future, 154
  of general information, 148
  international market for, 235, 236
  as prime display device, 80–81
  speech synthesizers with, 132
Television signal (*see* Vertical
  blanking intervals, systems using)
Telework services, 83
Telidon, 1, 20–23, 116, 155, 158, 209,
  270
  development of, 15, 16
  enhanced, 161
  services provided by, 20, 21
  standards for, 197
  structure of, 22–23, 25, 26, 28–30,
    32, 33
  text and graphics display on, 144
  U.S. trials with, 44
Telset, 33, 36
Terminal-to-terminal communication,
  62
Terminals:
  applications and characteristics of,
    64
  for DBS, 107
  dumb, 165
  foreign, 15, 19, 27
  information providers and, 193, 194
  international market for, 272
  for packet networks, 118
  portable, 119

Terminals (*Cont.*):
  for switched-telephone network-
    based systems, 116
  (*See also specific types of terminals,
    for example:* Intelligent
    terminals; User terminals)
Texas Instruments, 127, 131, 224
Texas Public Utilities Commission,
  179
Text and text format, 2, 3
  for basic services, 149
  display of, 29, 123–126, 144, 159
  with enhanced systems, 159
  input and, 132, 133
  for messaging services, 149
  standards for, 194
Ticketron-type services, 75
Time:
  transformative effects on power
    and use of, 267
  (*See also* Access time; Waiting time)
Time domain synthesizers, 130
Time Inc., 42, 47, 50, 224
Times/Mirror, 44, 48, 50–53, 224
Timesharing, 3, 17, 56, 89, 95, 149,
  150
  with enhanced services, 165
  sales of, 175
  U.S. trials with, 42, 43
Timing criterion, 63
Tokyo Juki Industrial Company, 129
Touch-sensitive displays, 132, 133,
  135, 136, 269
Touch-tone telephones, 132, 133, 136,
  153, 154
Transaction services, 3, 10, 59, 182,
  183, 187, 211, 269
  basic, 149, 151, 157
  case study of (*see* Telebanking
    services, case study of)
  consumer protection and, 221–222
  content regulation and, 205
  database for, 90, 93
  described, 56
  employment with, 231
  enhanced, 161–164
  equity of access to, 219
  European, 20
  industry structure and, 226, 228
  packet-switched network-based
    systems for, 118, 119
  policy options on, 212, 213
  popularity of, 85, 86
  potential market for, 64, 67–68,
    73–75, 79, 80
  privacy and, 216
  storage requirements of, 94
  technologies competing for, 63
  (*See also specific types of
    transaction services, for
    example:* Teleshopping
    services)
Transducers (sensors), 9, 132, 133,
  135, 136, 150, 153, 168
Transfer rate, storage and, 139–142
Transformative effects, 13, 253–254
  (*See also* Societal impacts)
Transmission:
  digital, 64

Transmission (*Cont.*):
  duration of, in European countries,
    17
  signal, 28, 29, 38, 196
  speed and rate of, 17, 50, 51, 120,
    121
  with basic services, 149, 150,
    156–157
  with coaxial cable, 113
  defined, 4
  effective, defined, 50*n.*
  with enhanced systems, 160
  as factor affecting application, 63
  to graphics terminals, 126
  with local distribution networks,
    100–102
  with MDS, 106
  with modems, 26
  with packet-switched networks,
    118
  to printers, 128
  of specific information, 148
  to speech synthesizers, 130, 131
  with switched-telephone net-
    works, 116, 117
  with telephone-based systems, 4–5
  to television sets, 123–125
  trade-off between access time
    and, 96, 97
  standards of, 195–196
  in U.S. trials, 45–47
  (*See also* One-way transmission;
    Two-way transmission)
Transpac, 25
Transponder power, defined, 107*n.*
Transponders, satellite, 43, 107
Travel reservations, 43, 44, 62
Treasury, Department of, 54, 177
Tree-searching (menu) searches,
  32–33, 36, 51, 61, 92–93
TRS-80 (computer), 43, 135
Truth-in-Lending Act, 178
Tudocarpe (firm), 129
Turnkey systems, 81, 89–90, 224, 235
TV (*see* Television)
Two-way transmission, 2, 102,
  111–112, 268
  access time with, 5–6, 96, 98
  with basic systems, 148–153, 155,
    156
  with computing services, 57–58
  future applications using, 64
  telemonitoring and, 58
  terms describing, 1
  U.S. trials with, 40–44
Tyler, 2
Tymnet (firm), 42, 224
TYMNET (system), 118, 119
Tymshare, Inc., 106
Type-'N-Talk, 130

Ultrahigh frequency (UHF), 104, 109,
  119
United American Bank, 44, 224
United American Services Corp., 243
United Video, 205
Upstream transmission (*see*
  Transmission)

Urbanet (system), 106
Usage:
  cost and, 119
  growth potential and, 87–88
User groups:
  closed, 3n., 19–20, 33–34, 82–83, 90
  foreign, 17–20
  standards setting and, 192
  in U.S. trials, 47
  (See also Business market; Home market)
User terminals, 122–145
  for basic systems, 151–153
  conclusion and findings on, 268–269
  for enhanced systems, 161, 163–165
  foreign, 25–32
  inputting and, 132–139
    (See also Input)
  processing and, 135–139
    (See also Processing)
  standards for, 193–194
  storage and, 139–144
    (See also Storage)
Users, 2, 6, 17, 18
  communication links with, 25–26
  identification of, 95, 96, 149, 161, 195, 196
  roles and tasks of, 45, 46
  service centers and monitoring of, 23
  standards setting and, 190, 196
  statistics on, 36
  system characteristics and technological options as viewed by, 49
  technological attributes important to, 47

V.3 (recommended standard), 26
Variable-format (asynchronous) signal transmission, 28, 29, 38, 196
VBI (see Vertical blanking intervals, systems using)
VCSI (very-large scale integration), 104–106, 122
Venezuelan telecommunications, 16

Vertical blanking intervals, systems using, 4, 5, 35, 64, 102, 120, 125, 176, 268
  access time and, 96, 97
  broadcast television-based, 102, 148
  centralized database and, 89
  closed captioning and, 41
  content regulation and, 205–207
  European, 17, 22, 25, 35
  FM radio-based, 109
  MDS-based, 106
  number of lines used with, 103
  policy issues related to, 186, 187
  transmission speed with, 50
  U.S. trials with, 41–43
Venture Development Corporation (VDC), 127, 138, 140
Verbung, 223
Very-high frequency (VHF), 104–106
Very-large scale integration (VCSI), 104–106, 122
VHF (very-high frequency), 104–106
Video cameras, 132, 135, 136
Video computer terminals, 101, 123–125, 165
Video encyclopedias, 43
Video games (see Games and entertainment)
Video input, 132, 133
Video Response System (VRS), 15
Video signals, 118
Videocassettes and recorders, 17, 80, 83–84, 140–143, 162, 178, 207
Videodial (firm), 90, 224
Videodiscs and players, 17, 140–144, 160
Vidiprinters, 129
Viewdata (see Teletext and videotex technology)
Viewdata Corp., 243
Viewtron, 1, 95, 158, 194, 203, 206, 224, 239
  U.S. trials with, 43–44, 46, 50–52, 54, 55
Viniprix (firm), 82
VIP (system), 78
Voice:
  analog model of human, 130
  display of, 123
  transmission of, 2, 167, 168

Voice (Cont.):
  voice input, 132, 133
  (See also Speech recognition devices; Speech synthesizers)
Votrax (firm), 130

Waiting time, 155
  with basic services, 148–151, 157
  database structure and, 96–99
Wall Street Journal, The (newspaper), 43, 54
Waller, 138
Warner Amex Cable Communications, 42, 56, 111, 200, 224, 239
Weather information (see News, sports, and weather information services)
Wessler, 118, 120
West German telecommunications, 15–17, 23–25, 33, 35, 37–39, 107
Westinghouse Broadcasting, 41
WETA (station), 47, 50, 54, 97
WFLD (station), 41, 47, 50, 51
Whitbreads (firm), 20
White, J., 81, 212
White House Office of Science and Technology, 193
Winchester disc drives, 140
Winsbury, 2, 225
Word processing (see Processing)
Work, blurring of boundaries between home and, 263
  (See also Employment)
World views, homogenization of, 262
WTBS (station), 42

Xerox Corp., 106
XTEN (network), 106

Yasuda, 15
Yurow, 208

Zenith (firm), 105, 192, 224
Zientara, 83